Nature-Based Technologies for Wastewater Treatment and Bioenergy Production

Edited by
Imran Ahmad and Norhayati Abdullah

IWA PUBLISHING

Published by **IWA Publishing**
Unit 104–105, Export Building
1 Clove Crescent
London E14 2BA, UK
Telephone: +44 (0)20 7654 5500
Fax: +44 (0)20 7654 5555
Email: publications@iwap.co.uk
Web: www.iwaponline.com

First published 2025

British Library Cataloguing in Publication Data
A CIP catalogue record for this book is available from the British Library

ISBN: 9781789064094 (paperback)
ISBN: 9781789064100 (eBook)
ISBN: 9781789064117 (ePub)

Doi: 10.2166/9781789064100

This eBook was made Open Access in February 2025.

Contents

To my beloved elder brother, Dr (Professor) Rizwan Ahmad (PhD)

Your unwavering support has been my constant pillar of strength. In the wake of our father's passing, you stepped in with a heart full of boundless love, offering guidance and light in moments when my world felt shrouded in darkness. This book is a testament to your belief in me and the strength and resilience you instilled, especially when I doubted myself. For every word penned and every challenge conquered, I owe it to your wisdom, kindness, and the incomparable bond that only a brother like you could share. You have been my anchor in the storm, my protector in times of uncertainty, and my greatest source of inspiration.

Thank you for standing by me, not just as a brother but as my unwavering guiding star, leading me through the darkest of nights and the brightest of days.

With all my love and gratitude,

Imran Ahmad

Authors' Biographies

Lead editors

Imran Ahmad

ProalgaeTech, Schwabbacher Str. 15, 70437 Stuttgart, Germany

Dr Imran Ahmad has valuable experience working as a Bioprocess Engineer with the innovative startup ProAlgaeTech, focusing on carbon sequestration and wastewater bioremediation. With over eight years of teaching experience in AICTE-accredited institutions, Dr Ahmad brings deep expertise in wastewater treatment, microalgal biotechnology, and nature-based technologies. He earned his PhD from the Malaysia-Japan International Institute of Technology (MJIIT). He previously served as a Research Scientist and Plant Operations Manager at Liquid Trees, where he contributed to cultivating diatoms for the bioremediation of the Hindon River. Dr Ahmad was pivotal in the prestigious Indo-European Horizon 2020 PAVITR project, advancing sustainable water treatment solutions. He has also led innovations in high-rate algal ponds and photobioreactor systems for bioenergy production and resource recovery. His contributions include numerous peer-reviewed publications and edited two books with IGI Publications and Springer. Additionally, he has presented his research at several international conferences, furthering advancements in green technologies and the circular bioeconomy. With a commitment to bridging the gap between research and real-world applications, Dr Ahmad continues to make meaningful contributions to environmental engineering and global sustainability efforts.

Norhayati Abdullah, Malaysia

Malaysia-Japan International Institute of Technology, Universiti Teknologi Malaysia, Jalan Sultan Yahya Petra 54100, Kuala Lumpur, Malaysia

Norhayati Abdullah is an associate professor in environmental engineering at the Malaysia-Japan International Institute of Technology, Universiti Teknologi Malaysia (UTM) Kuala Lumpur. Recently conferred the MWA Malaysia Water Award for Women Leadership 2024, Norhayati is chair of the KURITA Asia Research Grant Malaysia and secretary of the ASEAN Learning Network. She was formerly the associate director of UTM International Kuala Lumpur. Norhayati is the recipient of the 2019/2020 ASEAN Science and Technology Fellowship, which focuses on strategic recommendations for water sector transformation. She represented UTM and Malaysia at the University of Michigan Ann Arbor for the Fulbright US-ASEAN Visiting Scholar Initiatives in 2016–2017. In 2018, she received the L'Oréal-UNESCO for Women in Science Fellowship. She was the board director of the

International Water Association (IWA) from 2014 to 2018, an IWA Fellow and a council member of the Malaysian Water Association (MWA). At the global level, Norhayati has been actively involved in the IWA program and activities since 2000. She received the IWA Young Water Professional (IWAYWP) Award in 2012 for outstanding wastewater research achievements and active involvement in various professional development programs. Norhayati is an advanced adventure diver who loves reading and enjoys outdoor activities.

Authors

Nina Artanti, Indonesia

Research Center for Chemistry, Research Organization of Nanotechnology and Material, National Research and Innovation Agency (BRIN), KST B.J. Habibie, Tangerang Selatan, Banten 15314, Indonesia

Nina Artanti completed her bachelor honor's degree in agriculture engineering from IPB University (Indonesia), graduate diploma and master's degree in biochemistry and molecular genetics from the University of New South Wales (Australia), and doctorate degree in bioscience and biotechnology from Ehime University (Japan). She has been a researcher at the Indonesian Institute of Sciences (LIPI) since 1988. From 2021 until now, she has worked at BRIN, a merger of LIPI and several research institutions. She works at the Organic Chemistry Research Group in the Research Center for Chemistry BRIN.

A. Aswathy, India

Department of Food Science and Technology, Faculty of Ocean Science and Technology, Kerala University of Fisheries and Ocean Studies, Panangad, Kochi, Kerala 682506, India

Aswathy completed her bachelor's degree in food science and quality control from NSS Hindu College, Changanacherry and her master's degree in food science and technology from Kerala University of Fisheries and Ocean Studies, Kochi. She has also been a member of the only team to represent India and win the 2nd Best Project award in an International Student Project Competition conducted by ISEKI-Food STa in 2021. She has worked as a QC Executive at Jeevan Milk and Milk Products, Kottayam, and as a QC technician at Erie Meats, Ontario, Canada. She is currently working as a quality assurance technician at SucroCan, Canada.

Bhavesh Choudhary, India

Department of Aquaculture, College of Fisheries, Central Agricultural University (Imphal), Agartala 799210, Tripura (West), India

Bhavesh Choudhary completed his BFSc (2016–2020) from College of Fisheries, MPUAT, Udaipur. After BFSc, Bhavesh succeeded in the ICAR-All India Entrance Examination for Admission (ICAR-AIEEA (PG) and completed MFSc (aquaculture) from the College of Fisheries, Shirgaon, Ratnagiri (DBSKKV, Dapoli) in 2022.

Currently, Bhavesh is working in the Department of Aquaculture, College of Fisheries, Central Agricultural University (Imphal). He has cleared the ASRB-NET, SMS(T-6) exam. Bhavesh is an editorial member and reviewer of several international peer-reviewed, high-impact factor journals. He has attended several national and international training, conferences, seminars, workshops, etc. He is also an active volunteer in National and International programmes.

Tjandra Chrismadha, Indonesia

Research Center for Limnology and Water Resources, Research Organization of Earth and Maritime, National Research and Innovation Agency-the Rep. of Indonesia (PRLSDA-BRIN), KST Soekarno-Cibinong, Bogor 16911, Indonesia

Tjandra Chrismadha is currently an associate research professor (Peneliti Ahli Utama/IVe) at the Research Center for Limnology and Water Resources, Research Organization of Earth and Maritime, National Research and Innovation Agency Rep. of Indonesia (PRLSDA-ORKM-BRIN). He received his MPhil degree from the Department of Biology, Murdoch University, Australia in 1996. He is the team member of the Photobioreactor developer group for the development of a photobioreactor system for generating microalgae biomass for particular purposes (RIIM-Cost Effective Photobioreactor system) from 2022 to 2024, and the carbon dioxide biosequestration system based on microalgae cultivation (FCB configuration series for CO_2 biosequestration photobioreactor system) from 2020 to date. His main research is in biology, integrated multi-trophic aquaculture, microalgae acclimatization process, and outdoor photobioreactor systems. He has published 51 scientific papers, four book chapters, and three patents. He has engaged with various national, regional, and international collaborations.

Suparna Deb, India

Department of Aquaculture, College of Fisheries, Central Agricultural University (Imphal), Agartala 799210, Tripura (West), India.

Suparna Deb is a dedicated PhD research scholar in the Department of Aquaculture, College of Fisheries, Central Agricultural University (Imphal), Tripura, India. She holds a postgraduate degree in aquaculture and an undergraduate degree from the College of Fisheries, Central Agricultural University, Imphal, Tripura, India. Miss Deb pursued knowledge and academic excellence and has been recognized with an institutional fellowship award for her PhD research. Miss Deb's work reflects patience in advancing aquaculture research and contributing to the global scientific community.

Mostafa El-Sheekh, Egypt

Tanta University, Tanta 31527, Egypt

Mostafa El-Sheekh is a professor of phycology at Tanta University, Egypt. He has nearly 40 years of experience in research and teaching in the field of algae, microbiology, and its applications. Dr El-Sheekh served as the vice dean and dean of the Faculty of Science at Tanta University, Egypt, and as a cultural counsellor at the Egyptian Embassy in Yemen and Uzbekistan. He also served as vice president of Tanta University for postgraduate studies and research. He holds a PhD in phycology from Tanta University and Göttingen University, Germany, as a Chanel System fellowship. He is the recipient of fellowships in Germany, Hungary, India, Sweden, the Czech Republic, and Japan. He is the author of 312 papers, 36 book chapters, and seven books. He also serves as an editorial member and reviewer for more than 115 peer-reviewed journals. He was one of the World's Top 2% of Scientists List published by Stanford University in 2020, 2021, 2022, 2023, and 2024. Professor El-Sheekh initiated a research school at Tanta, Damanhur, Benha, and South Valley Universities. He holds the State Encouragement Award in Biological Sciences, the Tanta University Encouragement Award in Basic Sciences, the Tanta University Appreciation Award, the Tanta University Award for Excellence, the Tanta University Award for the Highest Scientific Citation, and the Best Institutional Performance Award at Tanta University as Dean of the Faculty of Science. He constructed research labs in phycology and renewable energy in the Botany Department, Faculty of Science, Tanta University. He was awarded several research projects from the Egyptian Academy for Science and Technology, Science, Technology and Innovation Funding Authority (STDF), and Tanta University. He is now chief of the Standing Committee for the Promotion of Professors and Associate Professors in Botany and Microbiology at the Supreme Council of Universities, Egypt.

Gunawan Gunawan, Indonesia

Research Center for Limnology and Water Resources, Research Organization of Earth and Maritime, National Research and Innovation Agency-the Rep. of Indonesia (PRLSDA-BRIN), KST Soekarno-Cibinong, Bogor 16911, Indonesia

Gunawan Gunawan is currently a senior engineer (Perekayasa Madya/IVb) at the Research Center for Limnology and Water Resources, Research Organization of Earth and Maritime, National Research and Innovation Agency Rep. of Indonesia (PRLSDA-ORKM-BRIN). He is the team member of the Photobioreactor developer group for the development of a photobioreactor system for generating microalgae biomass for special purposes (RIIM-Cost Effective Photobioreactor system) from 2022 to 2024, and the carbon dioxide biosequestration system based on microalgae cultivation (FCB configuration series for CO_2 biosequestration photobioreactor system) from 2020 to date. His main research is in biology and integrated multi-trophic aquaculture. He has published three scientific papers, two book chapters, and five patents.

Ardiyan Harimawan, Indonesia

Department of Chemical Engineering, Faculty of Industrial Technology, Institut Teknologi Bandung, Jl. Ganesa 10, Bandung 40132, Indonesia

Ardiyan Harimawan is an Associate Professor in the Department of Chemical Engineering, Institut Teknologi Bandung (ITB), Bandung, Indonesia. He received his PhD degree from the Department of Chemical and Biomolecular Engineering, National University of Singapore (NUS) in 2012. His main research area is bio-electrochemistry, mainly in microbial adhesion, bio corrosion, bioreactors using microalgae, and microbial fuel cells. He has published 35 scientific papers, three book chapters, and three patents.

Nair Chithra Harinarayanan, India

Department of Food Science and Technology, Faculty of Ocean Science and Technology, Kerala University of Fisheries and Ocean Studies, Panangad, Kochi, Kerala 682506, India

Chithra Harinarayanan graduated with the highest marks from St Teresa's College (Autonomous), Ernakulam, with a bachelor's degree in food processing technology and completed her master's in food science and technology from Kerala University of Fisheries and Ocean Studies, Kochi. She also holds a bachelor's degree in business administration from Annamalai University, India. She has been a member of the only team to represent India and win the 2nd Best Project award in an International Student Project Competition conducted by ISEKI-Food STa in 2021. She has also worked as a Senior Research Fellow in ICAR-Krishi Vigyan Kendra, Kottayam, and as an R&D and QC Executive in The Betel Leaf Co., Bangalore, India. She has published four research papers and four book chapters. She is pursuing her doctoral degree in food process engineering from Gyeongsang National University, South Korea.

Lamiaa H. S. Hassan, Egypt

Scientific Researcher in Applied Phycology, Faculty of Science, Menoufia University, Egypt

Lamiaa Hassan is a microbiology scientific researcher with phycology expertise at the Faculty of Science, Menoufia University. She was awarded MSc and PhD degrees from the Department of Botany and Microbiology, Faculty of Science, Menoufia University. She has several studies on microalgae cultivation. She is interested in applied phycology and participates in some studies related to bioethanol production from microalgae, in addition to other research focused on the biosynthesis of nanoparticles from microalgae and cyanobacteria. She also shared some studies regarding wastewater treatment by microalgae.

Kousar Banu Hoorzook, South Africa

University of Johannesburg, Process Energy Environment Technology Station (UJ, PEETS), South Africa

Dr Kousar Banu Hoorzook completed her post-doctoral fellow (PDRF) at the University of Johannesburg (UJ) in the Water and Health Research Centre (WHRC). She then joined the UJ Process

Energy Environment Technology station (PEETS) as a program manager in the Water and Sanitation focus area. She obtained her master's in biotechnology and her doctoral degree in biomedical technology from the University of Johannesburg, where she specialized in microbial and molecular biology techniques for detecting bacterial pathogens in water. Her current work at PEETS focuses on improving the competitiveness of industry and SMEs through applying specialized knowledge and technology and facilitating the interaction between industry (especially SMEs) and academia to enable innovation and technology transfer to grow the green economy. She is currently rated with SA NRF as a Y2-rated researcher.

Pangaman Jiyane, South Africa

University of Johannesburg, Process Energy Environment Technology Station (UJ, PEETS), South Africa

Pangaman Jiyane is a PhD candidate at the University of Johannesburg. He is currently a project manager at UJ PEETS in the Water and Sanitation focus area. He obtained his master's degree in chemistry from the University of Johannesburg specializing in biological activities of naturally occurring compounds from African medicinal plants. His current work focuses on implementing specialized knowledge and technologies in SMEs, various industry companies and communities in water-related projects.

Helen Julian, Indonesia

Department of Chemical Engineering, Faculty of Industrial Technology, Institut Teknologi Bandung, Jl. Ganesa no. 10, Bandung 40132, Indonesia

Helen Julian is an assistant professor at Institut Teknologi Bandung (ITB), Indonesia. After completing her undergraduate and graduate degrees (chemical engineering from ITB) in 2010 and 2012, she received her PhD in chemical engineering from the University of New South Wales in 2018, where her research was coordinated in the UNESCO Centre for Membrane Science and Technology. Her main research interests are membrane distillation for food applications and zero liquid discharge desalination.

Prasad Kaparaju, Australia

School of Engineering and Built Environment, Griffith University (Nathan Campus), Queensland, Australia

Prasad Kaparaju is a professor of environmental engineering at Griffith University. He holds a PhD in environmental engineering and an MSc in agricultural engineering. Professor Kaparaju has extensive experience in renewable energy, bioenergy production, and environmental technologies. Before joining Griffith, he worked in various academic and research roles across Europe and Australia, contributing to several international projects on sustainable energy solutions and waste management technologies. He has published 104 research outputs, including 78 peer-reviewed papers, with over 5,700 citations and an H-index 36 on Google Scholar. Professor Kaparaju has secured $5.86 million in research funding and was recognized in Stanford's top 2% of scientists in energy and biotechnology disciplines. He is the first researcher to demonstrate biogas production and upgrading the biogas to compressed biomethane (bio-CNG) for vehicle fuel use in Australia.

Selina Akhter Laizu, Bangladesh

Institute of Forestry and Environmental Sciences, University of Chittagong, Chattogram-4331, Bangladesh

Selina Akhter Laizu is an environmental science graduate from the University of Chittagong. Her research experience includes projects on phytoplankton bioremediation and integrated environmental management for coastal zones. She has also worked as an environmental consultant, monitoring wastewater treatment processes and contributing to sustainability initiatives. She has presented research papers at several conferences and received prestigious fellowships, including the National Science and Technology Fellowship from Bangladesh's Ministry of Science and Technology.

Miratul Maghfiroh, Indonesia

Research Center for Limnology and Water Resources, Research Organization of Earth and Maritime, National Research and Innovation Agency-the Rep. of Indonesia (PRLSDA-BRIN), KST Soekarno-Cibinong, Bogor 16911, Indonesia

Miratul Maghfiroh is a junior researcher (2010–present) and a government official at the Research Center for Limnology and Water Resources, National Research and Innovation Agency (BRIN), Indonesia. She received a master's degree from Humboldt State University with biology as her major under the US Government's Fulbright Scholarship scheme (2013–2015). She obtained a doctoral degree (2019–2022) from the University of Science and Technology – KICT School, South Korea. Her research interests include water treatment, environmental microbiology, and environmental engineering.

Hakiki Melanie, Indonesia

Research Center for Chemistry Research Organization of Nanotechnology and Material, National Research and Innovation Agency (BRIN), KST B.J. Habibie, Tangerang Selatan, Banten 15314, Indonesia

Hakiki Melanie is a senior researcher in the Organic Chemistry Research Group at the Research Center for Chemistry BRIN. In 2019, she obtained her doctoral degree in the Life Science Innovation Program from the School of Integrative and Global Majors, University of Tsukuba, Japan. Her research interests include organic chemistry and natural products, involving the extraction of bioactive components, chemical characterization, and bioactivity analysis. Her current research focuses on a comprehensive study of functional chemical compounds from algae.

Eva Nafisyah, Indonesia

Research Center for Limnology and Water Resources, Research Organization of Earth and Maritime, National Research and Innovation Agency-the Rep. of Indonesia (PRLSDA-BRIN), KST Soekarno-Cibinong, Bogor 16911, Indonesia

Eva Nafisyah is currently a junior engineer (Perekayasa Muda/IIIb) at the Research Center for Limnology and Water Resources, Research Organization of Earth and Maritime, National Research and Innovation Agency Rep. of Indonesia (PRLSDA-ORKM-BRIN). She is the team member of the Photobioreactor developer group for the development of a photobioreactor system for generating microalgae biomass for special purposes (RIIM-Cost Effective Photobioreactor system) from 2022 to 2024, and the carbon dioxide biosequestration system based on microalgae cultivation (FCB configuration series for CO_2 biosequestration photobioreactor system) from 2020 to recent. Her main research is in analytical chemistry and aquatic organic constituent analysis. She has published three scientific papers, three book chapters, and one patent.

Nofdianto Nofdianto, Indonesia

Research Center for Limnology and Water Resources, Research Organization of Earth and Maritime, National Research and Innovation Agency-the Rep. of Indonesia (PRLSDA-BRIN), KST Soekarno-Cibinong, Bogor 16911, Indonesia

Nofdianto Nofdianto is currently a senior researcher (Peneliti Madya/IVa) at the Research Center for Limnology and Water Resources, Research Organization of Earth and Maritime, National Research and Innovation Agency Rep. of Indonesia (PRLSDA-ORKM-BRIN). He received his doctoral degree from the University of Toulouse III, France. He is a team member of the Photobioreactor developer group for the development of a photobioreactor system for generating microalgae biomass for special purposes (RIIM-Cost Effective Photobioreactor system) from 2022 to 2024. His main research is in biology and microalgal–periphyton system implementation in integrated multi-trophic aquaculture. He has published two scientific papers, two book chapters, and one patent.

Maya Raman, India

Department of Food Science and Technology, Faculty of Ocean Science and Technology, Kerala University of Fisheries and Ocean Studies, Panangad, Kochi, Kerala 682506, India

Maya Raman graduated in Botany and received her MSc (industrial fisheries) and PhD degree from the School of Industrial Fisheries, Cochin University of Science and Technology, India in 2006. Her doctoral area of study was in fish biochemistry and fish processing technology. She is an associate professor and head of the Food Science and Technology Department at Kerala University of Fisheries and Ocean Studies. She also worked as a postdoc at Applied Nutrition and Food Chemistry, Centre for Chemistry and Chemical Engineering, Lund University, with a post-doctoral scholarship from the Erasmus Mundus External Cooperation Windows Lot 15 Programme. She also worked as a woman scientist at the Indian Institute of Technology Madras with the Women Scientist Scheme (WOS-A) funding from the Department of Science and Technology, India. She was selected as the Young Student Speaker IFCON 2003 by the Association of Food Scientists and Technologists, India. She also worked at a faculty of the Department of Food Science and Technology, MACFAST, as a certified technologist (Export Inspection Authority, India) at Koluthara Exports Limited, and as a research officer at NGIL, India. She has 34 peer-reviewed publications and has published three books and 16 book chapters. She has 37 abstracts/extended abstracts presented at national and international symposia/conferences. Her areas of expertise are fish biochemistry (collagen), fish processing technology, nutrition and health (cancer), nanotechnology, bioactive marine compounds, dietary fibers (cereal, vegetable, marine algae) and colon cancer (in vitro and in vivo), molecular modelling and dynamics, gut simulation techniques, drug encapsulation, hydrogels, antimicrobial biodegradable food wraps. She is a life member of AFST(I), SOFT(I), SASNET, a member of ACS and of several other renowned organizations. She qualified for NET conducted by ASRB in 2001. A patent on hydrogel from cyclic beta-glucan for cosmetics/food applications was granted in 2022. She offers two online courses through NPTEL-SWAYAM.

Rosidah Rosidah, Indonesia

Research Center for Limnology and Water Resources, Research Organization of Earth and Maritime, National Research and Innovation Agency-the Rep. of Indonesia (PRLSDA-BRIN), KST Soekarno-Cibinong, Bogor 16911, Indonesia

Rosidah Rosidah is currently a junior engineer (Perekayasa Muda/IIId) at the Research Center for Limnology and Water Resources, Research Organization of Earth and Maritime, National Research and Innovation Agency Rep. of Indonesia (PRLSDA-ORKM-BRIN). She is the team member of the Photobioreactor developer group for the development of a photobioreactor system for generating microalgae biomass for particular purposes (RIIM-Cost Effective Photobioreactor system) from 2022 to 2024, and the carbon dioxide biosequestration system based on microalgae cultivation (FCB configuration series for CO_2 biosequestration photobioreactor system) from 2020 to recent. Her main research is in biochemistry, water quality analysis, and chemical statistics. She has published nine scientific papers, two book chapters, and one patent. She has engaged with various national and regional collaborations.

Nilay Kumar Sarker, Australia

School of Engineering and Built Environment, Griffith University (Nathan Campus), Queensland, Australia

Nilay has a PhD in environmental engineering from Griffith University, QLD, Australia, focusing on sustainable and environmental solutions. He holds a master's in energy technology from AIT, Bangkok, and a BSc in chemical engineering from BUET, Bangladesh. His interdisciplinary academic and research background allows him to approach challenges from unique perspectives, as reflected in his publication record. Nilay is currently engaged in algal research, particularly on natural pigments. He receives the prestigious GUPRS and GUIPRS scholarships.

Awalina Satya, Indonesia

Research Center for Limnology and Water Resources, Research Organization of Earth and Maritime, National Research and Innovation Agency-the Rep. of Indonesia (PRLSDA-BRIN), KST Soekarno-Cibinong, Bogor 16911, Indonesia

Awalina Satya is currently an associate research professor (Peneliti Ahli Utama/IVd) at the Research Center for Limnology and Water Resources, Research Organization of Earth and Maritime, National Research and Innovation Agency Rep. of Indonesia (PRLSDA-ORKM-BRIN). She received her doctoral degree in Chemical Engineering Faculty of Industrial Engineering, Institut Teknologi Bandung in 2018. She is the team leader of the photobioreactor developer group for the development photobioreactor system for generating microalgae biomass for particular purposes (RIIM-Cost Effective Photobioreactor system) from 2022 to 2024 and the carbon dioxide biosequestration system based on microalgae cultivation (FCB configuration series for CO_2 biosequestration photobioreactor system) from 2020 to 2027. Her primary research is in the application of bioprocess engineering in treating various industrial wastewater, biosorbent production, aquacultural engineering, and other microalga-based products. She has published 45 scientific papers, eight book chapters, and three patents. She has engaged with various national, regional, and international collaborations.

Azalea Dyah Maysarah Satya, Malaysia

Department of Chemical and Environmental Engineering, University of Nottingham Malaysia, Semenyih, Selangor Daryl Ehsan 43500, Malaysia

Azalea completed her bachelor's degree in chemistry from Universitas Indonesia with her final project on the green synthesis of a metal-organic framework system for adsorption and desorption of aspirin in 2021. She worked on a research project under a research group in the National Research and Innovation Agency in Indonesia (NRIA) for alternative food products, aiding in analyzing nutrients from different microalgae and cyanobacteria species in different medium compositions. She is aiming to complete her MPhil in engineering and engineering trades (chemical and environmental engineering) in feasibility assessment of nature-based solutions for water reuse and the potential role of microalgae/cyanobacteria in its application in rural Malaysia. She has published nine scientific papers and five book chapters. She hopes to continue research for a broader application in the region, using the current research.

Ika Atman Satya, Indonesia

Research Center for Computation, Research Organization of Electronics and Informatics, National Research and Innovation Agency-the Rep. of Indonesia (PRLSDA-BRIN), KST Soekarno-Cibinong, Bogor 16911, Indonesia

Ika Atman Satya is a senior engineer (Perekayasa Madya/IVb) at the Research Center for Computation, Research Organization of Electronics and Informatics, National Research and Innovation Agency

Rep. of Indonesia (PRK-OREM-BRIN). He is the team member of the Photobioreactor developer group for the development of a photobioreactor system for generating microalgae biomass for particular purposes (RIIM-Cost Effective Photobioreactor system) from 2022 to 2024 and the carbon dioxide bio sequestration system based on microalgae cultivation (FCB configuration series for CO_2 bio sequestration photobioreactor system) from 2020 to recent. His main research is in electronic engineering, the Internet of Things, real-time monitoring, and automatization of photobioreactor systems. He has published five scientific papers, four book chapters, and two patents. He has been engaged with various national, regional, and international collaborations.

Tjandra Setiadi, Indonesia

Department of Chemical Engineering, Faculty of Industrial Technology, Institut Teknologi Bandung, Jl. Ganesa 10, Bandung 40132, Indonesia

Tjandra Setiadi is a full professor in the Department of Chemical Engineering, Institut Teknologi Bandung (ITB), Bandung, Indonesia. He was the director of the Centre for Environmental Studies (PSLH) from 2007 to 2021 and the Centre for Resource Efficient and Cleaner Production Indonesia (CRECPI) from 2014 to 2018. His primary research is applying bioprocess engineering to treat various industrial wastewater, biopolymer production, lignocellulosic enzymes, and bio-based products. He has published 75 scientific papers and eight book chapters. He is a fellow of the Faculty of Engineering (2019), University of Tokyo. He has engaged with various national, regional, and international collaborations and assignments.

Sindhu Subramanian, India

Department of Chemical Engineering and Materials Science, Amrita School of Engineering, Coimbatore, Amrita Vishwa Vidyapeetham, India

Sindhu completed her BTech with distinction in chemical engineering and MTech in biopharmaceutical technology from Anna University, Chennai. She is working as an assistant professor in the Department of Chemical Engineering and Materials Science at Amrita School of Engineering, Coimbatore. She completed her doctoral degree in chemical engineering and materials science from Amrita School of Engineering, Coimbatore, Amrita Vishwa Vidyapeetham, India. She has published 10 journal articles and worked as a reviewer for many reputed journals.

Amala Tony, India

Department of Food Science and Technology, Faculty of Ocean Science and Technology, Kerala University of Fisheries and Ocean Studies, Panangad, Kochi, Kerala 682506, India

Amala completed bachelor's and master's degree in food science and technology. She further majored in Future Food Sustainability from Cranfield University, UK. She works as a microbiologist at South East Water Scientific Services, UK. She was honoured with the Douglas Bomford Trust award for best utilizing science and technology for sustainable agriculture in 2024 from Cranfield University UK for working on colorimetric ethanol sensors for food packaging. She also secured 1st rank for MSc in food science and technology from KUFOS, India. She won the 2nd Best Project award in an International Student Project Competition conducted by the ISEKI food association and European Union for preserving food biodiversity in 2021.

Foreword

In the face of escalating global environmental challenges, the need for innovative and sustainable water and energy frameworks has never been more pressing. In these frameworks, wastewater management as a critical component of modern infrastructure should be addressed as a sustainable nexus with other elements. Thus, the applications of nature-based solutions (NBS) are a forward-thinking approach to treating wastewater efficiently and align with broader ecological goals. Integrating natural processes not only encourages establishing resilient systems that protect ecosystems but also instils hope for a more sustainable and resilient future, while addressing the growing demand for clean water, waste recovery and renewable energy.

This book, *Nature-Based Solutions for Wastewater Treatment and Bioenergy Production*, arrives at a pivotal time. It explores innovative techniques that leverage natural mechanisms–such as microalgae treatment, aquaculture systems, and constructed wetlands–to manage wastewater efficiently. These approaches highlight the potential for developing sustainable wastewater systems that work in harmony with the environment rather than against it.

Beyond wastewater treatment, this volume sheds light on another significant benefit of NBS–bioenergy generation. The world's transition to cleaner, more sustainable energy sources is essential, and NBS hold great promise in this area. By transforming waste into energy, these technologies open new doors for sustainable development, creating circular economies where resources are reused and environmental impact is minimized.

The breadth of case studies and expert contributions presented in this book underscores the global applicability of NBS. From low-income communities to industrialized regions, these approaches can be tailored to meet local needs while providing scalable models for a more sustainable future. The combination of technical insights and real-world applications is a vital resource for practitioners, policymakers, and researchers alike, offering a roadmap toward a more sustainable, resilient future.

I congratulate the authors for their dedication in assembling this valuable resource. Their work demonstrates that when properly harnessed, natural resources offer some of the most efficient, sustainable solutions to the world's pressing environmental challenges. I hope this book will inspire and guide readers in rethinking how we manage water, energy, and waste in innovative and environmentally sound ways.

Zaini Ujang, PhD, PE (Malaysia), CEng (UK)
Vice-Chancellor and Professor of Environmental Sustainability
Universiti Teknologi Malaysia (2008–2013)

Preface

Our world is currently grappling with a pressing issue: The escalating need for clean water and sustainable energy, coupled with the imperative to safeguard and rehabilitate our natural ecosystems. This book, *Nature-Based Solutions for Wastewater Treatment and Bioenergy Production*, is a response to this critical challenge. In the wake of rapid technological progress over the past few decades, it has become increasingly clear that nature harbors the solutions to some of our most intricate problems.

Within the pages of this book, we have assembled a wealth of pioneering research and practical insights that illuminate the transformative potential of NBS in the fields of wastewater treatment and bioenergy production. Drawing from a diverse array of ecosystems and resources, the chapters in this volume not only delve into the theoretical underpinnings but also showcase real-world success stories of sustainable approaches. Each chapter offers a unique perspective on wastewater treatment, be it the functional dynamics of wetlands, bio-aided water filtration, or the potential of photobioreactors powered by microalgae.

The concept behind this book is straightforward: natural systems have been efficiently treating water and generating energy for billions of years. From wetlands serving as biofilters to the utilization of forest resources for bioenergy, these methods operate in perfect harmony with our environment, offering cleaner, more resilient solutions. Biodiesel and bioenergy generation, coupled with advanced strategies like integrated multi-trophic aquaculture, demonstrate how nature-inspired approaches can reduce environmental impact while creating sustainable energy sources.

Our book also tackles the contemporary challenges that we face, such as the current state of wastewater treatment plants in Gauteng and the urgent issue of effluent discharge quality into water bodies. By presenting case studies from various parts of the world, including the bioenergy potential of forests, this book underscores the interconnectedness of ecosystems and the pivotal role they can play in alleviating the environmental strains caused by industrial and human activities.

As we confront the twin crises of climate change and resource depletion, *Nature-Based Technologies for Wastewater Treatment and Bioenergy Production* serves as a timely reminder that many of the solutions to our challenges are rooted in the natural systems that have evolved over time. This book is not only for academics and practitioners but also for policymakers, environmentalists, and anyone with a stake in building a sustainable future. I hope the ideas and solutions presented here will inspire innovative applications and further research into NBS as we strive toward a cleaner, greener world.

Imran Ahmad

Acknowledgments

First and foremost, we extend our deepest gratitude to Allah, the Almighty, for His divine guidance and blessings, which have been instrumental in preparing this book. The strength, knowledge, and perseverance he has given us have been our constant companions throughout this journey, and without His mercy, this work would not have been possible.

We would also like to sincerely thank IWA Publishing for providing us with the platform to present this book to a global audience. Their support and commitment to advancing knowledge in water and environmental sciences have been invaluable in bringing this project to fruition.

A special thanks goes to the esteemed reviewers and authors whose insightful feedback and contributions have been instrumental in shaping this book. Their expertise and dedication have significantly enriched its content, ensuring its relevance and quality for readers across various disciplines.

We hope this book will serve as a beacon of hope and inspiration for researchers, academics, and policymakers in the field of environmental sustainability. By addressing the need for nature-based technologies, we believe this work can help fill the existing knowledge gap and inspire innovative solutions for wastewater treatment and bioenergy production, paving the way for a more sustainable future.

Lastly, we thank our family for their unwavering support throughout this journey. Their love, patience, and encouragement have been our most significant source of strength and motivation. This book would not have been possible without their constant belief in our work.

Imran Ahmad
Norhayati Abdullah

doi: 10.2166/9781789064100_0001

Chapter 1

Introduction to nature-based solutions for wastewater treatment

I. Ahmad

ProalgaeTech, Schwabbacher Str. 15, 70437 Stuttgart, Germany
Corresponding author: mustafwibinqamar@gmail.com

ABSTRACT

Nature-based solutions (NBS) for wastewater treatment present an environmentally sustainable and cost-effective alternative to conventional treatment systems. These approaches harness natural processes involving plants, microorganisms, and soil systems to treat and purify wastewater, reducing the reliance on energy-intensive and chemical-based methods. Wetlands, biofilters, and vegetated swales are examples of NBS that utilize natural filtration, adsorption, and microbial degradation to remove contaminants from wastewater, such as nutrients, heavy metals, and pathogens. In addition to wastewater purification, these solutions offer co-benefits, including biodiversity enhancement, carbon sequestration, and improved water security. NBS are particularly advantageous in decentralized and rural contexts where traditional infrastructure may be limited, promoting resilience against climate change and providing ecosystem services. However, the effectiveness of NBS can vary based on site-specific factors, such as climate, pollutant load, and land availability, necessitating careful design and monitoring. This chapter will briefly introduce the NBS and its advantages and challenges, which provide a platform for the upcoming chapters incorporating various technologies. Future research is needed to optimize these systems, quantify long-term benefits, and scale their application in urban settings. Overall, NBS represent a promising and versatile approach to achieving sustainable water management and addressing global wastewater challenges.

Keywords: Nature-based solutions, wastewater treatment, pollutants, sustainability

1.1 INTRODUCTION

Globally, 2.4 billion individuals lack improved sanitation (i.e., facilities that hygienically isolate human excreta from human touch). At the same time, an additional 2.1 billion have inadequate sanitation (i.e., systems where wastewater discharges directly into surface waterways). Notwithstanding advancements in recent decades, the hazardous management of faecal waste and wastewater remains a significant threat to public health and the environment. The United Nations World Water Assessment Programme estimates that 80% of wastewater is released untreated (Jones *et al.*, 2021).

There is a growing interest in cost-effective treatment techniques that harness natural systems. However, the effective integration of conventional infrastructure, such as wastewater treatment facilities, with natural alternatives like wetlands, is a challenge that often falls on the shoulders of wastewater utility managers. This chapter delves into the potential of nature-based solutions (NBS) for wastewater treatment and its societal co-benefits. The International Union for the Conservation of Nature defines NBS as 'actions to protect, sustainably manage, and restore natural or modified ecosystems that effectively and adaptively address societal challenges while simultaneously providing benefits for human well-being and biodiversity' (Cohen-Shacham *et al.*, 2016).

NBS for wastewater treatment offer an innovative and eco-friendly alternative to traditional methods. These approaches rely on natural systems such as wetlands, biofilters, and microbial processes to treat and remediate wastewater. These approaches reduce chemical and energy input and provide co-benefits such as biodiversity enhancement and ecosystem restoration. Recent studies have demonstrated the effectiveness of constructed wetlands in removing contaminants, including antibiotics and antimicrobial resistance genes (ARGs), with success rates significantly higher than conventional methods like UV and chlorination treatments (Pastor-Lopez *et al.*, 2024).

Furthermore, advancements in constructed wetland designs have shown promising results in meeting water reuse standards and reducing antimicrobial resistance, making them a sustainable solution to growing water scarcity (Gentile *et al.*, 2024). The development of modelling frameworks for NBS systems also highlights the potential for further optimization, offering guidance on their design and implementation for improved performance (Dehghani Tafti *et al.*, 2024).

Microalgae-based solutions for wastewater treatment provide an effective and sustainable approach to removing pollutants such as nitrogen, phosphorus, heavy metals, and organic compounds. Microalgae not only help treat wastewater but also offer the potential for biomass production, which can be utilized in agriculture, biofuels, and other industries (Ahmad *et al.*, 2021). The use of microalgae in wastewater treatment has been demonstrated to reduce pollutants without the emission of harmful by-products like methane, which is often associated with anaerobic treatments (Kaparapu & Krishna Prasad, 2024). Additionally, microalgae can bioremediate contaminants while producing high-value biomass, making the process economically viable (Nayana *et al.*, 2024). A UK pilot-scale study has shown promising results, particularly in removing phosphates and emerging contaminants from wastewater using microalgal systems integrated into existing infrastructure for enhanced efficiency (Ekins-Coward *et al.*, 2024). Despite these benefits, challenges such as low biomass density and the need for improved separation techniques remain. Advances in microalgae biogranulation show potential for overcoming these issues, creating denser, more easily harvested biomass that enhances wastewater treatment effectiveness (Ahmad *et al.*, 2024).

Overall, NBS presents a viable option for addressing the global wastewater crisis while promoting environmental sustainability and resilience. This introductory chapter gives a brief overview of the advantages and impacts of some of the NBS.

1.2 UTILIZATION OF NBS FOR WASTEWATER TREATMENT

NBS can be utilized in a constructed or grey infrastructure wastewater treatment system and is applicable for several wastewater types, including municipal, agricultural, and industrial wastewater, leachates, and stormwater. The application of NBS in wastewater treatment seeks to create engineered systems that replicate and utilize effective ecosystems while minimizing reliance on mechanical components.

NBS employs flora, soil, permeable substrates, bacteria, and various natural elements and processes to eliminate contaminants in wastewater, including suspended particles, organic matter, nitrogen, phosphorus, and pathogens (Kadlec & Wallace, 2008). NBS possesses the capability to eliminate new pollutants, including steroid hormones and biocides (Chen *et al.*, 2019), personal care products (Ilyas & van Hullebusch, 2020), and insecticides (Vymazal & Březinová, 2015).

Various forms of NBS can be integrated to attain the required treatment efficacy. The application of NBS in wastewater treatment can foster healthier ecosystems by enhancing water quality and improving the natural environment and adjacent habitats. Natural spaces and NBS can foster physical and emotional well-being, improve air and water quality, and contribute to overall human health. Moreover, NBS can enhance aesthetic value and offer healing benefits, fostering social cohesion

Nature-Based Solutions for Wastewater Treatment

Advantages

- Reliable process with quality effluent
- Can be used in different climates and site locations
- Ease of construction: local materials and plants can be used
- Reduced operational, labour, chemical and energy requirements compared with conventional treatment
- Can be applied for decentralised treatment
- Sustainable and environmentally friendly & Diverse microbial communities
- Multi-purpose functionality, which can reduce the impacts of water scarcity

Challenges

- Multi-stage and hybrid schemes can be required to fulfil stringent limits on nutrient removal
- High area demand compared with conventional technological solutions
- Proper operation and maintenance also of the primary treatment step (regular removal of settled sludge)
- Lack of standard guidelines on design and sizing for recently developed types of NBS
- Require accurate design according to local conditions
- Accumulation of phosphorus and metals in soil or other compartments of NBS

Figure 1.1 Advantages and challenges of using NBS for wastewater treatment.

and reinforcing communal bonds. The economic advantages encompass diminished water treatment expenses, decreased flood damage charges, enhanced fishery health, improved recreational prospects, and augmented tourist and economic growth. A comprehensive cost–benefit analysis is required to incorporate such benefits when evaluating NBS solutions (Connor, 2015).

Investing in NBS can assist wastewater treatment operators in reducing operational expenses, accessing new revenue sources, enhancing customer engagement, and delivering public environmental goods and services (European Investment Bank, 2019). Operational and maintenance expenses and initial expenditures are frequently reduced compared to typical activated sludge (CAS) systems, contingent upon land costs, used technology, and resource availability (Vymazal, 2010). Some of the advantages and challenges of using NBS for wastewater treatment are summarized in Figure 1.1.

1.3 CHRONICLE OF NBS UTILIZATION FOR WASTEWATER TREATMENT

NBS has historically facilitated wastewater treatment; ancient Egyptian and Chinese civilizations utilized wetlands for wastewater disposal (Brix, 1995). Wastewater was directly released into surface water, accumulating biosolids and nutrients, facilitating the formation of natural wetlands and subsequently leading to flora proliferation. The wastewater would undergo natural treatment, and the ecology was preserved despite a modest discharge volume (Brix, 1995).

As people began to increase, contamination of ecosystems, including aquatic environments, also escalated. Technologies were developed to address excessive pollution levels without harming aquatic ecosystems. This resulted in the proliferation of conventional wastewater treatment facilities, which employ physical, chemical, and biological processes to eliminate particles, organic matter, and, where necessary, nutrients from wastewater. Since the 1950s, NBS, including treatment wetlands (TWs), have developed into a dependable wastewater treatment technology capable of processing substantial wastewater volumes to achieve the requisite effluent quality while preserving the surrounding environment (Vymazal, 2010).

This is accomplished by managing several elements of the treatment wetlands, including macrophytes and soil components in surface flow systems or using appropriately selected infill mediums such as sand and gravel in subsurface flow wetlands. Comparable considerations apply to waste stabilization ponds, which have been extensively utilized in developing nations. Treatment wetlands and ponds are currently regarded as appropriate NBS for wastewater treatment and eliminating hazardous pathogens, serving as effective alternatives to conventional technology methods.

Innovative methodologies for using NBS in wastewater treatment are expanding. Living walls and green roofs recycle greywater for applications such as toilet usage or landscape irrigation while providing additional benefits of cooling, air filtration, and enhancing urban aesthetics (Boano *et al.*, 2020; Pradhan *et al.*, 2019). Another example is willow systems, which utilize wastewater for irrigation and provide woody biomass for various purposes, including firewood for local heating, soil amendment, landscaping, and branches for riverbank stabilization. This technique is called zero discharge, as all effluent is evaporated or utilized for plant growth. These examples illustrate how NBS for wastewater treatment can integrate into a circular economy framework that seeks to eradicate waste and promote the perpetual utilization of resources (Masi *et al.*, 2018; Nika *et al.*, 2020).

1.4 IMPACT ON THE UN SUSTAINABLE DEVELOPMENT GOALS

NBS are increasingly recognized as innovative approaches to address water-related risks, playing a significant role in the 2030 Agenda for Sustainable Development. They offer many benefits, such as enhancing human health and livelihoods, ensuring food and energy security, promoting sustainable economic growth, and aiding in ecosystem rehabilitation (Martin *et al.*, 2020). Various services offered by NBS can aid in reaching diverse sustainable development goal (SDG) targets, such as minimizing greenhouse gases and environmental toxins, preserving stable groundwater levels, and contributing to planetary cooling (Seifollahi-Aghmiuni *et al.*, 2019).

The connection between NBS for wastewater treatment and SDG 6 on Clean Water and Sanitation is evident. The advantages provided by NBS can differ across different spatial and temporal dimensions and among various societal groups. This indicates that the role of these solutions in contributing to diverse SDGs will be specific to the context (Martin *et al.*, 2020). Wetlands have a significant impact on ecosystem processes that relate to multiple SDGs, including those addressing poverty, hunger, clean water and sanitation, responsible production and consumption, and climate action, along with their specific targets (Seifollahi-Aghmiuni *et al.*, 2019). The location and application of the NBS may yield contributions to various SDGs, including SDG 3 (Good Health and Well-being), SDG 7 (Affordable and Clean Energy), SDG 11 (Sustainable Cities and Communities), SDG 14 (Life Below Water), and SDG 15 (Life on Land) (Seifollahi-Aghmiuni *et al.*, 2019).

1.5 CONCLUSION

NBS for wastewater treatment represent a transformative approach that merges ecological sustainability with effective pollutant removal. By mimicking and enhancing natural processes, NBS, including wetlands, biofilters, and microalgae-based systems, offer a multi-functional alternative to traditional chemical and mechanical treatment methods. This chapter highlighted the efficacy of these systems in removing contaminants such as nitrogen, phosphorus, heavy metals, and pathogens and mitigating the risks of emerging pollutants like antimicrobial resistance genes (ARGs), personal care products, and insecticides.

Microalgae, in particular, play a critical role in tertiary wastewater treatment by efficiently absorbing nutrients and producing biomass that can be repurposed for bioenergy and agricultural applications, thus contributing to circular economy principles. Their capacity for bioremediation, without producing harmful by-products, makes them a promising candidate for sustainable wastewater management. However, as demonstrated, implementing NBS is contingent upon site-specific factors such as climate, pollutant load, and available land, necessitating careful system design and optimization. Advances in microalgal biogranulation and engineered wetland configurations address scalability and operational challenges, making these systems more feasible for large-scale applications, particularly in decentralized and rural settings. Importantly, NBS also contribute to broader environmental goals by enhancing ecosystem services, promoting biodiversity, and aligning with key United Nations SDGs, especially in areas such as clean water and sanitation, climate action, and sustainable cities.

Integrating NBS into existing grey infrastructure will require continued research, particularly in developing models to predict long-term benefits and optimize performance. By fostering collaborations between ecological, technological, and social dimensions, NBS, including microalgae, can be pivotal in addressing global wastewater challenges while promoting environmental and economic resilience.

REFERENCES

Ahmad I., Abdullah N., Koji I., Yuzir A. and Mohamad S. E. (2021). Potential of microalgae in bioremediation of wastewater. *Bulletin of Chemical Reaction Engineering & Catalysis*, **16**(2), 413–429, https://doi.org/10.9767/bcrec.16.2.10616.413-429

Ahmad K., Faraheen S., Kanadasan G., Lee K. T. and Vadivelu V. M. (2024). Insight into recent advances in microalgae biogranulation in wastewater treatment. *Critical Reviews in Biotechnology*, **44**(8), 1–16, https://doi.org/10.1080/07388551.2024.2317785

Boano F., Caruso A., Costamagna E., Ridolfi L., Fiore S., Demichelis F., Galvão A., Pisoeiro J., Rizzo A. and Masi F. (2020). A review of nature-based solutions for greywater treatment: Applications, hydraulic design, and environmental benefits. *Science of the Total Environment*, **711**, 134731, https://doi.org/10.1016/j.scitotenv.2019.134731

Brix H. (1995). Treatment wetlands: an overview. Oczyszczalnie hydrobotaniczne. *Miedzynarodowa Konferencja Naukowo-Techniczna, Gdansk*, **167**, 176.

Chen J., Liu Y. S., Deng W. J. and Ying G. G. (2019). Removal of steroid hormones and biocides from rural wastewater by an integrated constructed wetland. *Science of the Total Environment*, **660**, 358–365.

Cohen-Shacham E., Walters G., Janzen C. and Maginnis S. (2016). Nature-based solutions to address global societal challenges. *IUCN: Gland, Switzerland*, **97**, 2016–2036.

Connor R. (2015). The United Nations World Water Development Report 2015: Water for A Sustainable World. UNESCO Publishing, Paris, Vol. **1**.

Dehghani Tafti A., Houweling D., Perron J.-M., Bencsik D., Johnson T., Vanrolleghem P. A. and Comeau Y. (2024). Towards a modelling framework for nature-based solutions in wastewater treatment. *Water Science & Technology*, **90**(3), 758–776, https://doi.org/10.2166/wst.2024.235

Ekins-Coward T., Mei Ho F., Whittle I. and Viñuela R. (2024). Microalgae as a multibenefit natural solution for the wastewater industry: A UK pilot-scale study. *Water and Environment Journal*, **38**(4), 628–639.

European Investment Bank (Ed.). (2019). Climate Solutions. European Investment Bank.

Gentile A., Piccolo P., Iannece P., Cicatelli A., Castiglione S. and Guarino F. (2024). Reduction of antimicrobial resistance: Advancements in nature-based wastewater treatment. *Journal of Hazardous Materials*, **471**, 134330, https://doi.org/10.1016/j.jhazmat.2024.134330

Ilyas H. and van Hullebusch E. D. (2020). Performance comparison of different constructed wetlands designs for the removal of personal care products. *International Journal of Environmental Research and Public Health*, **17**(9), 3091.

Jones E. R., Van Vliet M. T., Qadir M., and Bierkens M. F. (2021). Country-level and gridded estimates of wastewater production, collection, treatment and reuse. *Earth System Science Data*, **13**(2), 237–254.

Kadlec R. H. and Wallace S. (2008). Treatment Wetlands. CRC Press, Boca Raton.

Kaparapu J. and Krishna Prasad M. (2024). Microalgae for wastewater treatment: A review. *Algal Biotechnology*, 107–124, https://doi.org/10.1201/9781003219194-8

Martin E. G., Costa M. M. and Máñez K. S. (2020). An operationalized classification of Nature Based Solutions for water-related hazards: From theory to practice. *Ecological Economics*, **167**, 106460.

Masi F., Rizzo A. and Regelsberger M. (2018). The role of constructed wetlands in a new circular economy, resource oriented, and ecosystem services paradigm. *Journal of Environmental Management*, **216**, 275–284, https://doi.org/10.1016/j.jenvman.2017.11.086

Nayana K., Vidya D., Srinivasa Rao A., Periyasamy C., Suresh Kumar K., Arun Kumar K. and Subba Rao P. V. (2024). Valorization of wastes using microalgae: A glance. In: Algal Biotechnology, CRC Press, Boca Raton, pp. 173–180.

Nika C. E., Gusmaroli L., Ghafourian M., Atanasova N., Buttiglieri G. and Katsou E. (2020). Nature-based solutions as enablers of circularity in water systems: A review on assessment methodologies, tools and indicators. *Water Research*, **183**, 115988, https://doi.org/10.1016/j.watres.2020.115988

Pastor-Lopez E. J., Casas M. E., Hellman D., Müller J. A. and Matamoros V. (2024). Nature-based solutions for antibiotics and antimicrobial resistance removal in tertiary wastewater treatment: Microbiological composition and risk assessment. *Water Research*, **261**, 122038, https://doi.org/10.1016/j.watres.2024.122038

Pradhan S., Al-Ghamdi S. G. and Mackey H. R. (2019). Greywater recycling in buildings using living walls and green roofs: A review of the applicability and challenges. *Science of The Total Environment*, **652**, 330–344.

Seifollahi-Aghmiuni S., Nockrach M. and Kalantari Z. (2019). The potential of wetlands in achieving the sustainable development goals of the 2030 agenda. *Water*, **11**(3), 609, https://doi.org/10.3390/w11030609

Vymazal J. (2010). Constructed wetlands for wastewater treatment. *Water*, **2**(3), 530–549.

Vymazal J. (2011). Enhancing ecosystem services on the landscape with created, constructed and restored wetlands. *Ecological Engineering*, **37**(1), 1–5.

Vymazal J. and Březinová T. (2015). The use of constructed wetlands for removal of pesticides from agricultural runoff and drainage: a review. *Environment International*, **75**, 11–20, https://doi.org/10.1016/j.envint.2014.10.026

doi: 10.2166/9781789064100_0007

Chapter 2

Bio-aided water filtration: an approach to food processing by-product utilization

Nair Chithra Harinarayanan, Maya Raman, Amala Tony and A. Aswathy*

Department of Food Science and Technology, Faculty of Ocean Science and Technology, Kerala University of Fisheries and Ocean Studies, Panangad, Kochi 682506, Kerala, India

**Corresponding author: Dr. Maya Raman, Associate Professor & Head, Department of Food Science and Technology, Faculty of Ocean Science and Technology, Kerala University of Fisheries and Ocean Studies, Kochi. E-mail address: mayaraman@kufos.ac.in (M. Raman)*

ABSTRACT

In the wake of industrialization, urbanization, and population growth, the scarcity of readily available freshwater resources has become a great concern and there is an urgent need for energy-efficient, cost-effective, and chemical-free water purification systems. The central treatment plants are considerably expensive and inefficient in certain cases; therefore, low-income countries prefer onsite filtration systems. Chemocoagulants such as aluminium sulphate and ferric chloride work well in water treatment, but they usually require long mining and processing of raw materials and expensive disposal of sludge; further, these may be toxic and could be associated with various health issues, and, thus, cannot be used indefinitely. Coagulation with natural coagulants is a sustainable alternative, which is readily available, economical, convenient, atoxic, less sludge generation, eco-friendly, and effective. In addition to removing contaminants, natural materials can also add nutrients to the water. Hence, these could be used as an alternative for water treatment assuring health security and sustainability. However, natural coagulants are still inadequately accepted; are not commercialized, and are widely applied throughout the world, despite their known benefits. In order to promote a circular economy, more intensive research is needed on natural coagulants to determine their mechanism of action, how they are adopted, and how they can be commercialized.

Keywords: Bio-aided filtration, by-products, underutilized, valorizing

2.1 INTRODUCTION

Water pollution is a rising concern these days and has attracted a lot of attention from various groups. Dumping, poor waste management, and so on, which are some of the contributing elements to this crisis. Providing safe water as an industry has boomed. Though Earth is covered by three-fourths of water, it is estimated that only 3% of the world is equipped with fresh water from which 2.5% is unavailable as it is locked up in glaciers, ice caps, the atmosphere, and in the soil. India has only 4% of the world's freshwater but has about 16–17% of the world's population (ide-india.org). As an agrarian nation, the majority of India's water reserves are used for irrigation (78%), while the domestic and industrial utilization accounts for 6% and 5%, respectively (Press Information Bureau, 2013). Approximately half of the Indian population does not have access to safe drinking water because of chemical contamination from industrial and farming effluents with arsenic and fluoride (UNICEF, 2019). In developed countries, chemical contamination is the most significant cause of water pollution, while in developing countries, including India, agricultural discharges are the most significant (Sharma & Bhattacharya, 2017). As a result, the safety and quality of water is becoming a greater concern since it is linked to several diseases such as neuronal damage, cancer, diabetes, cardiovascular diseases, kidney injuries, and many infectious diseases such as typhoid, hepatitis, and cholera (Rehman *et al.*, 2018; Shabiimam *et al.*, 2018). Having clean, safe, and fresh drinking water is fundamental to human and animal health, and water filtration may be one of the best techniques for achieving this goal.

Several water filtration techniques were explored. Water clarification using conventional coagulants has become an increasingly difficult task due to quality, reliability, economy, and sustainability demands. Historically, sand filtration and rapid sand filtration were the most common filtration methods; however, with the advancement in technology, many alternative methods have been developed such as bank filtration, membrane filtration, disinfection, and solar purification. There are several techniques for treating water that have become popular in recent years, including reverse osmosis (RO), ultraviolet radiation, distillation, microporous filtration, and ion exchange (Khulbe *et al.*, 2012). However, some of these techniques have limitations such as retaining toxic residues, altering the taste, larger residence time, being expensive, and so on. The most basic is the use of alum, which though easy to use, has several drawbacks such as a high carbon footprint (He *et al.*, 2013), improper disposal of raw alum sludge leading to environmental pollution (Gutierrez *et al.*, 1998), impacts of bauxite mining on health and the environment (Abdullah *et al.*, 2016), and so on. Moreover, trihalomethanes (THMs) are produced when chlorine combines with synthetic organic matter and other natural organic ingredients found in water, and these possess carcinogenic properties. In recent years, increased urbanization and human interventions have placed tremendous pressure on freshwater resources. As a result, new resources and sustainable alternatives are given thrust and repurifying and recycling water is gaining momentum. The reuse of greywater would result in a significant amount of wastewater being transformed into valuable resources (Friedler & Hadari, 2006). If treated, this water can be reused for other purposes than drinking, for example, gardening, flushing toilets, and irrigation. As a result of the high cost and insufficient supply of various components used in filtration systems, natural substitutes are explored, chiefly the underutilized or by-products of agriculture or manufacturing processes, available in abundance. Hence, the use of natural substitutes derived from plants and animal resources could be a significant step. In addition to being excellent coagulants, flocculants, and antimicrobials, these by-products are underutilized and may have least impact on the environment (Kurniawan *et al.*, 2020).

To make water suitable for various applications, it needs to be treated properly. Also, the parameters indicating water quality can differ from place to place depending on the source and the treatment provided. In terms of public health, the quality of water is determined by three major parameters: physical, chemical, and microbiological. In a nutshell, the drinking water should be:

(1) Clear
(2) Tasteless or odourless
(3) Pathogen-free
(4) Free from other harmful compounds like heavy metals, and
(5) Harmful chemicals that may corrode the materials being washed in it.

So, in order to ensure safe water for the population, detailed water quality standards have been proposed by different countries and international organizations such as European Economic Community (EEC), World Health Organisation (WHO), and so on.

2.2 TYPES OF POLLUTANTS SEEN IN WATER

Water pollutants are generally classified into three categories:

(a) Based on source of water: (1) surface water pollution, (2) ocean water pollution, (3) underground water pollution.
(b) Based on nature of pollutants: (1) natural sources (land destruction, gas, etc.), (2) effluents (sewage water, oil, etc.), (3) anthropogenic source (domestic, industries, agriculture, etc.).
(c) Based on type of pollution: (1) physical, (2) chemical, and (3) biological.

The classification of pollutants based on the type is more commonly followed and gives a better understanding of the pollutants. Majority of the water pollutants are biological (bacteria, viruses), inorganic (chemical discharge, fertilizers) and organic (hydrocarbon) in nature (Hesham *et al.*, 2021).

2.2.1 Physical, chemical, and biological pollutants

The physical, chemical, and biological pollutants can be broadly classified into the following:

- Organic and inorganic materials – drugs, hair, pharmaceuticals, human faeces, food waste, sand, tier, metal, rubber and so on.
- Different pollutant paints, adhesives, gases, and pesticides.
- Macro solids – sanitary napkins, diapers, dead organic substances
- Heavy metals
- Pathogenic microorganisms – *Salmonella, Shigella, Vibrio cholerae*, viruses, and so on (Godbole *et al.*, 2023).

Water contaminants generally in the range of less than 0.001 to over 100 μm are of great interest in research. Those of size smaller than 0.1 μm are generally viruses, macromolecules (proteins, lipids, nucleic acids), cell debris, and so on, which are present in water supplies and water infiltration. Those smaller than 10^3 amu are chlorophyll, vitamins, fatty acids, and so on. These contaminants are separated into four categories based on size: settleable (>100 μm), supracolloidal (1–100), colloidal (0.08–1.00) and soluble (<0.08) (Levine *et al.*, 1985). Micro- (measuring <5 mm) and nanoplastics (<0.1 μm) are yet another group of pollutants that are dominating the pollutants due to their enormous impact. The plastics discarded are broken down to micro- to nano-sizes leading to toxicity to the environment and humans. The plastics find entry into the human body through various means and pose as threat to health. Several recent studies are undertaken to understand their breakdown pathway and behaviour in animals, humans, and environment (Yee *et al.*, 2021). Hence, it is also a serious concern and need to be addressed at the earliest.

Human interventions and increasing urbanization have created immense pressure on freshwater resources; henceforth, it is necessary to identify new alternatives and other techniques that will emphasize on three important criteria such as, reduce, reuse, and recycle. One such alternative is to recyle and reuse grey water. Greywater is the water collected from the household activities such as, dishwashing, laundry, and kitchen sinks, except from toilet water. Approximately 75% of domestic wastewater is disposed; with high-income countries and EU countries producing up to 150 L per person per day and low-income countries produce smaller volume but considerably in same proportion (Ghaitidak & Yadav, 2013). Grey water normally contains wide range of soluble and insoluble materials such as hair, saliva, skin, detergent, oil, grease, fats, and so on. There has been some evidence in recent research that greywater also contain pharmaceuticals, health and beauty products, aerosols, pigments, (Eriksson *et al.*, 2003); and toxic heavy metals such as, lead, nickel, cadmium, copper, and chromium (Aonghusa & Gray 2002; Eriksson *et al.*, 2010).

The physical constituents of grey water mainly include turbidity, temperature, suspended solids and electrical conductivity. The temperature of greywater is usually in the range of 18–35°C (Oteng-Peprah *et al.*, 2018). This high temperature may even favour microbiological growth. The total soluble solids (TSS) of greywater can range between 190 and 537 mg/l and is mainly high in water from the kitchen and laundry (Edwin *et al.*, 2014; Oteng-Peprah *et al.*, 2018); whereas the electrical conductivity ranges between 14 and 3000 μS/cm (Ciabattia *et al.*, 2009; Prathapar *et al.*, 2005). The major chemical contaminants found in greywater usually are surfactants that are used for washing and cleaning activities. Other constituents such as nitrates, phosphates, and sodium, are also present in appreciable amounts. The biodegradability of greywater is determined by the ratio of biological oxygen demand (BOD)/chemical oxygen demand (COD). And in conventional greywater, it can be observed that COD has slight dominance over BOD. To treat greywater, various physical, chemical, and biological methods are currently used; though these have limitations, these are still preferred. The conventional processes include:

(a) *Physical treatment*:
  (1) *Equalization*: A properly equalized industrial effluent can give a more consistent and homogenous flow to the flotation machine. It also lowers hydraulic surging which can harm the system's performance. In some circumstances, an equalization tank can be sized to allow flotation units to operate for specific periods (Baluev, 2021).
  (2) *Screening*: Removal of relatively larger solids of 0.7 mm or above can be removed through this process and minimize the pressure on the following treatment such as flotation. This is one of the most cost-effective and straightforward methods available for wastewater treatment. The advantage of this method is that they are less prone to clogging and self-cleaning mechanisms are available nowadays. The trapped materials are washed out of the screen into a collector and can eliminate about 30–80% of the solids from seafood wastewater from the industries (Netsol Water Solutions, 2022).
  (3) *Flotation*: The fats, oils, and grease form a fatty cake on the screens of the removal system, trapping the low-density organic suspended solids, making it easier to remove them from the water. As the particulate matter is suspended in aqueous solutions, hydrophobic suspended particles attach themselves to bubbles and flow in the opposite direction of gravity towards the surface of the solution. Bubble formation is accomplished using two types of techniques: one is dispersed-air flotation (or froth flotation, in mineral processing), which incorporates electroflotation, and another is dissolved-air flotation (DAF), which works by attaching small bubbles of air to suspended solids. The coagulation–flotation process is used to remove the antibiotics process in water, while ion flotation is used to remove the surface active ions in the wastewater. Recent studies have shown that the flotation process is efficient in removing from the wastewater microplastics that are less than 5 mm.

(b) *Biological processes*: Biological treatment involves the use of microorganisms to remove the nutrients from the wastewater. The presence of nitrogenous and other organic compounds can serve as nutrients for the microbes. Usually, there are three different mechanisms depending upon the availability of oxygen:
  (i) Aerobic
  (ii) Anaerobic, and
  (iii) Facultative conditions.

Aerobic and facultative predominate in the aerobic phase, while anaerobic organisms are used in anaerobic conditions only. Microorganisms, when they are suspended in wastewater for treatment, are called the 'suspended growth process', while those attached to the surface on which they grow are called as 'attached growth process' (Sachit, 2018). Biological treatments are used only on a continuous basis, for instance, 365 days/year. The excess biomass produced at the end needs to be properly disposed.

  (i) *Aerobic treatment*: The most commonly used techniques in aerobic treatment include aerated lagoons, activated sludge, trickling filters, and rotating disc contractors. An adequate amount of oxygen must be available or provided during this process.
    - *Activated sludge system*: An attuned, mixed biological growth of microorganisms interacts with organic materials in the wastewater in the presence of excess oxygen

and nutrients. The microbes convert these materials into CO_2 and cellular materials. By applying air, adequate mixing is ensured and oxygen is obtained. The effluent is settled to separate the biological solids and a portion of the sludge is recycled, the excess is wasted for further treatment such as dewatering (Shmeis, 2018).

- *Aerated lagoons*: Used where there is insufficient land or money is not available for activated sludge systems. Efficient treatment can be obtained using an aerated lagoon system. In lieu of a steel or concrete tank, these earthen basins are used to treat wastewater biologically and solids are not recycled (Tay *et al.*, 2004). In naturally aerated lagoons, cyanobacteria and algae are the main photosynthetic microorganisms. Mechanically aerated lagoons (Fig. 2.1) do not have the depth requirements as the oxygen is provided artificially instead of by algal photosynthesis (Zhang, 2001). Microorganisms oxidize the organic matter into CO_2, ammonia, inorganic radicals, and new organisms as wastewater flows through the system. A BOD removal efficiency of 90–95% can be achieved for seafood processing wastewater (Parvathy *et al.*, 2017).
- *Trickling filters*: One of the most common attached biofilm processes is biomass attached to a support media on which they grow and an under-drainage system that captures any microbial solids that come off the media along with the treated wastewater. The primary mechanism involves diffusion and microbial assimilation. As the water flows through the filter bed, the microbes on the gravel bed absorb organic matter (Grady *et al.*, 1999). A process known as 'sloughing' occurs, where the microbial slime that grows thicker detaches off from the supporting media and leaves the filter bed as the inner portions of the biomass get deprived of oxygen and nutrients (Gonzalez 1996). Depending on the filter medium, hydraulic and BOD loading rates, recirculation ratio, and depth, trickling filters can be classified as low-rate, intermediate-rate, high-rate, super high-rate, roughing, and two-stage (Metcalf *et al.*, 1991).
- *Rotating biological contractors*: The necessity to develop innovative, cost-effective wastewater treatment alternatives led to the development of aerobic rotating biological contractors. It is an attached growth process that combines the advantages of biological fixed film and partial stir. It usually involves a corrugated, light plastic disc that rotates at 1–3 rpm when immersed about 40% in wastewater and the

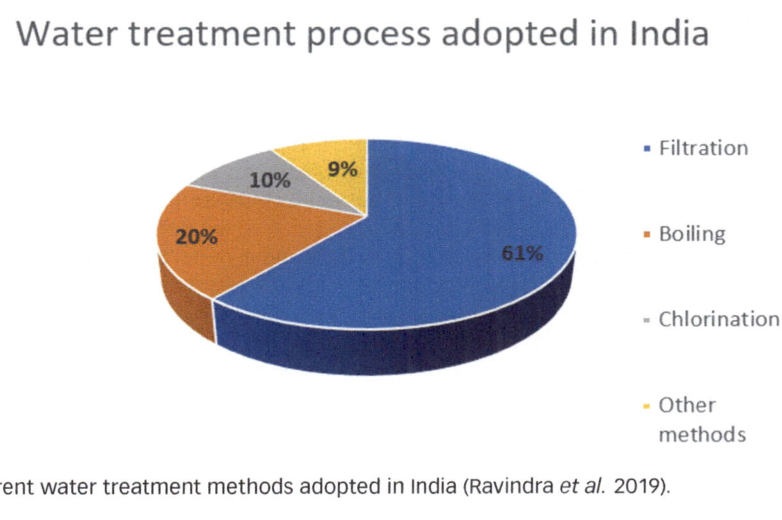

Figure 2.1 Different water treatment methods adopted in India (Ravindra *et al.* 2019).

biomass attached to it absorbs the air when exposed and absorbs the organic load when immersed (Reynolds & Richards, 1996). The advantages of this system include low energy cost, short hydraulic retention time, and insensitivity to toxic substance shock loads, high biomass concentration, and easy operation. Hence, these reactors can be widely used for both domestic and industrial wastewater. A multistage RBC will be required for treating wastewater from fish processing industries (Chowdhury *et al.*, 2010).

(ii) *Anaerobic processes*: This process also helps in converting the organic dissolved solids to organic suspended solids and also into methane and CO_2 (Clanton *et al.*, 1990). Several reactions occur during anaerobic treatment: (1) Water is converted into soluble organic material as a result of the acid-producing bacteria that produces CO_2, hydrogen, and volatile fatty acids. (2) The conversion of these products into CO_2 (30–40%), methane (60–70%), and other gases by methane-producing bacteria (Metcalf & Eddy, 2002; Nielsen, 1996). In comparison with aerobic systems, anaerobic systems can remove a high percentage of BOD from fish processing wastewater at a much lower cost, and produce a smaller quantity of highly stabilized, easily dewatered sludge. Due to the temperature-sensitive nature of the anaerobic process, the digester may need to be heated to 30°–35°C in some cases. This type of process is used where lagoon construction or expansion is not available.

(iii) *Facultative processes*: In municipal and industrial settings, facultative waste stabilization ponds are commonly used for wastewater treatment. Usually, 1.2–2.4 metres deep, these earthen lagoons do not have mechanical mixing or aeration. The lagoon bottom contains sludge deposits that support anaerobic bacteria. For facultative ponds to function effectively, algae must be present in the aerobic and facultative zones (USEPA, 2022). FPs are generally categorized into secondary or primary FPs based on the amount of settled wastewater they receive from the primary treatment stage (usually anaerobic pond effluent). After screening and grit removal, primary FPs receive raw wastewater whereas secondary FPs receive settled wastewater from the primary stage. Anaerobic purple sulphide-oxidizing photosynthetic bacteria may occasionally cause FPs to appear red or pink.

(c) *Chemical processes*:

- *Coagulation and flocculation*: Coagulation, flocculation, and flotation are the three most efficient methods for removing fats, oils, grease, suspended particles, and colloidal components. This method can significantly minimize the pressure on industries and its impact on nearby water bodies such as rivers, seas, lakes, and so on. Ferric sulphate is used as the coagulant to remove the traces of blood colour from fish wastewater. The dosage is kept low (20 mg/l) so as to avoid bulky wet sludge production. The dosage of anionic flocculant is kept constant at 10 mg/l. Removal of TSS and FOGs (fats, oils, and greases) also significantly reduce odour problems. In crab processing, the conductivity of wastewater was high and also varied between 6000 and 22 000 μΩ/cm. The dissolved sodium, chloride, calcium, and sulphate ions interfere with the activation and flocculation performance of polyacrylamide flocculants. A new generation of ultrahigh molecular weight anionic polyacrylamide was used to flocculate particulates in such streams. The ferric and cationic polyamine coagulants were used to overcharge initially negatively charged particulates from wastewater, prior to adding anionic polyacrylamide (Colic *et al.*, 2007). Aluminium sulphate or alum is also used as a flocculant to remove colour and turbidity from water, especially during the rainy season. This removal depends on pH, initial turbidity of water, and the dosage of alum. It can reduce the total hardness of water by reducing the levels of Ca, Mg, Mn, Fe, and other metallic ions in the water (Ahirrao, 2008).

- *Disinfection*: Most of the chemicals used in wastewater treatment is for the purpose of disinfection, in which the disease-causing organisms are either rendered inactive or killed. In microorganisms, disinfection occurs when their cell wall corrosion causes the contents to leak, or when the membrane permeability, protoplasm changes, or enzyme activity changes due to structural modification of the proteins. According to Gross and Farrell-Poe (2005), it is important to note that the effectiveness of wastewater disinfection depends on several factors, namely: (1) pH, (2) contact time, (3) flow rate, (4) concentration of disinfecting agents, (5) wastewater demand, (6) wastewater temperature, (7) concentration of interfering substances. Disinfection is mainly achieved by the use of bactericidal agents such as ozonation, chlorination, UV radiation, and so on.
  (i) *Ozonation*: It is one of the highly effective techniques in the treatment of water, where ozone is generated in situ in an ozone generator, and the gas dissolves in water and kills the microorganisms present as well as removes both inorganic and organic pollutants. It is effective in removing dyes from textile wastewater. High-voltage alternating current (6–20 kV) is applied across a dielectric discharge gap containing oxygen to generate ozone in wastewater treatment plants (USEPA, 1999). The ozone dissolving in water attaches to the microorganisms and pollutants; oxidizes and destroys them. As ozone decomposes in the water, free radicals of hydrogen peroxy ($-HO_2$) and hydroxyl (OH) ions are formed, which play an important role in the disinfection process due to its great oxidizing capacity. Ozonation is normally used as the final step in the water treatment processes, as it reduces the COD and BOD in the discharge water. This is a short contact process (10–30 minutes). Also, the ozone decomposes rapidly, hence no harmful products or effects are seen (Sasáková *et al.*, 2021).
  (ii) *Chlorination*: It is a commonly used disinfection method in municipal and domestic water treatments. This step is generally carried out just before the final discharge into the water bodies. Generally used are hypochlorite salts such as calcium hypochlorite [$Ca(OCl)_2$] or sodium hypochlorite (NaOCl) as it is easy to handle and use, and sometimes, in the form of chlorine gas. The filtered clear water in a chlorination tank should be given at least 20 min detention with a baffle to prevent short-circuiting. On the other hand, chlorination solutions are added through a diffuser at the inlet in turbulent conditions. In recent years, researchers have found out that the use of chlorine can cause undesirable effects and other hazards and this can be overcome with the selection of proper alternatives, such as ozone, chloramine, boiling, potassium permanganate, chlorine dioxide, and so on (USEPA, 1999).

However, there are certain demerits involved in these methods of water treatment. The most critical ones are high installation and maintenance costs, large space for the equipment, and accumulation of dirt and other particles that may lead to bacterial growth, causing odours, and may turn out to be hazardous to human health. The improper handling of these may have a negative impact on the environment (UNICEF, 2008). Biological methods have poor decolorization capacity, require maintenance and management, and a favourable environment to work, sometimes producing gas and unwanted odour (Hussain *et al.*, 2021). Sludge bulking, foaming, and bioreactor membrane fouling are some other demerits associated with this technique (Meng *et al.*, 2009).

Chemical methods create undesirable hazards and effects as some of the chemicals used possess carcinogenic properties such as trihalomethanes (THMs) which are produced when chlorine reacts with synthetic organic matter and other natural organics present in water. Sometimes, it produces an off-taste (NRCSDWC, 1980). Moreover, it has high initial capital and energy costs, the more chemical dosage will be required to remove the metal ions, handling and disposal problems, non-reusable chemicals, and needs constant monitoring (Crini & Lichtfouse, 2019). Alum has a negative impact if not used within limits. Salinity and electrical conductivity may increase beyond the values prescribed by WHO. Also,

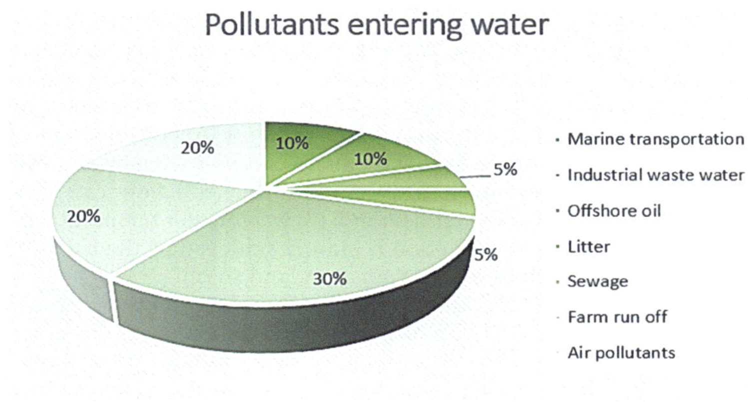

Figure 2.2 Sources of water pollution (courtesy: Ocean Graph).

it causes digestive troubles, irritation to the respiratory tract, redness of the eyes, irritation, and itching of the skin (ACAAI, 2023). If used in higher amounts, it can impart a metallic taste to the water being treated (Oo *et al.*, 1993). Figure 2.2 shows different sources from where pollutants enter in water.

2.3 INNOVATIONS IN WATER FILTRATION

To overcome the limitations of these techniques, new innovative techniques were developed. The main objective of these techniques is to minimize further waterborne illness. Filtered water generally tastes better compared to unfiltered water as it removes pesticides, heavy metals, bacteria, and so on. Generally, in wastewater treatment, conventional processes occasionally fail to achieve the required standards depending on the purpose of use. This makes it necessary to use advanced water filtration methods. The filtration methods usually involve using a physical barrier and chemical or biological processes (Fig. 2.3).

(i) *Distillation*: Distillation is one of the oldest methods of water purification. In a distillation unit, water is heated and vapourized in a boiling chamber, cooled in a condensing chamber, and stored in a purification tank (Fig. 2.4). It is the liquid-phase vessel that holds the contaminants. The disadvantages are that it cannot remove certain contaminants like herbicides and pesticides that have boiling points lower than 100°C, the equipment required is costly and a large amount of energy as well as the water is required for operation. It is time-consuming as well (Hulett, 1896). Also, there are chances that the organic compounds that are removed may re-contaminate the water if the purified product is not removed before the condensation process. This process requires continuous or regular maintenance. Unevaporated compounds if left in the system may contaminate or cause fouling. Operation and equipment costs are higher, generally ranging between 300 and 1200 USD.

(ii) *Ion-exchange*: This method was invented back in the 1850s by Harry Stephen Meysey Thompson and John Thomas Way (Association of Water Technologies). The first phenolic ion exchange resin was developed by English chemists, Basil Albert Adams and Eric Leighton Holmes in 1935 (WCP, 2014). A water treatment method called ion exchange involves exchanging undesirable ions with less objectionable or non-objectionable ions in a solution. In water treatment, it percolates water through bed-like spherical ion-exchange resin materials. The ions present in water are exchanged for other ions which are fixed to the beads. The most commonly used method for ion exchange is softening and deionization. Softening is used to reduce the water hardness (generally caused by the presence of chlorides of calcium and

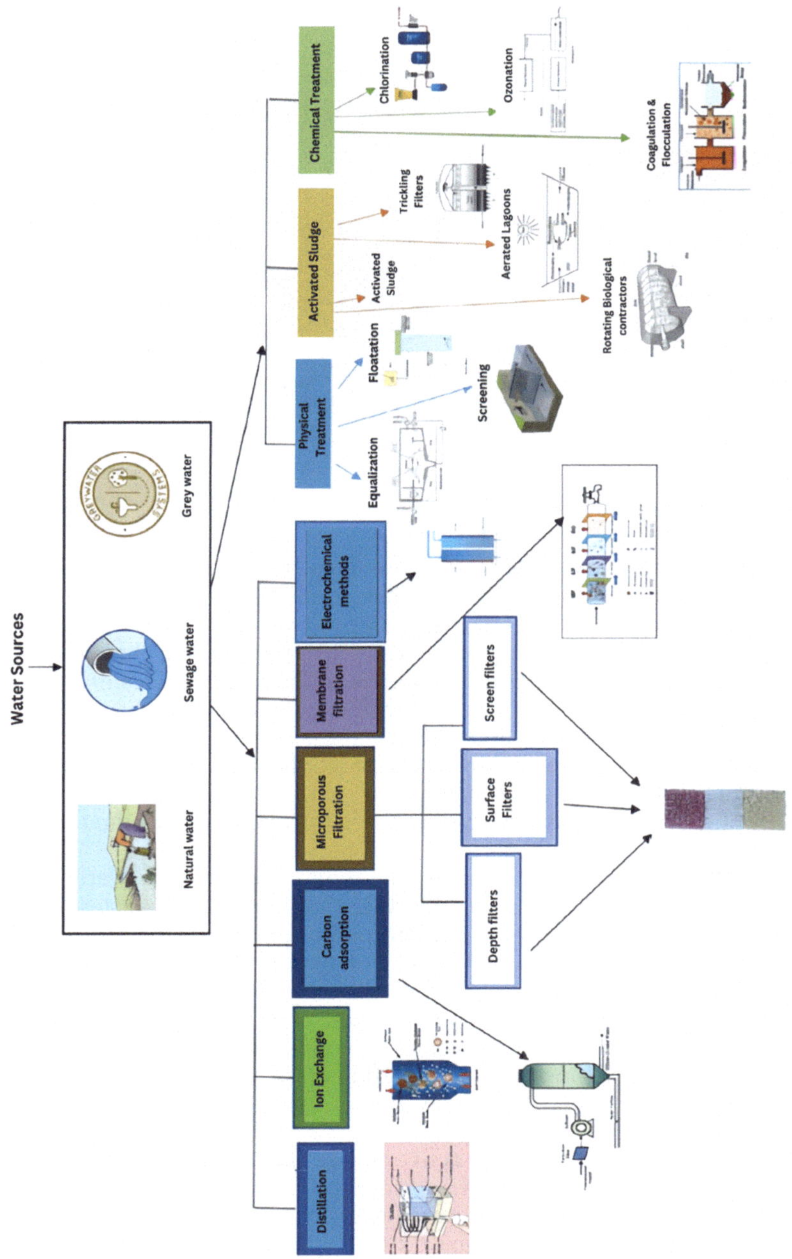

Figure 2.3 Different methods of water treatment.

Figure 2.4 Lab-scale filter developed at KUFOS.

magnesium, bicarbonates, and sulphates) before RO processing, while deionization uses beads that exchange hydrogen ions for cations or hydroxyl ions for anions (Fig. 2.5). After a particular time, the beds become saturated and the exchange resin needs to be regenerated/ recharged. It also effectively reduces dissolved organic matter, nitrate, and phosphate (Gregory & Dhond, 1972). Currently, India has nine major suppliers of ion-exchange resins, such as ResinTech Inc., Felite Resin Technology, Evoqua Water Technologies, Resindion, Thermax, Lanxess, DuPont Water Solutions, Purolite, and Mitsubishi Chemicals.

Cation and anion exchange resins are the commonly used ones. The cation exchange is generally used to remove iron and magnesium, while anion exchange resins are used for removing arsenic, nitrate, and bicarbonates. Figure 2.6 shows estimated markets of ion-exchange resins

(iii) *Carbon adsorption*: One of the widely used methods used in home water filtration method because of its ability to remove odours, chlorine, and unfavourable tastes. Carbon activation was discovered in the early 1800s. Activated carbon technology was first used in potable water treatment in the 1960s and 1970s to enhance organic contamination removal (Jin *et al.*, 2013). The structure and distribution of pores, surface area, molecular weight, pH, presence of aliphatic or aromatic compounds, and so on are critical factors in determining the effectiveness of these filters in absorbing various contaminants. Activated carbon filters have been found effective in reducing the concentration of toxic organic compounds such as DDT, malathion, dieldrin, and so on to a low level in wastewater (Whitehouse, 1967). By filtering the water through granular carbon, impurities are removed methodically from the water; the water becomes less saturated and more active as it gradually clears itself of its pollution. Thus, as the water becomes more active, activated carbon fractions become less and less saturated.

(iv) *Microporous basic filtration*: There are three types of microporous filtration; depth, surface, and screen. Depth filters consist of a matrix that retains particles by random adsorption or entrapment. It can entrap 98% of the suspended solids and protect the elements. Screen filters have uniform structures that retain particles larger than the controlled pore size on the surface. Surface filters are made from multiple layers of media and when fluid passes through the filter, it retains the larger particles. It can remove 99.9% of suspended solids and

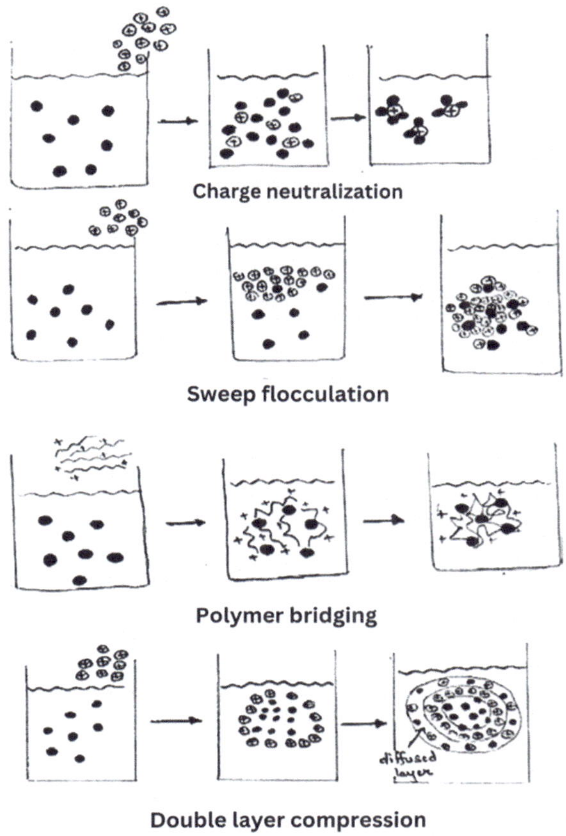

Figure 2.5 Mechanisms involved in natural coagulation processes (courtesy: Koul *et al.*, 2022).

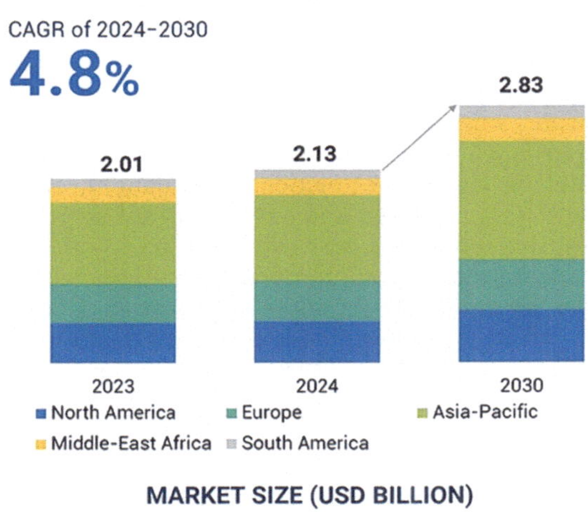

Figure 2.6 Estimated market of ion-exchange resins globally (Markets & Markets, 2020).

can be used as a clarifying filter. Also, certain charged microporous filters like electronegative microporous filters composed of epoxy fibreglass can absorb a greater percentage of coliphages and enteroviruses compared to untreated filters (Preston *et al.*, 1988). Materials used for microporous filters are carbonaceous materials such as activated carbon, carbide derivatives, zeolite, conjugated polymer nanoparticles (CPNs), and so on.

(v) *Membrane filtrations*: Membrane filtration is a separation method wherein a liquid is forced through a membrane filter with a high surface area. Filters are available in different pore size, based on which is categorized into microfiltration, RO, ultrafiltration (UF), and nanofiltration. In microfiltration, a particle size of 0.05–10 μm can be filtered. Membranes have a nominal pore size of 0.1–1 μm and operate between the pressure of 0.5–5 bar. The feed flows tangentially to the surface of the membrane which prevents fouling (Charcosset, 2012). This membrane can retain colloids, suspended solids, and microorganisms. It can operate at low pressure and consumes low energy compared to nanofiltration and RO. Few manual actions are required and create less impact on the environment.

This membrane technology is sensitive to oxidative chemicals such as nitric acid, sulphuric acid, peroxides, and persulphates in high concentrations. Often membrane fouling occurs due to suspended solids and the degree of purification is lower than in other membrane processes. Also if pressure exceeds, it can damage the membrane. It is also not effective against sodium, iron, and volatile organic acids (VOCs).

RO is an economical method of removing 90–99% of the contaminants. The pores of RO are much packed or tighter than UF membranes. The membranes are typically composed of a thin film composite membrane with three membranes: a polyester support web, a microporous polysulphone interlayer, and an ultrathick polyamide barrier layer on the top surface (Rashed *et al.*, 2020). RO membranes have a pore size of around 0.0001 μm and the uniqueness of the RO among other filtration systems is that the residence time is very short (a few seconds), and are capable of rejecting all particles, bacteria, and organics that have molecular weight higher than 300 Da. This is the leading technology used in most water-bottling plants. It is the osmotic pressure that drives water through the semi-permeable membrane, the water dilutes the more concentrated solution and the result is equilibrium. There are certain limitations for RO as well. Cellulose acetate membranes have limited pH tolerance, hence, they degrade at temperatures greater than 35°C. These are vulnerable to bacteria and can easily hydrolyse. Polyamide membranes are intolerant to a temperature greater than 35°C and have poor tolerance to free chlorine. Thin-film composites are intolerant to chlorine. It is necessary to soften or deionize the feed water before using polysulphones with high flux.

Nanofiltration is a pressure-driven or chemical-free process. It has an average pore size of 1 nm. It removes most of the organic molecules such as salts, and viruses, and is often used with water that has low TDS. Water softening is achieved as it involves the removal of hardness-causing multivalent ions, such as calcium and magnesium. The advantage of nanofiltration over RO is that it has a higher flux rate. The membrane functionality of nanofiltration depends on three layers: an active layer, a porous supporting layer, and a macroporous structure underneath (Abdel-Fatah, 2018). The active layer determines the permeability of a certain component and hence, helps in selecting the membrane for the separation process. The supporting layer helps in modifying the mechanical properties. This process does not require heating or cooling as in distillation and can handle high volumes of feed at a stable flow rate and in a continuous manner.

UF is a membrane technology that purifies, separates, and concentrates solution between micro and nanofiltration. The pore size of the membranes ranges from 0.001 to 0.1 μm. It can intercept particles with a molecular weight of 0.5–500 kDa. The diameter of the separated components is about 0.005–10 μm (Liuqin, 2011). The UF process has a higher removal rate of particulate matter and turbidity (up to 99.9%). It is also efficient in removing pathogenic bacteria and viruses such as *Giardia*, *Cryptosporidium*, and so on in the water (Laîne *et al.*,

2000). UF can completely remove microorganisms from water, improving its biological quality. It is also an excellent way to remove pyrogen contamination from water. Figure 2.1 shows the proportion of different water treatment methods adopted in India.

(vi) *Electrochemical methods*: Electrochemical treatment of water is another method that is gaining popularity nowadays. In the electrochemical unit, anodic electrolysis is used to combine with simultaneous cation removal; electrocatalytic and chemical additional oxidation takes place in an intermediate vortical reaction tank, and catalytic purification takes place in a catalytic reactor. The advantages of these methods are that disinfection agents can be produced on-site, are effective in removing dyes, oils, grease, and suspended solids, have coagulation and separation of Cu and other ions, are rapid and effective, rapid biodegradability and the ability to adapt to different flow rates and pollutant loads.
Electrochemical treatment will require technological and economic improvement before it can be used widely. This means that investment and machinery costs are high, especially in processes like electrocoagulation. Moreover, there will be the formation of sludge and anode passivation, which will affect the continuous operations of the system. It also requires the addition of chemicals for carrying out flocculation and post-treatment requirements to remove the ion concentrations.

Most of these techniques, although have advantages, faces certain disadvantages too. The disadvantage of almost all the existing techniques is that these are expensive and/or need expensive chemicals. In the case of the ion exchange method, specialized resins will range between USD $500 and $2000 per cubic foot. If the turbidity of water is greater than 10 ppm, then it will reduce the process efficiency. Some organic matter or Fe^{3+} fouls the resins, which means that untreated water from lakes and rivers contains dissolved organic matter derived from decaying vegetation which imparts a yellow or brown colour is adsorbed by the anion beads and reduces its exchange capacity, leading to a reduction in treated water quality (Patel, 2016). Sometimes, it is possible that the treated water may be contaminated with organic contamination from the resin. The carbon adsorption technique generally does not remove total dissolved solids (TDS), heavy metals, or hardness. Only certain carbon filters remove lead, coliform, asbestos, and so on (Siong *et al.*, 2013). Exhausted carbon must be regenerated or disposed of as of hazardous waste. They also have a limited service life and these filters are not effective in removing any pathogenic bacteria or viruses. Hence, frequent filter changes may be required. In processes such as UV, continuous electrical supply, regular inspection, and maintenance is necessary. If turbidity is higher in the water being treated, it reduces the process efficiency as well. In the case of membrane filtration techniques, the major drawback is that it reduces essential minerals that are required for good health. Along with these advantages, there are also certain disadvantages to this system such as those consuming RO-treated water as a primary source for drinking may suffer from muscle cramps, weakness, cardiovascular diseases in severe cases, and tiredness. System cost is also higher. RO membranes require high-pressure pumps to extract clear water through the membrane. High TDS in water may require the membrane to be replaced in a shorter time than the actual lifetime of the filter. It has limited retention of salts and univalent ions and is sensitive to free chlorine. Organic and inorganic solutes, biological solids, and colloids can cause membrane fouling at the nano level (Van Der Bruggen *et al.*, 2003). Formation and accumulation of exopolymeric substances is the real cause of flux decline when biofouling occurs (Ivnitsky *et al.*, 2007; Kim *et al.*, 2006).

As most of the currently used water treatments/technologies are expensive, especially in rural areas, and attempts are being made to find a cost-effective, sustainable alternative technique to treat water, and the use of natural plant extracts could be one of the finest options. As a result of the financial and technical challenges of filtration systems in developing countries including India, researchers contemplated using natural substitutes that are either underutilized or are by-products of agricultural or manufacturing processes that are available in abundance for filtration purposes. Moreover, the use of chemical coagulants has resulted in drastic effects on human health, such as convulsions, constipation,

Table 2.1 Difference between natural and chemical coagulants.

Parameters	Natural Coagulants	Chemical Coagulants
Cost	Abundantly available, less cost	High cost of treatment and procurement
Toxicity	Non-toxic	Highly toxic (some are even carcinogens)
pH	Hardly any change to the pH	Significant change due to the formation of metallic salts
Dosage	Less dosage required	A larger dosage is required to bring coagulation
Heavy metal removal	Settles along with the coagulation process	Certain heavy metals such as Cd, Zn, and so on cannot be easily settled.
Effectiveness	More stable and no formation of dissolved solids	Extended usage leads to the formation of dissolved solids and less effectiveness
Sludge formation	Biodegradable sludge formed	Sludge contains hazardous chemicals.

abdominal colics' and loss of energy (Rasool *et al.*, 2016; Saravanan *et al.*, 2017; Subramonian *et al.*, 2014) The use of natural plant-based materials such as coir, pith, and so on, in water purification, has been reported earlier (Swan, 2008). Also, due to their less impact on health and the environment, there is a tendency to utilize plant-based natural substitutes in removing chemical and biological contaminants from water. A few disadvantages prevent natural coagulants from being widely used:

(1) The supply of natural coagulants is influenced by seasonal production since they are mostly derived from plants (Freitas *et al.*, 2018).
(2) By mixing rapidly during the coagulation process, cells rupture, resulting in organic matter loading that reacts with disinfectants during subsequent treatment, resulting in disinfectant by-products (Oladoja, 2015; Šćiban *et al.*, 2009).
(3) Coagulants made from natural sources can decompose when stored for long periods of time (Saleem & Bachmann, 2019).

Eureka Forbes and Aquaguard had introduced Ayurfilter, wherein they used ayurvedic plant parts, especially spices for their disinfection and health-improving properties.

For better clarity, the table below shows the difference between natural and chemical coagulants (Mathuram *et al.*, 2018) (Table 2.1).

2.4 NATURAL SUBSTITUTES IN WATER PURIFICATION

Natural substitutes are mainly obtained from plants, animals, and microbes. Some of the commonly used ones are sand filters, diatomaceous earth, and oysters, since long ages, but only a few are effective in removing undesirable particles from water. These substances work by bridging molecules, adsorbing, and neutralizing charges (Koul *et al.*, 2022). Because of their natural origin, it is considered safe for use. A confirmed environmental safety assessment is needed for organic coagulants, as their toxicity in humans and the environment is currently unclear. They are readily available, and robust. Nevertheless, they are susceptible to biodegradation by microbial or environmental factors, which compromises their long-term storage and commercialization (Abidin *et al.*, 2019; Ho *et al.*, 2014), and the toxic sludge formed post-treatment creates a huge problem during disposal. Hence, an efficient method of utilizing them is using these in combination with other coagulants available to increase their efficiency and reduce the utilization and the overall cost of the treatments. In low- and middle-income countries, natural wastewater treatment systems provide sustainable sanitation services at low maintenance and operation costs, as well as energy consumption and mechanical complexity. There is very little or no electricity required for bio-wastewater treatment systems, as they utilize natural processes and factors, such as sunlight, sedimentation, temperature, adsorption, filtration, and biodegradation.

2.4.1 Plant-based filtering aids

Water filtering systems developed using plants as base material have been explored as natural alternatives to synthetic filters. Purifying and extracting natural coagulants in an effective manner increases their efficacy, and therefore, their coagulative properties improve. The use of plants in water filtration has been widely documented. The fruit peels, cortexes, seeds, or rinds, such as bananas, oranges, mango, pineapple, oranges, and lemons, which have been researched for their capacity to purify water. Not only that, several other plants such as *Moringa oleifera*, *Aloe vera*, cactus (*Opuntia ficus-indica*), tea and coffee wastes, coconut husks, *Cassia fistula*, tamarind seeds, date seeds, and hibiscus are all promising alternatives. Figure 2.7 shows the production of different nature-based materials in India.

(i) *Banana peels*: According to Food and Agricultural Organization (FAO), the World banana production in 2018 was 115.74 million metric tonnes where India ranked first accounting for 30.808 million metric tons. Approximately, 30–40% of total banana production goes as waste; of which banana peel contributes to 34.72–46.29 million metric tons. Basically, banana peels have nitrogen, sulphur, carboxylic acid, and other phytoconstituents that are capable of attracting heavy metals. Heavy metals are positively charged and carboxylic acid is negatively charged, which attract each other (Verma, 2020). Banana peel has the capacity to absorb 97% of metals (especially Pb and Cu) in 1 h and can be used up to 11 times without losing its metal binding capacity (Castro *et al.*, 2011). Other metals such as, Cr, As, Fe, Ni, Co, Zn, and S can also be removed (Hanafiah *et al.*, 2022; Negroiu *et al.* 2021). It has also got an excellent capacity to absorb dyes and pesticides (Mane & Bhusari 2012; Silva *et al.*, 2013). It has also been proven to remove cyanide (Abbas *et al.*, 2014) as well as Co^{2+} and Ni^{2+} from water (Abbasi *et al.*, 2013). Processed dry banana peels are shown in Figure 2.8.

(ii) *Citrus peels*: Citrus peels are another class of by-product whose waste generated accounted for approximately 12.68–15.85 million metric tons in 2019 (40–50% of the fruit is contributed by the peel) (El Barnossi *et al.*, 2021). The citrus peels were reported to be used as a low-cost coagulant, which helped in reducing the turbidity along with maintaining a constant pH throughout the process. It reduces the biological oxygen demand (BOD) up to 89–96% (Subashree *et al.*, 2017). In addition, it is effective at removing calcium and magnesium hardness, chlorides, and total hardness of water. Wastewater treated with orange peels removed 85% of arsenic (Khaskheli *et al.*, 2011).

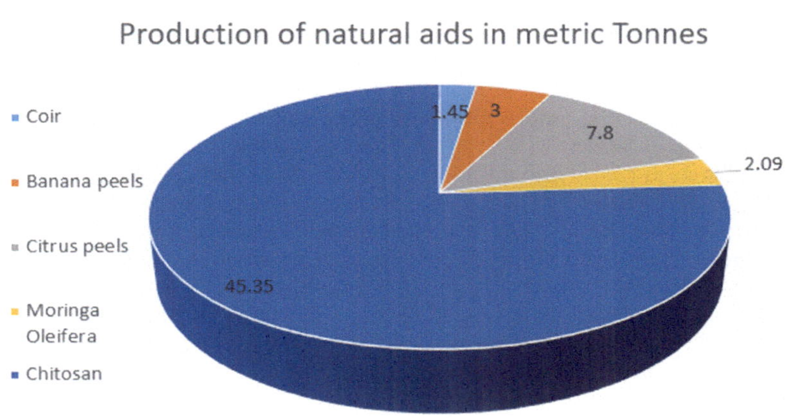

Figure 2.7 Production of different nature-based materials in India.

Figure 2.8 Processed, dried banana peels.

(iii) *Moringa oleifera*: India is the prime producer of moringa (Fig. 2.9), with an annual production of 2.20–2.40 million tonnes (Sekhar *et al.*, 2018). India, Hawaii, Philippines, Pakistan, and many other Asian countries use its leaves, flowers, fruits, and immature pods for their nutritious vegetable diets. The roots, fruits, and seeds are used in medicine as they lower cholesterol, protect the liver, and have antitumour and antipyretic properties, provide antioxidants, fight infections, and prevent the development of cancer (Anwar *et al.*, 2007). Still, its potency in purifying water is still unknown. *Moringa oleifera* seeds have positively charged proteins called *Moringa oleifera* cationic proteins (MOCP), which inhibit the growth of *E. coli*, *Shigella*, *Salmonella typhi* and reduce turbidity without altering the pH (Delelegn *et al.*, 2018). A proton is taken up by the amino acids in moringa protein, which are then released as a hydroxyl group, which makes the basic solution. There are six amino acids in this protein, including glutamic acid, methionine, and arginine, and the molecular weight varies between 6000 and 16,000 Da (Mangale Sapana *et al.*, 2012). Seed extracts contain a protein called niaziminin, which provides its antiviral and antibiofilm formation properties (Nasr-Eldin *et al.*, 2017).

(iv) *Coir and coconut shell husk*: The coir production in India is approximately 11.63 lakh tonnes (Coir Board of India, 2014). Coir has traditionally been used in water filtration and as an adsorbent for oil spills on water surfaces. They are normally used as mats that can efficiently remove sediment from drainages (Swan, 2008). Coconut shells have the potential to be utilized more effectively in the form of activated carbons. Raw and Fe-modified coconut shell biochar were compared for their ability to adsorb Cr (VI) from wastewater, with Fe-modified biochar adsorbing 94.89% more than raw biochar (Bu *et al.*, 2021); and when modified with Mn, its capacity to absorb Pb and Cd increased by 20 and 30 times, respectively (Wu *et al.*, 2020). The use of biochar in domestic wastewater reduced turbidity (99.9%), TDS (77.5%), tannery wastewater turbidity (99%), and TDS (83%) (Vidya *et al.*, 2019).

Figure 2.9 *Moringa oleifera* leaves and seeds.

(v) *Cassia fistula*: *Cassia fistula* seeds come from the *Indian Laburnum* tree. It is used only for its flower (golden shower) (Fig. 2.10) on a festive occasion (*Vishu*) in India, specifically in Kerala, where this tree has been declared as the State tree. The seeds usually fall on the ground and get dumped away with the garbage. Recent studies have shown that they have proven to have antibacterial (Seyyednejad *et al.*, 2004; Sharma, 2017) and antifungal properties. It is reported that it contains proteins called napins that exhibit antifungal and antiviral activities. The extracts of these seeds have been proven to be effective against *E. coli* and *K. pneumoniae* (Maqsood *et al.*, 2020). The gum from the seeds can be used to treat textile wastewater with a removal efficiency of 93.83% (Dao *et al.*, 2021). The positive charge of the pollutants interacts with the negatively charged carboxylate group of the coagulant via adsorption/bridging interactions (Koul *et al.*, 2022). Using 100% ethanol extracts of seeds, membrane

Figure 2.10 *Cassia fistula* flower and pods.

filters can be developed with chitosan, which exhibited antibacterial activity against *E. coli*. Water treated with these membrane filters showed a reduction in pH, turbidity, TDS, and alkalinity. According to FTIR and SEM images, chitosan encapsulates bioactive molecules of seed extracts, possibly modifying the membrane filters' antimicrobial properties, thereby increasing water BOD (Harinarayanan & Raman, 2022).

(vi) *Papaya seeds*: According to the Food and Agriculture Organization (FAO) of the United Nations. India and Brazil are the major producers of papaya with 5.7 and 1.4 million tons, respectively. Industrial papaya utilization mainly contributes to the production of papain and the manufacturing of food products, which disposes approximately 25% w/w of processed fruit as waste, particularly seeds (Desai & Wagh, 1995). These days, the global production of agro-industrial papaya waste is estimated around 3 million tons/year of which 70% are seeds (FAO, 2016). The seeds of papaya have coagulant properties because they contain positively charged proteins that attract negatively charged particles, such as clay, bacteria, toxins, and slit, through adsorption and charge neutralization (Yimer & Dame, 2021). 91.92% of the total hardness was removed from the water (Chandran & George, 2018). As a polyelectrolyte, papaya seed protein is primarily composed of C = O, C–O, and –OH groups. The removal efficiency for cadmium was 60–85%, 52–90% for lead, and 67–87% for chromium (Hussain & Sheriff, 2014). Coagulants derived from deshelled *Carica papaya* seeds were capable of removing 57.6 and 62.1% of total coliforms and *E. coli* (Amran *et al.*, 2021).

(vii) *Tamarind seeds*: Study results showed the seed coat is rich in fibre (21.6%) but the seed is rich in protein (13.6%). It also contains a high content of copper, calcium, potassium, magnesium, phosphorus, zinc, and manganese compared to that of pulp (El-Gindy & Youssif, 2015). The tamarind seed and other parts of the plant also exhibit antibacterial, antiviral, and antifungal properties, making them an effective tool in the treatment of water and purification of wastewater. Tamarind seed extracted using distilled water and 1 M NaCl effectively removed 99.2% turbidity and 77.2% of COD, while extract prepared using 1 M NaOH removed 99.4% turbidity and 73.8% COD (Zainol & Fadli, 2020). However, this can be used in water with low turbidity levels, than those with high turbidity, because of mechanisms like hindered settling and Brownian movement (Ali *et al.*, 2021). Effendi *et al.* (2017) reviewed that tamarind seed (4 g/l) can reduce turbidity by (75%), TSS (77.14%), COD (68.18%), sulphate ion (71.42%) and mercury ion concentration (82.73%) (Kamalakannan *et al.*, 2019).

(viii) *Tea and coffee waste*: Waste tea residue can be utilized for the treatment of contaminated wastewater and can be utilized for fluoride removal. Brewed tea waste was found to be an efficient adsorbent in removal of lead, nickel, cadmium, and zinc from wastewater in the order Pb (97.97%) >Ni (82%) >Zn (76%) >Cd (84.74%) (Çelebi *et al.*, 2020). Methylene blue (MB) was removed from contaminated wastewater using raw green tea waste (GTW) and thermally treated green tea waste (TTGTW). The % removal was found to be 96.58% and 98.07% using RGTW and TTGTW, respectively. RGTW showed mesopores and micropores, while TTGTW showed micropores, according to Brunauer–Emmett–Teller (BET). (El-Azazy *et al.*, 2021). Spent coffee grounds with or without treatment can be used for dye removal. Without treatment, it gave a removal efficiency of 23.40 mg/g for methylene blue, while those treated, gave an efficiency of 986.80 mg/g (Jung *et al.*, 2016).

(ix) *Water hyacinth (Eichhornia crassipes)*: It is a vascular plant and thrives well in polluted water (Fig. 2.11). It significantly reduces turbidity and removes flocs, reducing the organic matter, fecal coliform, suspended particles, toxic metals, and other impurities present in the water. Its root system is capable of absorbing toxic and heavy metal pollutants like Mn and Strontium-90 and also harbours bacteria that help in removing impurities in wastewater (Dash, 2022). The removal efficiency for heavy metals like As, Pb, Cu, Cd, and Zn reached 59–92% from industrial wastewater (Huynh *et al.*, 2021). According to Adewumi and Ogbiye (2009), a floating and running reed bed system of water hyacinth could reduce 99% of faecal coliform and total coliform from raw wastewater.

Figure 2.11 Water hyacinth (courtesy: Aquatic Biologists Inc).

(x) *Date seeds*: It was observed that date seeds could be used as a natural coagulant to reduce the turbidity and COD, and to increase the efficiency of removal, a combination can be used with chemical coagulants such as alum and ferric chloride. So, for a turbidity removal efficiency of 95%, a combination of 70 mg/l date seeds +50 mg/l of alum can be used, and for 90% COD removal, a combination of 40 mg/l date seeds +70 mg/l of ferric chloride can be used (Jaeel, 2016). Modified carbon produced from the date seed without activation had a methyl orange adsorption capacity of 66.26 mg/g and was a temperature-dependent activity (Arunachalam *et al.*, 2018).

(xi) *Aloe vera*: Successfully used as a natural coagulant, the turbidity was reduced from 6.0 nephelometric turbidity units (NTU) at pH 6 to 1.42 NTU at pH 7.5 (Benalia *et al.*, 2021). *Aloe vera* leaves are crushed, powdered and sieved and modified with carboxylated carbon nanotubes to remove cadmium (II), and a high level of adsorption (98%) was obtained. Leaves when treated with H_2SO_4 allowed high removal of 98.66% for Cr (VI) (Gupta *et al.*, 2019; Mahmoodi *et al.*, 2020; Prajapati *et al.*, 2020). *Aloe vera* gel is mainly composed of polysaccharides with low protein and lipid contents (L-arabinose, D-galactose, L- rhamnose, D-xylose, and galacturonic acid) have been reported to be implicated as coagulant agent (Vijayaraghavan *et al.*, 2011). It also has polysaccharide functional groups such as carboxyl, amine, and hydroxyl and hydrogen bonds useful for flocculation (Sepúlveda *et al.*, 2007). Also, dye biodegradation is possible because of the presence of enzymatic systems such as glutathione peroxidase, carboxypeptidase, and isoenzymes of superoxide dismutase (Katubi *et al.*, 2021).

2.4.2 Animal-based coagulants

(i) *Chitosan*: In most cases, coagulants can be extracted from shellfish, shrimp, lobster, molluscs, diatoms, and sponge exoskeletons. Chitin is a naturally derived mucopolysaccharide that is produced abundantly by the process of biosynthesis. It is a nitrogenous polysaccharide that is hard, white, and non-elastic (Muzzarelli, 1977). It is similar to cellulose, but has acetamide groups, while chitosan is deacetylated product of chitin. Chitosan is a linear polymer formed by α (1 4)-linked 2-amino-2-deoxy-β-D-glucopyranose and derived by *N*-deacetylation, characterized by the degree of deacetylation, which is the copolymer of N-acetylglucosamine and glucosamine (Periayah *et al.*, 2016). Chitosan, prepared from the shells of crustaceans has excellent antibacterial, flocculation, and filtration properties (Ferrero *et al.*, 2014; Hahn & Zibek, 2018; Kangama *et al.*, 2018). India is one of the major producers and exporters of prawns and the conversion of shrimp shells into chitin and chitosan by demineralization and deacetylation has led to economic gain and waste management (Renuka *et al.*, 2019). Due to

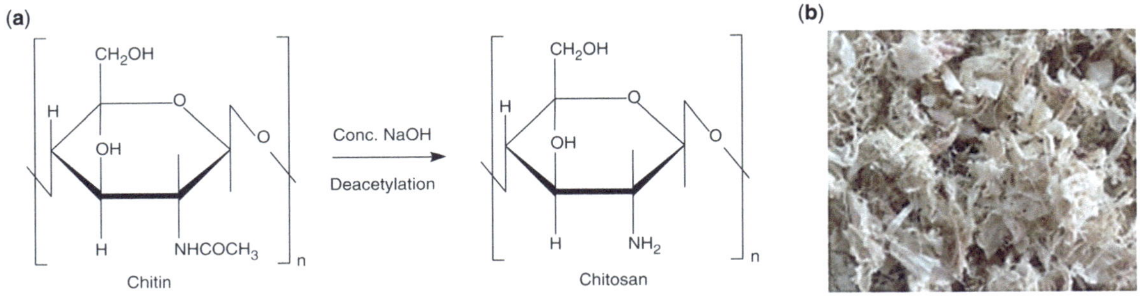

Figure 2.12 (a) Structure of chitin and chitosan (Dutta *et al.*, 2012). (b) Commercially available chitosan.

their flocculation properties, these agents are used in water filtration to act as chelators, which attach to various metal ions by donating electrons to nitrogen and oxygen atoms to the metals and forming stable compounds. It can therefore be used to trap heavy metals, radioactive metals, and dyes (Bassi *et al.*, 2000; Mouzdahir *et al.*, 2010). The polycationic structure of chitosan makes it an antimicrobial agent as well (Másson, 2021). Generally, it is believed that chitosan works by binding negatively charged cell walls of bacteria, resulting in their disruption, followed by inhibition of DNA replication, and subsequent cell death (Nagy *et al.*, 2011). Figure 2.12 shows the figure of chitin and chitosan and commercially available chitosan.

(ii) *Eggshells*: Eggshells are an undervalued waste and represent an environmental and public health concern when improperly disposed of. Microorganisms, some of which are pathogenic, can multiply in eggshells that have been incorrectly discarded, attracting disease vectors and predators (Ayodele *et al.*, 2021). Eggshells have been efficiently used to remove dyes and metallic ions in water. In soils and water contaminated by metallic ions, eggshells are an excellent adsorbent material due to their calcium carbonate content (Lee *et al.*, 2022). Eggshell has a high adsorptive capacity as a result of its expanse of surface area and the presence of several pores, ranging in number from 7000 to 17 000 (Rajoriya *et al.*, 2021). And another advantage is that it can be used without any surface modification using chemical methods due to its high porosity. It was found that 1.0 g of eggshell was sufficient to adsorb 99% of Cd^{2+}; 97% of Direct Black 22 (DB); 96% of Reactive Blue 198 (RB); and 95% of Cr^{3+} (de Oliveira Zonato et al., 2022).

2.4.3 Microbial-based coagulants

Aerobic and anaerobic bacteria are used for this purpose. The metabolism of aerobic bacteria is much higher than that of anaerobic bacteria. Hence, 90% fewer organisms will be required for the metabolism compared to the anaerobic processes and the process is accomplished in less time. Anaerobic bacteria live and reproduce in absence of oxygen and utilize compounds such as, nitrates and sulphates to produce energy. So, for this process to occur, the organic matter must be exposed to a high quantity of bacteria compared to the aerobic process for a longer period of time (Climate Policy Watcher, 2023). Microalgae are also one of the important sources used for wastewater treatment, as they have the photosynthetic capability to convert solar energy to biomass and absorb nutrients such as, phosphorous and nitrogen in a short period of time. The general species used are *Golenkinia*, *Euglena*, *Ankistrodesmus*, *Micractinium*, and so on (Dadrasnia *et al.*, 2017). Microbial fuel cells (Fig. 2.13) are also utilized for this purpose. Microbial fuel cell system uses microorganisms' metabolic activity in order to generate electricity. It can be utilized to break down the organic matter in wastewater. Microbial fuel cell technology has become an innovation in renewable energy technology by degrading organic pollutants in wastewater (Mustakeem, 2015; Wang *et al.*, 2015). The bacteria on the anode decompose the organic matter and release H^+ ions and electrons. These electrons are transferred to anode and flow to the cathode linked by the conductive material containing a resistor;

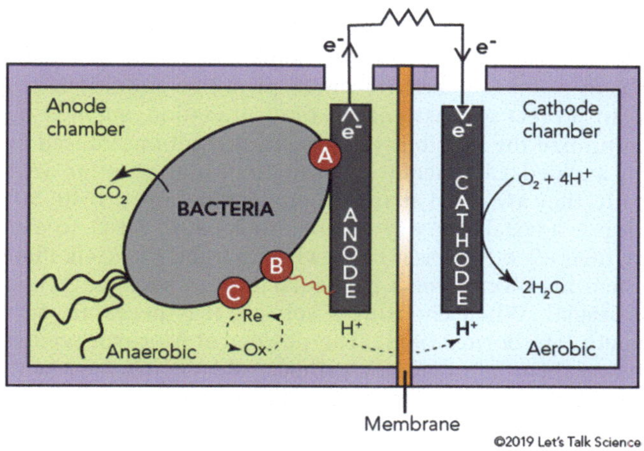

Figure 2.13 Microbial fuel cell technology (www.lotusArise.com).

thus, producing electricity (Hernandez-Flores *et al.*, 2015; Ramadan & Purwono, 2017). The H^+ ions flow through the semi-permeable membrane and cathode. The electrons from the cathode may also combine with the dissolved oxygen and H^+ ions to form pure water (Brown, 2012).

2.4.4 Miscellaneous materials

Apart from the above-mentioned natural materials, there are other substances on which very few studies have been conducted. These materials are listed in Table 2.2.

Table 2.2 Miscellaneous items used in water treatment.

Scientific Name	Family	Genera	Plant Part	Habit	Uses	References
Phaseolus vulgaris L.	Fabaceae	Phaseolus	Seed	Herb	Coagulation	Šćiban *et al.* (2006)
Dolichos Lablab L.	Fabaceae	Lablab	Fruit	Herb	Disinfection	Zhang *et al.* (2006)
Mangifera Indica L.	Anacardaceae	Mangifera	Fruit	Tree	Coagulation	Qureshi *et al.* (2011)
Cicer arietinum L.	Fabaceae	Cicer	Seeds	Herb	Coagulation and disinfection	Choubey *et al.* (2012)
Phoenix spp.	Arecaceae	Phoenix	Seeds	Tree	Coagulation	Al-Sameraiy (2012)
Luffa cylinderca M. Roem	Cucurbitaceae	Luffa	Fruit	Vine	Coagulation	Sowmeyan *et al.* (2011)
Coccinia indica (L.) Voight	Cucurbitaceae	Coccinia	Fruits	Vine	Coagulation	Patale and Pandya (2012)
Hibiscus sabdarifa L.	Malvaceae	Hibiscus	Calyx	Herb	Coagulation and disinfection	Sowmeyan *et al.* (2011)
Parkinsonia aculeata L.	Fabaceae	Parkinsonia	Seed	Tree	Coagulation	Marobhe and Gunaratna (2012)

2.5 PROCESSING OF THE NATURAL FILTER AIDS

The preparation of the filter aids generally involves simple processing. For preserving plant materials, open-air drying at ambient room temperature under primitive conditions is advised to avoid heat and moisture and also to protect against enzymatic degradation, such as hydrolysis of glucosides (Harborne, 1998). To minimize the possibility of chemical reactions caused by ultraviolet radiation, direct sunlight is usually avoided. Chemical transformations may generate artefacts after exposure to ultraviolet light. Generally, they are dried at an optimum temperature of 40–50°C, and if thermolabile compounds are present, it is advisable to dry the materials below 30°C, to avoid degradation. In the case of leaves and seeds, their effectiveness is more when extracts are used. Plant materials are usually extracted by soaking them in solvents for a particular time. Choosing the right solvent for solvent extraction is therefore crucial. When choosing a solvent, it is necessary to consider its solubility, selectivity, cost, and safety (Harborne, 1998). The powder of Moringa seeds is usually obtained by shelling the seeds, grinding them in an electric grinder, and sieving them with a sieve. This powder is then suspended in water and kept in contact for a definite amount of time. Moringa seed powder reduces turbidity and coliform count in treated water, while the extracts exhibit good antibacterial activity, especially against *Salmonella typhii* (Delelegn *et al.*, 2018). The extracts prepared out of leaves and seeds are filtered and stored in a refrigerator for further use. Cactuses are generally peeled, and the gel part is oven-dried, powdered, and sieved until further use. Seed extracts of tamarind and dates are commonly used in water treatment studies.

Now in case of preparation of filter film, it is done by using chitosan. Chitosan is used to prepare membrane filters by using a casting solution method (Rupiasih *et al.*, 2016). A plain chitosan sheath is prepared by dissolving the desired quantity of chitosan (generally between 1 and 2% for the purpose of water treatment) in 1% acetic acid at room temperature. Another sheath with extracts can be prepared by adding the extracts in a 1:1 ratio. Then they are cast onto a muslin cloth or Petri dish without any air bubbles. The membranes are then dried in a hot air dryer at 40°C for 6–7 h. Apart from chitosan, other polysaccharides are also used in water remediation. Cellulose for instance, has an enhanced adsorption capacity for Cr (III) and Fe (III) (Díaz Bukvic *et al.*, 2023), Cd (II), Cu (II), Zn (II), and Pb (II) (Zhou *et al.*, 2020). Its adsorption capacity can be improved by introducing charged groups into the structure of cellulose. For example, carboxymethylation introduces carboxylic groups into cellulose. Using cellulose derivatives from CMC and a freeze–thaw treatment without organic crosslinkers, cellulose sponges were obtained without organic crosslinking with different concentrations of sulphuric acid (Lin *et al.*, 2021). It also effectively removed dyes such as Acid Blue 92 dyes, Acid Red 9, and Acid Red 13 (Kono, 2015). Alginate is a polysaccharide abundant in marine brown algae such as *Ascophyllum nodosum, Macrocystis pyrifera, Phaeophyceae*, and so on. Cationic pollutants are well absorbed by calcium alginate beads. The adsorption capacities for these beads increased with increasing cation size: Pb (II) > Cu (II) > Cd (II). This suggests that larger ions fit better in binding sites (He & Chen, 2014; Papageorgiou *et al.*, 2006). Polyaniline/sodium alginate composite was used for the removal of both cationic and anionic dyes such as Methylene Blue, rhodamine B, orange II, and methyl orange (Majhi & Patra, 2020). Starch is another polysaccharide focused on by researchers. Amylose and amylopectin are its two homopolysaccharides. There are several common sources of starch, including maize, potatoes, wheat, and rice. Native starch generally shows poor adsorption capacity (Awokoya *et al.*, 2021), hence requiring chemical modifications. A maximum adsorption capacity value almost three times higher than native starch was obtained for Methylene Blue after crosslinking with epichlorohydrin and hydrolysis with amylase (Guo *et al.*, 2013). Esterification of starch with maleic acid and itaconic acids helped increase the adsorption capacity values for Pb (II), Zn (II), Cd (II), and Ni (II) (Soto *et al.*, 2015, 2016).

An attempt to develop a simple cartridge (Fig. 2.4: a lab-scale filter) was done using food by-products such as peels, seeds, chitosan, and so on inside a plastic bottle with one end cut off in the food science laboratory of KUFOS. Citrus, banana peels and papaya seeds were cleaned and then dried in a tray dryer at 40°C for 2–3 days, then powdered and washed to remove the colour and starch. Clean coir were separated into threads, washed, dried, and cut into uniform sizes. Banana peels are washed, ground coarsely, and soaked in a slaked lime solution for 2 h to remove soluble starch. Then they were dried in a tray dryer at 50°C. Coconut shells are burned at a high temperature (500–800°C), soaked in $ZnCl_2$ for activation for 24 h, dried, and then used. Cassia seeds are powdered, suspended in solvents, and stirred for 2 days on a magnetic stirrer. Similarly, Moringa seed extract is prepared by dissolving dry seed powder insolvents. Moringa leaves are macerated and suspended in solvent. Dried chitosan sheets were cut to fit the bottle diameter and stacked in the bottle in such a way that the chitosan sheet with extract is on top of that of the plain chitosan sheet. Over the chitosan sheath, coarsely ground dried banana peels were stacked, followed by dried coir. An absorbent cotton was also placed at the bottom outlet of the bottle.

The sample water is then treated with fruit peel or seed powders for a definite period of time, and filtered using muslin cloth or Whatman No. 1 filter paper to remove coagulants. This water is then passed through the developed filter. Following filtration, this water is passed through coconut shell-activated charcoal.

2.6 EFFECT OF FILTERING AIDS ON PHYSICAL, CHEMICAL, AND MICROBIOLOGICAL WATER QUALITY

Physical factors that influence water quality include colour, smell, turbidity, pH, and taste. Water should be colourless, but if the water has colour, it is usually due to industrial colorants or humic substances dissolved in it. The turbidity is generally caused by the presence of microorganisms or colloidal particles. It is important to note that both organic and inorganic aspects are included in the chemical aspects. Disinfection by-products and disinfection action are considered as the major causes of chemical pollutants in water. These compounds possess significant risks to human health when exposed to prolonged levels (UNICEF, 2008). The presence of some of these substances in the food chain can lead to biomagnification, including cadmium, chromium, mercury, and lead, which can accumulate and cause health hazards. Different governmental organizations and international groups have proposed detailed water quality standards, which include Bureau of Indian Standards (BIS), the European Economic Community (EEC) and the World Health Organization (WHO) (Table 2.3).

Water filtration is essential for maintaining the health and safety of all living organisms and the sustainability of the environment. The use of a technique or methodology to filter or purify water, aims to eliminate pathogens and prevent diseases or illnesses caused by water bodies. The natural plant-based treatment techniques aid in water treatment/purification with an aim to valorize biodiversity as well as to establish safe and cost-effective treatment methods. A wide range of plant parts including roots, leaves, stems, pith, husk, and seeds have been tested for their flocculating capacity, coagulation efficiency, TSS, and COD properties, and their capacity to bring down/alter these parameters to the desired standards specified by International/national bodies. The probable mechanism of action of natural coagulants is shown in Figure 2.5.

Natural coagulants have a similar mechanism of action to polyelectrolytes and contain a variety of functional groups such as $-NH_2$, $-OH$, and $-COOH$ (Alazaiza *et al.*, 2022). A natural coagulant acts by combing and flocculating, neutralizing charge, compressing double layers, and bridging antiparticles. As a result of hydrolysis, amorphous metal hydroxide precipitates are formed by sweep flocculation/coagulation. The process reduces colloids by entrapping and enmeshing them within a net-like structure. The flocs produced are generally smaller in size with good settling capacity (Jiao *et al.*, 2017). Despite the larger size of their smaller particles, flocs formed by the charge neutralization method are stronger than those formed by sweep flocculation. By infiltrating the double layer covering colloidal particles with ions with a charge opposite to theirs, double-layer compression can be achieved. The counter ions will thin and shrink the double layer (Ghernaout *et al.*, 2020).

Table 2.3 Standards for physical and chemical quality (Bureau of Indian Standards, 2012; Sayre, 1988).

Characteristic	EEC	U.S.	BIS	WHO
Physical quality				
Turbidity (NTU)	0–4	1–5	1–5	<5
Colour (TCU)	20 mg Pt.Co/L	15	5–15	15
Odour (TON)	0–2 dilution Number at 12°C	3	Agreeable	No standard
Taste	2 dilution Number at 12°C	–	Agreeable	–
pH	6.5–8.5	6.5–8.5	6.5–8.5	6.5–8
Chemical quality				
Arsenic	0.05	0.05	0.01	0.01
Barium	0.1	1.0	0.7	0.7
Cadmium	0.005	0.01	0.003	0.003
Chromium	0.05	0.05	0.05	0.05
Fluoride	1.5	4.0	1.0	1.5
Lead	0.05	0.05	0.01	0.01
Mercury	0.001	0.002	0.001	0.001
Nitrate	50 (as NO_3^-)	10 (as N)	45 (as NO_3^-)	50 (as NO_3^-) 3 (as NO_2^-)
Selenium	0.01	0.01	0.01	0.01
Silver	0.01	0.05	0.1 (as Ag)	No standard
Aluminium	–	–	0.03	0.2
Chloride	250	250	250	25[a]
Hardness (as $CaCO_3$)	–	–	200	–
Copper	1	1	0.05	0.1
Total dissolved solids (TDS)	500	500	500	–
Iron	0.3	0.3	0.3	0.3
cyanide	–	–	0.05	–
Manganese	0.05	0.05	0.1	0.05

[a]Maximum admissible concentration not reported.

As a result of inter-particle bridge, polymers are formed as long chains with highly reactive groups dangling from them in water, and certain regions of the chain will attach to the colloids. In a complex colloid–polymer–colloid structure, the polymer is used as a bridge between the unattached polymeric chain and other colloidal particles. A floc formed through this mechanism is flaky with irregular void spaces between network structures and not complex in nature (Nimesha *et al.*, 2022). This mechanism was also reported in case of Cassia seeds and cactus mucilage. There is an electrostatic attraction between contaminant particles and coagulants, which leads to the adsorption, charge reversal, and neutralization of contaminant particles, which ultimately makes them easier to remove (Bolto & Gregory, 2007).

M. oleifera seed powder were also observed to effectively reduce turbidity of water samples by 83–87% at a concentration of 0.016 g/l and concentration above this resulted in higher turbidity. The acetone extract of the seed also exhibited excellent activity against pathogens, implying that compounds such as, polyphenols and tannins which portray antibacterial activity are extracted using acetone (Delelegn *et al.*, 2018). Another study revealed that compared to alum at 50 mg/l, hexane extract of *M. oleifera* leaves significantly improved the quality parameters of water such as, fluoride, turbidity, hardness, TDS, pH and *E. coli* reduction (Aziz *et al.*, 2023). In the case of papaya seeds, its defatted salt extract revealed superior total coliform removal (96.3%) and residual turbidity of 4.25 NTU, which are acceptable based on WHO standards (Yimer & Dame, 2021). When papaya seed powder was used, it showed 89.1% turbidity removal efficiency along with a 90.3% reduction in TDS at a dosage level of 0.6 g/l (Chandran & George, 2018). The powder obtained from papaya seeds contains functional groups such as −OH, −NH, and C = O that indicate good potential as natural coagulants. It also removed 84.8% of drimarene dark red (DDR) in textile wastewater at a dosage of 0.57 g/l and pH 2 (Kristianto *et al.*, 2018). The orange peel treated with acetic acid was effective (98.5%) in removing antimony from water with a contact time of 150 min (Hasan *et al.*, 2021). Though the use of *Cassia fistula* seeds in water treatment is not much explored, the chloroform extracts of *Cassia fistula* seeds were reported to exhibit high sensitivity against *Streptococcus pyogenes* and *Streptococcus aureus* than *E. coli* and *Pseudomonas aeruginosa*, and antifungal activity against *Candida albicans* than *Aspergillus niger* and *Aspergillus clavatus* (Bhalodia *et al.*, 2011). The methanolic seed extract of *C. fistula* did not produce any significant toxic effects when administered orally to mice, suggesting its potential for pharmaceutical use (Jothy *et al.*, 2011). *Cassia fistula* seed gum was recently proven to be effective in removing dye from wastewater with an efficiency of 93.8% at a dosage of 1.17 mg/L (Dao *et al.*, 2021). Another study wherein the Cassia seeds were converted into activated carbon modified with HNO_3 showed good absorption efficiency for zinc (78.7%) (Trung & Luu, 2021).

An attempt using rice husk, charcoal, *M. oleifera* seed and fine sand to treat sewage water was proven to be successful and when analysed for pH, turbidity, COD and BOD, the water quality parameters were within limits (Akshay & Jeyaprakash, 2020). It was revealed that the developed method had a filtration efficiency of 85%. The pH of the treated water was found to be 7, turbidity – 12.36 NTU, BOD – 10.62 mg/l and COD of 15.04 mg/l.

An attempt was made to develop the water filtering aid using the natural ingredients at the Food Science laboratory, Kerala University of Fisheries and Ocean Studies, Kochi. The ingredients included *Moringa oleifera* leaves and seed extracts, *Cassia fistula* seed extract, banana peel, orange peel, papaya seeds, chitosan and coir. The filtering aid thus developed showed significant improvement in the water quality parameters of aquaculture pond water (Harinarayanan *et al.*, 2021). Though, this unit needs further modifications and streamlining, these ingredients were proven to be highly efficient natural filtering sources that could be used to develop a sustainable and cost-effective water treatment techniques. The water quality results are shown in Table 2.4.

Iron content was below the detectable level in water from backwaters but was found to be increased in filtered water. This may have occurred since the water has a pH above 5, where the tannic acid in the plant extract dissociates and the ferrous irons combine with tannin anions and form ferrous-tannate compounds which when oxidized, release ferric ions (Hem, 1960). Biochemical oxygen demand was reduced in filtered water, which indicates better water quality. The most probable number (MPN) of both filtered water samples acknowledged a reduced microbiologic effect of the membrane filter, which may be because the antimicrobial activities of the plant extracts are inhibited due to the encapsulation of bioactive phenols and flavonoids in chitosan during membrane development. The flow rate (under gravity) was determined to be 0.07 ml/s, and this slow rate may be due to the increased concentration of chitosan. Overall, it is understandable that ingredients have given their significant contributions in purifying/altering the properties of treated water. However, more research needs to be done in this area, so that an improved, safe water treatment technology can be developed.

It is important to note that water in daily life is significant and unfathomable. With limited resources, it may not be possible to meet the increasing demand. Filtration and reuse could be sustainable options. However, the high cost, availability of the raw materials, disposal, and so on could be of great concern and limiting factors even for developing countries including India. Therefore, it is urgent to develop a simple and cost-effective solution by utilizing natural sustainable resources including by-products of agriculture and the food industry. As an alternative to conventional water filtration

Table 2.4 Physicochemical properties of water before and after filtration (unpublished data).

Parameters	Results	
	Before Filtration	**After Filtration**
Odour	Disagreeable	Agreeable
Taste	Disagreeable	Disagreeable
Turbidity (NTU)	37	12.5
Total dissolved solids (mg/L)	1190	437
Conductivity at 25°C (μS/cm)	1983	729
pH at 25°C	5	7
Total chloride (as Cl) (mg/L)	257	176
Total alkalinity (as $CaCO_3$) (mg/L)	151	33.1
Total hardness (as $CaCO_3$) (mg/L)	170	68
Calcium hardness (as $CaCO_3$) (mg/L)	60	28
Calcium (as Ca) (mg/L)	24	11.2
Magnesium (as Mg) (mg/L)	26.7	9.72
Sulphate (as SO_4) (mg/L)	62.3	28.9
Nitrate (as NO_3) (mg/L)	28.1	BDL
Fluoride (as F) (mg/L)	BDL	BDL
Iron (as Fe) (mg/L)	BDL	1.27
Copper (as Cu) (mg/L)	BDL	BDL
Manganese (as Mn) (mg/L)	BDL	BDL
Sodium (as Na) (mg/L)	125	88.9
Potassium (as K) (mg/L)	68.9	25.6
Biochemical oxygen demand (BOD) (mg/L)	80	4

BDL: below detectable level.

systems, plant material, and chitosan membrane filters offer rich bioactive components that could be utilized in water purification. It is possible for innovations in water filtration to be achieved through the use of novel plant-based bioactive components and other biopolymers (such as alginates, starch, etc.). Agriculture by-products exhibit greater flexibility and are less sensitive to flow fluctuations than other adsorption methods. It would also reduce the environmental impact of wastewater effluents through this innovation.

Plant extracts may have effects on quality parameters such as colour, taste, and odour (Ndabigengesere & Narasiah, 1998). These may be enhanced by the microbial decomposition of its organic compounds; therefore, the shelf stability of such treated water is not more than 24 h (Jahn, 1986). It was found that many plant-based flocculants in the laboratory showed significant flocculation behaviour, and their pollutant removal efficiency often exceeded 90% depending on the type of plant material and the characteristics of the water being treated.

2.7 CONCLUSION

It is important to note that water in daily life is significant and unfathomable. With growing population and limited water resources, it may not be possible to meet the increasing demand. The three mantras, reduce recycle, and reuse are the only sustainable options. The reduction in water usage is inevitable with the growing population. Nevertheless, the recycle and reuse are the existing options. However, the high cost, availability of the raw materials, disposal, and so on could be of great concern and limiting factors even for developing countries including India. Therefore, it is urgent to develop a simple and cost-effective solution by utilizing natural sustainable resources including by-products of agriculture and the food industry. As an alternative to conventional water filtration systems, plant materials, chitosan and other biopolymers like starch, alginates, and so on offer novel possibilities of using these in water purification. It is possible for innovations in water filtration through the use of novel plant-based bioactive components and other biopolymers. Agriculture by-products exhibit greater flexibility and are less sensitive to flow fluctuations than other adsorption methods. It would also reduce the environmental impact of wastewater effluents through this innovation.

2.8 FUTURE PERSPECTIVES

Plants can potentially be used as cheap, readily available alternatives to water treatment agents or disinfectants. With these natural materials, processes such as chlorination of water can be avoided without causing harm to the environment. There is a need for follow-up research to standardize the use of medicinal plants in water disinfection. For plant extracts to be used as a coagulant in the purification of drinking water for human consumption, further testing, including the assessment of toxicity, is necessary. To meet guideline values, faecal coliforms need to be reduced to the maximum extent possible. A greater number of plant extracts should be tested for their ability to purify water. To improve the quality of their water, rural villages in India need to develop a localized water purification system using plant extracts. Phytochemicals need to be refined for industrial use to overcome the chemical complexity, poor water solubility, lack of standardization, and slow pace of working, purification, and extraction intricacies. Finally, an evaluation of the techno-economic effects of green flocculants should be assessed on a large scale, in real-world situations, with a wide range of wastewater treatment systems (Harinarayanan *et al.*, 2021).

REFERENCES

Abbas N. M., Abbas S. F. and Ibrahim A. S. (2014). Cyanide removal from waste water by using banana peel. *Journal of Asian Scientific Research*, **4**(5), 239–247.

Abbasi Z., Alikarami M., Nezhad E. R., Moradi F. and Moradi V. (2013). Adsorptive removal of Co^{2+} and Ni^{2+} by peels of banana from aqueous solution. *Universal Journal of Chemistry*, **1**(3), 90–95, https://doi.org/10.13189/ujc.2013.010303

Abdel-Fatah M. A. (2018). Nanofiltration systems and applications in wastewater treatment. *Ain Shams Engineering Journal*, **9**(4), 3077–3092, https://doi.org/10.1016/j.asej.2018.08.001

Abdullah N. H., Mohamed N., Sulaiman L. H., Zakaria T. A. and Rahim D. A. (2016). Potential health impacts of bauxite mining in Kuantan. *The Malaysian Journal of Medical Sciences: MJMS*, **23**(3), 1.

Abidin Z. Z., Norhafizah M., Robiah Y. and Aishah D. (2019). Effect of storage conditions on *Jatropha curcas* performance as biocoagulant for treating palm oil mill effluent. *Journal of Environmental Science and Technology*, **12**(2), 92–101, https://doi.org/10.3923/jest.2019.92.101

ACAAI (American College of Allergy, Asthma and Immunology) (2023). Chlorine allergy. Retrieved on 15th October 2023 from https://acaai.org/allergies/allergic-conditions/chlorine-allergy/

Adewumi I. and Ogbiye A. S. (2009). Using water hyacinth (Eichhornia crassipes) to treat wastewater of a residential institution. *Toxicological and Environmental Chemistry*, **91**(5), 891–903.

Ahirrao S. D. (2008). Effect of a common water purifier potash alum on drinking water quality and human health. *Nature Environment and Pollution Technology*, **7**(1), 79.

Akshay J. R. and Jeyaprakash S. (2020). Recycling of wastewater into useable water by using natural materials. *International Journal of Engineering Research & Technology (IJERT)*, **09**, 16–18, https://doi.org/10.17577/IJERTV9IS070014

Alazaiza M. Y., Albahnasawi A., Ali G. A., Bashir M. J., Nassani D. E., Al Maskari T., Abu Amr S. S. and Abujazar M. S. S. (2022). Application of natural coagulants for pharmaceutical removal from water and wastewater: a review. *Water*, **14**(2), 140, https://doi.org/10.3390/w14020140

Ali S. K. H., Vrandana K. G., Shetty K. H. and Deepika B. V. (2021–22). Treatment of sullage water using tamarind seed powder as coagulant using coagulant process. Karnataka State 45th Series Student Project Programme (SPP). Project ref no: 45S_BE_0015, https://www.kscst.org.in/spp/45_series/SPP45S/01_Seminar_Projects/040_45S_BE_0015.pdf

Al-Sameraiy, M. 2012. A novel water pretreatment approach for turbidity removal using date seeds and pollen sheath. *Journal of Water Resource and Protection*, **4**, 79–92.

Amran A. H., Zaidi N. S., Syafiuddin A., Zhan L. Z., Bahrodin M. B., Mehmood M. A. and Boopathy R. (2021). Potential of *Carica papaya* seed-derived bio-coagulant to remove turbidity from polluted water assessed through experimental and modeling-based study. *Applied Sciences*, **11**(12), 5715, https://doi.org/10.3390/app11125715

Anwar F., Latif S., Ashraf M. and Gilani A. H. (2007). *Moringa oleifera*: a food plant with multiple medicinal uses. *Phytotherapy Research: An International Journal Devoted to Pharmacological and Toxicological Evaluation of Natural Product Derivatives*, **21**(1), 17–25, https://doi.org/10.1002/ptr.2023

Aonghusa C. N. and Gray N. F. (2002). Laundry detergents as a source of heavy metals in Irish domestic wastewater. *Journal of Environmental Science and Health, Part A*, **37**(1), 1–6, https://doi.org/10.1081/ESE-100108477

Arunachalam A., Chaudhuri R. G., Iype E. and Prakash Kumar B. G. (2018). Surface modification of date seeds (*Phoenix dactylifera*) using potassium hydroxide for wastewater treatment to remove azo dye. *Water Practice & Technology*, **13**(4), 859–870, https://doi.org/10.2166/wpt.2018.095

Awokoya K. N., Oninla V. O. and Bello D. J. (2021). Synthesis of oxidized *Dioscorea dumentorum* starch nanoparticles for the adsorption of lead (II) and cadmium (II) ions from wastewater. *Environmental Nanotechnology, Monitoring & Management*, **15**, 100440, https://doi.org/10.1016/j.enmm.2021.100440

Ayodele O., Olusegun S. J., Oluwasina O. O., Okoronkwo E. A., Olanipekun E. O., Mohallem N. D., … and Duarte H. A. (2021). Experimental and theoretical studies of the adsorption of Cu and Ni ions from wastewater by hydroxyapatite derived from eggshells. *Environmental Nanotechnology, Monitoring & Management*, **15**, 100439, https://doi.org/10.1016/j.enmm.2021.100439

Aziz H., Ahmad A., Zahra A., Ali H. U., Qaisrani T. B., Razzaq W., … and Bilal M. (2023). Evaluating the purification potential of *Moringa oleifera* seed in drinking water. *Jammu Kashmir Journal of Agriculture*, **3**(2), 59–69.

Baluev I. (2021). Dairy Farm – Biogas production. Project report. Technische Hochschule Ingolstadt. Retrieved on 11th October 2023, from https://www.researchgate.net/publication/351330013

Bassi R., Prasher S. O. and Simpson B. K. (2000). Removal of selected metal ions from aqueous solutions using chitosan flakes. *Separation Science and Technology*, **35**(4), 547–560, https://doi.org/10.1081/SS-100100175

Benalia A., Derbal K., Khalfaoui A., Bouchareb R., Panico A., Gisonni C., … and Pizzi A. (2021). Use of *Aloe vera* as an organic coagulant for improving drinking water quality. *Water*, **13**(15), 2024, https://doi.org/10.3390/w13152024

Bhalodia N. R., Nariya P. B., Acharya R. N. and Shukla V. J. (2011). Evaluation of in vitro antioxidant activity of flowers of *Cassia fistula* Linn. Coden (USA).

Bolto B. and Gregory J. (2007). Organic polyelectrolytes in water treatment. *Water Research*, **41**(11), 2301–2324, https://doi.org/10.1016/j.watres.2007.03.012

Brown J. L. (2012). Fuel cells treat wastewater, generate electricity. *Civil Engineering Magazine Archive*, **82**(7), 34–35, https://doi.org/10.1061/ciegag.0000643

Bu J., Li W., Niu N., Guo N., Zhou H., Chen C. and Ding A. (2021). Adsorption of Cr (VI) from wastewater by iron-modified coconut shell biochar. E3S Web of Conferences. EDP Sciences, Vol. **248**, p. 01059, https://doi.org/10.1051/e3sconf/202124801059

Bureau of Indian Standards. (2012). Drinking water- Specification (Second Revision). Retrieved on 6th February 2023.

Castro R. S., Caetano L., Ferreira G., Padilha P. M., Saeki M. J., Zara L. F., … and Castro G. R. (2011). Banana peel applied to the solid phase extraction of copper and lead from river water: pre-concentration of metal ions with a fruit waste. *Industrial & Engineering Chemistry Research*, **50**(6), 3446–3451, https://doi.org/10.1021/ie101499e

Çelebi H., Gök G. and Gök O. (2020). Adsorption capability of brewed tea waste in waters containing toxic lead (II), cadmium (II), nickel (II), and zinc (II) heavy metal ions. *Scientific Reports*, **10**(1), 17570, https://doi.org/10.1038/s41598-020-74553-4

Chandran A. and George D. (2018). Use of papaya seed as a natural coagulant for water purification. *International Journal of Scientific Engineeering and Research (IJSER)*, **6**(3), 41–46.

Charcosset C. (2012). *Membrane Processes in Biotechnology and Pharmaceutics*. Elsevier. https://doi.org/10.1016/C2011-0-04261-8

Choubey S., Rajput S. K. and Bapat K. N. (2012). Comparison of efficiency of some natural coagulants – bioremediation. *International Journal of Emerging Technology and Advanced Engineering*, **2**(10), 429–434.

Chowdhury P., Viraraghavan T. and Srinivasan A. (2010). Biological treatment processes for fish processing wastewater – a review. *Bioresource Technology*, **101**(2), 439–449, https://doi.org/10.1016/j.biortech.2009.08.065

Ciabattia I., Cesaro F., Faralli L., Fatarella E. and Tognotti F. (2009). Demonstration of a treatment system for purification and reuse of laundry wastewater. *Desalination*, **245**(1–3), 451–459, https://doi.org/10.1016/j.desal.2009.02.008

Clanton C. J., Goodrich P. R., Fox E. J., Backus B. D. and Morris H. A. (1990). Anaerobic digestion of cheese whey. Proceedings 1987 Food Processing Waste Conference, Atlanta, Georgia.

Climate Policy Watcher. (2023). Characteristics of Agricultural and Food Wastewater - Wastewater Treatment. Retrieved from https://www.climate-policy-watcher.org/wastewater-treatment/characteristics-of-agricultural-and-food-wastewater.html

Coir Board, Ministry of MSME, Government of India. (2014). Coirboard.gov.in (accessed 05 August 2023)

Colic M., Morse W., Hicks J., Lechter A. and Miller J. D. (2007). Case study: fish processing plant wastewater treatment. In Industrial Wastes (IW) Conference 2007 (pp. 1–27). Water Environment Federation.

Crini G. and Lichtfouse E. (2019). Advantages and disadvantages of techniques used for wastewater treatment. *Environmental Chemistry Letters*, **17**, 145–155, https://doi.org/10.1007/s10311-018-0785-9ff.ffhal-02082890f

Dadrasnia A., Usman M. M., Lim K. T., Velappan R. D., Shahsavari N., Vejan, P., Mahmud A. F. and Ismail S. (2017). Microbial aspects in wastewater treatment-a technical review. *Environmental Pollution and Protection*, **2**(2), 75–84, https://doi.org/10.22606/epp.2017.22005

Dao M. T., Nguyen V. C. N., Tran T. N., Nguyen X. D., Vo D. T., Nguyen V. K. and Hoang L. T. T. T. (2021). Pilot-scale study of real domestic textile wastewater treatment using *Cassia fistula* seed-derived coagulant. *Journal of Chemistry*, **2021**, 1–8, https://doi.org/10.1155/2021/7608856

Dash P. (2022). Water hyacinth proves to be a boon for wastewater treatment. Ecodeaz. Retrieved on 7th March 2023 from https://www.ecoideaz.com/innovative-green-ideas/water-hyacinth-proves-to-be-a-boon-for-wastewater-treatment

de Oliveira Zonato R., Estevam B. R., Perez I. D., dos Santos Ribeiro V. A. and Boina R. F. (2022). Eggshell as an adsorbent for removing dyes and metallic ions in aqueous solutions. *Cleaner Chemical Engineering*, **2**, 100023, https://doi.org/10.1016/j.clce.2022.100023

Delelegn A., Sahile S. and Husen A. (2018). Water purification and antibacterial efficacy of *Moringa oleifera* Lam. *Agriculture & Food Security*, **7**(1), 1–10, https://doi.org/10.1186/s40066-018-0177-1

Desai U. T. and Wagh A. N. (1995). Papaya. In: Handbook of Fruit Science and Technology: *Production, Composition, Storage and Processing*, D. K. Salunkhe and S. S. Kadam (eds), CRC Press, Boca Raton, pp. 313–330, https://doi.org/10.1201/9781482273458

Díaz Bukvic G., Rossi E. and Errea M. I. (2023). Polysaccharides as economic and sustainable raw materials for the preparation of adsorbents for water treatment. *Polysaccharides*, **4**(3), 219–255, https://doi.org/10.3390/polysaccharides4030016

Dutta J., Tripathi S. and Dutta P. K. (2012). Progress in antimicrobial activities of chitin, chitosan and its oligosaccharides: a systematic study needs for food applications. *Food Science and Technology International*, **18**(1), 3–34, https://doi.org/10.1177/1082013211399195

Edwin G. A., Gopalsamy P. and Muthu N. (2014). Characterization of domestic gray water from point source to determine the potential for urban residential reuse: a short review. *Applied Water Science*, **4**, 39–49, https://doi.org/10.1007/s13201-013-0128-8

Effendi H., Hydayah R. and Hariyadi S. (2017). Tamarindus indica Seed as Natural Coagulant for Traditional Gold Mining Wastewater Treatment. *World Applied Sciences Journal*, **35**(3), 330–333.

El Barnossi A., Moussaid F. and Housseini A. I. (2021). Tangerine, banana and pomegranate peels valorisation for sustainable environment: a review. *Biotechnology Reports*, **29**, e00574, https://doi.org/10.1016/j.btre.2020.e00574

El-Azazy M., El-Shafie A. S. and Al-Shaikh Yousef B. (2021). Green tea waste as an efficient adsorbent for methylene blue: structuring of a novel adsorbent using full factorial design. *Molecules*, **26**(20), 6138, https://doi.org/10.3390/molecules26206138

El-Gindy A. and Youssif M. (2015). Chemical studies and utilization of *Tamarindus indica* and its seeds in some technological application. *Egyptian Journal of Nutrition and Health*, **10**(1), 1–15, https://doi.org/10.21608/ejnh.2015.4825

Eriksson E., Auffarth K., Eilersen A. M., Henze M. and Ledin A. (2003). Household chemicals and personal care products as sources for xenobiotic organic compounds in grey wastewater. *Water SA*, **29**(2), 135–146, https://doi.org/10.4314/wsa.v29i2.4848

Eriksson E., Srigirisetty S. and Eilersen A. M. (2010). Organic matter and heavy metals in grey-water sludge. *Water SA*, **36**(1), 130–142, https://doi.org/10.4314/wsa.v36i1.50921

FAO-Food and Agriculture Organization of the United Nations FAOSTAT. (2016). http://www.fao.org/faostat/es/#data/QC/visualize (accessed 05 August 2023)

Ferrero F., Periolatto M., Vineis C. and Varesano A. (2014). Chitosan coated cotton gauze for antibacterial water filtration. *Carbohydrate Polymers*, **103**, 207–212, https://doi.org/10.1016/j.carbpol.2013.12.037

Freitas T. K. F. S., Almeida C. A., Manholer D. D., Geraldino H. C. L., de Souza M. T. F. and Garcia, J. C. (2018). Review of utilization plant-based coagulants as alternatives to textile wastewater treatment. In: Detox Fashion: Textile Science and Clothing Technology, S. Methu (ed.), pp. 27–79, Springer, Singapore.

Friedler E. and Hadari M. (2006). Economic feasibility of on-site greywater reuse in multi-storey buildings. *Desalination*, **190**(1–3), 221–234, https://doi.org/10.1016/j.desal.2005.10.007

Ghaitidak D. M. and Yadav K. D. (2013). Characteristics and treatment of greywater – a review. *Environmental Science and Pollution Research*, **20**(5), 2795–2809, https://doi.org/10.1007/s11356-013-1533-0

Ghernaout D., Elboughdiri N. and Ghareba S. (2020). Fenton technology for wastewater treatment: dares and trends. *Open Access Library Journal*, **7**(1), 1–26.

Godbole A., Wadetwar R. N., Lawal T. O., Mahady G. B. and Raut N. A. (2023). Microbiology of waste treatment. In: 360-Degree Waste Management, Volume 1 Fundamentals Agricultural and Domestic Wase, and Remediation, N. A. Raut, B. A. Bhanvase, S. J. Dhoble, D. M. Kokare and K. R. Randive (eds.), Elsevier, Vol. **1**, pp. 185–211.

Gonzalez J. F. (1996). Wastewater Treatment in the Fishery Industry (No. 355). Food & Agriculture Org.

Grady C. L. Jr, Daigger G. T., Love N. G. and Filipe C. D. (1999). Biological Wastewater Treatment. 2nd edn. Marcel Dekker Inc. New York. Retrieved on 13th October 2023, from https://rb.gy/tnrwxe

Gregory J. and Dhond R. V. (1972). Wastewater treatment by ion exchange. *Water Research*, **6**(6), 681–694, https://doi.org/10.1016/0043-1354(72)90183-2

Gross M. A. and Farrell-Poe K. (2005). University Curriculum Development for Decentralized Wastewater Management. National Decentralized Water Resources Capacity Development Project, University of Arkansas, Fayetteville, AR, pp. 1–35.

Guo L., Li G., Liu J., Meng Y. and Tang Y. (2013). Adsorptive decolorization of methylene blue by crosslinked porous starch. *Carbohydrate Polymers*, **93**(2), 374–379, https://doi.org/10.1016/j.carbpol.2012.12.019

Gupta S., Sharma S. K. and Kumar A. (2019). Biosorption of Ni (II) ions from aqueous solution using modified *Aloe barbadensis* Miller leaf powder. *Water Science and Engineering*, **12**(1), 27–36, https://doi.org/10.1016/j.wse.2019.04.003

Gutierrez R., Delvasto S. and Mesa E. (1998). Problem of alum sludges disposal in Cali, Colombia. 14th International Conference on Solid Waste Technology and Management, Philadelphia, USA, pp. 1–6.

Hahn T. and Zibek S. (2018). Sewage polluted water treatment via chitosan: A. Review. In *Chitin-Chitosan: Myriad Functionalities in Science and Technology*, R. Dongre (ed). IntechOpen, 120–133.

Hanafiah S. F. B. M. (2022). Adsorption Of Chromium, Cadmium and Manganese Ions Using Nanomagnetic Alumina Composites From An Aqueous Solution (Masters thesis, Universiti Sains Malaysia). Retrieved from eprints.usm.my

Harborne J. B. (1998). Methods of extraction and isolation. *Phytochemical Methods*, **3**, 60–66.

Harinarayanan N. C. and Raman M. (2022). Utilization of chitosan and cassia (*Cassia fistula*) seed extract based filters in efficient wastewater management. *Journal of Water Chemistry and Technology*, **44**(6), 458–466, https://doi.org/10.3103/S1063455X22060066

Harinarayanan N. C., Tony A. and Aswathy A. (2021). Bioaided water filtration: an approach to food processing byproduct utilization. Masters project, Kerala University of Fisheries and Ocean Studies, Kochi, Kerala, India.

Hasan M. B., Al-Tameemi I. M. and Abbas M. N. (2021). Orange peels as a sustainable material for treating water polluted with antimony. *Journal of Ecological Engineering*, **22**(2), 25–35, https://doi.org/10.12911/22998993/130632

He J. and Chen J. P. (2014). A comprehensive review on biosorption of heavy metals by algal biomass: materials, performances, chemistry, and modeling simulation tools. *Bioresource Technology*, **160**, 67–78, https://doi.org/10.1016/j.biortech.2014.01.068

He C., Liu Z. and Hodgins M. (2013). Using life cycle assessment for quantifying embedded water and energy in a water treatment system. *Water Research Foundation*, Vol. **1**, pp. 1–110.

Hem J. D. (1960). Complexes of Ferrous Iron with Tannic Acid (No. 1459). In Geological Survey Water-Supply Paper 1459-D, US Government Printing Office.

Hernández-Flores G., Poggi-Varaldo H. M., Solorza-Feria O., Ponce-Noyola M. T., Romero-Castañón T., Rinderknecht-Seijas N. and Galíndez-Mayer J. (2015). Characteristics of a single chamber microbial fuel cell equipped with a low cost membrane. *International Journal of Hydrogen Energy*, **40**(48), 17380–17387, https://doi.org/10.1016/j.ijhydene.2015.10.024

Hesham A., Awad Y., Jahin H., El-Korashy S., Maher S., Kalil H. and Khairy G. (2021). Hydrochar for industrial wastewater treatment: an overview on its advantages and applications. *J. Environ. Anal. Toxicol.*, **11**, 2021.

Ho Y. C., Norli I., Alkarkhi F. M. A. and Morad N. (2014). New vegetal biopolymeric flocculant: a degradation and flocculation study. *Iranian (Iranica) Journal of Energy & Environment*, **5**(1), 26–33, https://doi.org/10.5829/idosi.ijee.2014.05.01.05

Hulett G. A. (1896). Purification of water by distillation. *Journal of Physical Chemistry*, **1**(2), 91–95, https://doi.org/10.1021/j150584a004

Hussain A. Z. and Sheriff K. M. (2014). Removal of heavy metals from wastewater using low cost adsorbents. *Archives of Applied Science Research*, **6**(6), 52–54.

Hussain A., Kumari R., Sachan S. G. and Sachan A. (2021). Biological wastewater treatment technology: advancement and drawbacks. In: Microbial Ecology of Wastewater Treatment Plants, Elsevier, pp. 175–192, https://doi.org/10.1016/B978-0-12-822503-5.00002-3

Huynh A. T., Chen Y. C. and Tran B. N. T. (2021). A small-scale study on removal of heavy metals from contaminated water using water hyacinth. *Processes*, **9**(10), 1802, https://doi.org/10.3390/pr9101802

Ivnitsky H., Katz I., Minz D., Volvovic G., Shimoni E., Kesselman E. … and Dosoretz C. G. (2007). Bacterial community composition and structure of biofilms developing on nanofiltration membranes applied to wastewater treatment. *Water Research*, **41**(17), 3924–3935, https://doi.org/10.1016/j.watres.2007.05.021

Jaeel A. J. (2016). Investigation the optimum combined dosages of date seeds powder as natural coagulant with chemical coagulants in domestic wastewater pretreatment. *Journal of Engineering*, **22**(6), 76–86, https://doi.org/10.31026/j.eng.2016.06.06

Jahn S. A. A. 1986. Monitored water coagulation with *Moringa oleifera* seeds in village household. *Journal of Analytical Science*, **1**, 40–41.

Jiao Y. N., Chen H., Gao R. X., Zhu Y. G. and Rensing C. (2017). Organic compounds stimulate horizontal transfer of antibiotic resistance genes in mixed wastewater treatment systems. *Chemosphere*, **184**, 53–61, https://doi.org/10.1016/j.chemosphere.2017.05.149

Jin P., Jin X., Wang X., Feng Y. and Wang X. C. (2013). Biological activated carbon treatment process for advanced water and wastewater treatment. *Biomass Now-Cultivation and Utilization*, **1**, 153–192.

Jothy S. L., Zakaria Z., Chen Y., Lau Y. L., Latha L. Y. and Sasidharan S. (2011). Acute oral toxicity of methanolic seed extract of *Cassia fistula* in mice. *Molecules*, **16**(6), 5268–5282, https://doi.org/10.3390/molecules16065268

Jung K. W., Choi B. H., Hwang M. J., Jeong T. U. and Ahn K. H. (2016). Fabrication of granular activated carbons derived from spent coffee grounds by entrapment in calcium alginate beads for adsorption of acid orange 7 and methylene blue. *Bioresource Technology*, **219**, 185–195, https://doi.org/10.1016/j.biortech.2016.07.098

Kamalakannan S., Kumar S. C. R., Bose R. S. C., Simon M. S. M. and Sreekumar N. R. S. (2019). Turbid flood water treatment using tamarind by products. *Journal for Innovative Development in Pharmaceutical and Technical Science*, **2**. ISSN (O): 2581-6934, 129–134, https://jidps.com/wp-content/uploads/2019/11/TURBID-FLOOD-WATER-TREATMENT-USING-TAMARIND-BYPRODUCTS.pdf

Kangama A., Zeng D., Tian X. and Fang J. (2018). Application of chitosan composite flocculant in tap water treatment. *Journal of Chemistry*, **2018**, 1–9, https://doi.org/10.1155/2018/2768474

Katubi K. M., Amari A., Harharah H. N., Eldirderi M. M., Tahoon M. A. and Ben Rebah F. (2021). *Aloe vera* as promising material for water treatment: a review. *Processes*, **9**(5), 782, https://doi.org/10.3390/pr9050782

Khaskheli M. I., Memon S. Q., Siyal A. N. and Khuhawar M. Y. (2011). Use of orange peel waste for arsenic remediation of drinking water. *Waste and Biomass Valorization*, **2**, 423–433, https://doi.org/10.1007/s12649-011-9081-7

Khulbe K. C., Feng C. Y., Matsuura T. and Ismail A. F. (2012). Progresses in membrane and advanced oxidation processes for water treatment. *Membrane Water Treatment*, **3**(3), 181–200, https://doi.org/10.12989/mwt.2012.3.3.181

Kim A. S., Chen H. and Yuan R. (2006). EPS Biofouling in membrane filtration: an analytic modeling study. *Journal of Colloid and Interface Science*, **303**(1), 243–249, https://doi.org/10.1016/j.jcis.2006.07.067

Kono H. (2015). Preparation and characterization of amphoteric cellulose hydrogels as adsorbents for the anionic dyes in aqueous solutions. *Gels*, **1**(1), 94–116, https://doi.org/10.3390/gels1010094

Koul B., Bhat N., Abubakar M., Mishra M., Arukha A. P. and Yadav D. (2022). Application of natural coagulants in water treatment: a sustainable alternative to chemicals. *Water*, **14**(22), 3751, https://doi.org/10.3390/w14223751

Kristianto H., Kurniawan M. A. and Soetedjo J. N. (2018). Utilization of papaya seeds as natural coagulant for synthetic textile coloring agent wastewater treatment. *International Journal on Advanced Science, Engineering and Information Technology*, **8**, 2071–2077, https://doi.org/10.18517/ijaseit.8.5.3804

Kurniawan S. B., Abdullah S. R. S., Imron M. F., Said N. S. M., Ismail N. I., Hasan H. A., Othman A. R. and Purwanti I. F. (2020). Challenges and opportunities of biocoagulant/bioflocculant application for drinking water and wastewater treatment and its potential for sludge recovery. *International Journal of Environmental Research and Public Health*, **17**(24), 9312, https://doi.org/10.3390/ijerph17249312

Laîné J. M., Vial, D. and Moulart P. (2000). Status after 10 years of operation-overview of UF technology today. *Desalination*, **131**(1-3), 17–25.

Lee J. I., Kim J. M., Yoo S. C., Jho E. H., Lee C. G. and Park S. J. (2022). Restoring phosphorus from water to soil: using calcined eggshells for P adsorption and subsequent application of the adsorbent as a P fertilizer. *Chemosphere*, **287**, 132267, https://doi.org/10.1016/j.chemosphere.2021.132267

Levine A. D., Tchobanoglous G. and Asano T. (1985). Characterization of the size distribution of contaminants in wastewater: treatment and reuse implications. *Journal (Water Pollution Control Federation)*, **57**(7), 805–816.

Lin X., Jin J., Guo X. and Jia X. (2021). All-carboxymethyl cellulose sponges for removal of heavy metal ions. *Cellulose*, **28**, 3113–3122, https://doi.org/10.1007/s10570-021-03685-1

Liuqin S. (2011). Application of Ultrafiltration in Purification of Drinking Water and Advanced Treatment of Municipal Wastewater. Tsinghua University, Beijing.

Mahmoodi Z., Aghaie H. and Fazaeli R. (2020). Application of response surface methodology for optimization of cadmium removal by *Aloe vera*/carboxylated carbon nanotubes nanocomposite-based low-cost adsorbent. *Materials Research Express*, **7**(6), 065015, https://doi.org/10.1088/2053-1591/ab9cee

Majhi D. and Patra B. N. (2020). Polyaniline and sodium alginate nanocomposite: a pH-responsive adsorbent for the removal of organic dyes from water. *RSC Advances*, **10**(71), 43904–43914, https://doi.org/10.1039/D0RA08125F

Mane R. S. and Bhusari V. N. (2012). Removal of color (dyes) from textile effluent by adsorption using orange and banana peel. *International Journal of Engineering Research and Applications*, **2**(3), 1997–2004.

Mangale Sapana M., Chonde Sonal G. and Raut P. D. (2012). Use of *Moringa oleifera* (drumstick) seed as natural absorbent and an antimicrobial agent for ground water treatment. *Research Journal of Recent Sciences*. ISSN, 2277, 2502 **10**(1), 136–140.

Másson M. (2021). Antimicrobial Properties of Chitosan and Its Derivatives. In: R. Jayakumar, M. Prabaharan (eds). *Chitosan for Biomaterials III*. Volume **287**. Springer; Cham, Switzerland. pp. 131–168.

Maqsood A., Munir A. and Shahid S. (2020). A phytopharmacological evaluation of *Cassia fistula*. A comprehensive review. *International Journal of Pharmaceutical Sciences Review and Research*, **62**(2), 45–53.

Markets and Markets. (2020). Ion exchange market by type (cationic resins, anionic resins), application (water, non-water), end use industry (power, chemical & petrochemical, food & beverage, and region) global forecast to 2025. http://surl.li/mhiss Retreived on 18th October 2023

Marobhe N. J. and Gunaratna R. K. (2012). Effect of coagulant protein from *Parkinsonia aculeata* and citrus juice on bacteria isolated from Ruvu river in Tanzania. *International Journal of Applied Sciences and Engineering Research*, **1**(5), 714–724.

Mathuram M., Meera R. and Vijayaraghavan G. (2018). Application of locally sourced plants as natural coagulants for dye removal from wastewater: a review. *Journal of Materials and Environmental Science*, **2508**, 2058–2070.

Meng F., Chae S. R., Drews A., Kraume M., Shin H. S. and Yang F. (2009). Recent advances in membrane bioreactors (MBRs): membrane fouling and membrane material. *Water Research*, **43**(6), 1489–1512, https://doi.org/10.1016/j.watres.2008.12.044

Metcalf and Eddy. (2002). Wastewater Engineering-Treatment and Reuse. 4th edn. McGraw-Hill Publishing Company, New York. ISBN13:978-0070418783.

Metcalf L., Eddy H. P. and Tchobanoglous G. (1991). Wastewater Engineering: Treatment, Disposal, and Reuse. McGraw-Hill, New York, Vol. **4**.

Mouzdahir Y., Elmchaouri A., Mahboub R., Gil A. and Korili S. A. (2010). Equilibrium modeling for the adsorption of methylene blue from aqueous solution on activated clay minerals. *Desalination*, **250**, 335–338 https://doi.org/10.1016/j.desal.2009.09.052

Mustakeem M. (2015). Electrode materials for microbial fuel cells: nanomaterial approach. http://doi.org/10.1007/s40243-015-0063-8

Muzzarelli R. A. A. (1977). Chitin. Pergamon Press, Oxford.

Nagy A., Harrison A., Sabbani S., Munson R. S. Jr, Dutta P. K. and Waldman W. J. (2011). Silver nanoparticles embedded in zeolite membranes: release of silver ions and mechanism of antibacterial action. *International Journal of Nanomedicine*, **6**, 1833.

Nasr-Eldin M. A., Abdelhamid A. and Baraka D. (2017). Antibiofilm and antiviral potential of leaf extracts from *Moringa oleifera* and rosemary (*Rosmarinus officinalis* Lam.). *Egyptian Journal of Microbiology*, **52**(1), 129–139.

Ndabigengesere A. and Narasiah K. S. (1998). Quality of water treated by coagulation using *Moringa oleifera* seeds. *Water Research*, **32**(3), 781–791, https://doi.org/10.1016/S0043-1354(97)00295-9

Negroiu M., Țurcanu A. A., Matei E., Râpă M., Covaliu C. I., Predescu A. M., ... and Predescu C. (2021). Novel adsorbent based on banana peel waste for removal of heavy metal ions from synthetic solutions. *Materials*, **14**(14), 3946, https://doi.org/10.3390/ma14143946

Netsol Water Solutions (2022). Detergents ending up in our rivers prone fatal. Netsol Water Solutions Pvt. Ltd. https://www.netsolwater.com/detergents-ending-up-in-our-rivers-prove-fatal.php?blog=2250

Nielsen V. C. (1996). Treatment and disposal of processing wastes. In: Processing of Poultry, G. C. Mead (ed.), Chapman & Hall Publishing Company, New York, 361–411.

Nimesha S., Hewawasam C., Jayasanka D. J., Murakami Y., Araki N. and Maharjan N. (2022). Effectiveness of natural coagulants in water and wastewater treatment. *Global Journal of Environmental Science and Management*, **8**(1), 101–116.

NRCSDWC. (1980). An evaluation of activated carbon for drinking water treatment. *Drinking Water and Health*.

Oladoja N. A. (2015). Headway on natural polymeric coagulants in water and wastewater treatment operations. *Journal of Water Process Engineering*, **6**, 174–192.

Oo K. N., Aung K. S., Thida M., Knine W. W., Soe M. M. and Aye T. (1993). Effectiveness of potash alum in decontaminating household water. *Journal of Diarrhoeal Diseases Research*, **11**(3), 172–174.

Oteng-Peprah M., Acheampong M. A. and DeVries N. K. (2018). Greywater characteristics, treatment systems, reuse strategies and user perception – a review. *Water, Air and Soil Pollution*, **229**(8), 255, https://doi.org/10.1007/s11270-018-3909-8

Papageorgiou S. K., Katsaros F. K., Kouvelos E. P., Nolan J. W., Le Deit H. and Kanellopoulos N. K. (2006). Heavy metal sorption by calcium alginate beads from *Laminaria digitata*. *Journal of Hazardous Materials*, **137**(3), 1765–1772, https://doi.org/10.1016/j.jhazmat.2006.05.017

Parvathy U., Jeyakumari K. R. A. and Zynudheen A. A. (2017). Biological treatment systems for fish processing wastewater – a review.

Patale V. and Pandya J. (2012). Mucilage extract of *Coccinia indica* fruit as coagulant-flocculent for turbid water treatment. *Asian Journal of Plant Sciences*, **2**(4), 442–445.

Patel J. (2016). Ion Exchange Resins. Ion exchange India Ltd. Retrieved on 17th October 2023 from https://www.researchgate.net/publication/305658612_ION_EXCHANGE_RESINS

Periayah M. H., Halim A. S. and Saad A. Z. M. (2016). Chitosan: a promising marine polysaccharide for biomedical research. *Pharmacognosy Reviews*, **10**(19), 39, https://doi.org/10.4103/0973-7847.176545

Prajapati A. K., Das S. and Mondal M. K. (2020). Exhaustive studies on toxic Cr (VI) removal mechanism from aqueous solution using activated carbon of *Aloe vera* waste leaves. *Journal of Molecular Liquids*, **307**, 112956, https://doi.org/10.1016/j.molliq.2020.112956

Prathapar S. A., Jamrah A., Ahmed M., Al Adawi S., Al Sidairi S. and Al Harassi A. (2005). Overcoming constraints in treated greywater reuse in Oman. *Desalination*, **186**(1–3), 177–186, https://doi.org/10.1016/j.desal.2005.01.018

Press Information Bureau. (2013). Withdrawal of Fresh Water. Ministry of Water Resources, Government of India (accessed 6th February, 2023). https://pib.gov.in/newsite/PrintRelease.aspx?relid=101519

Preston D. R., Vasudevan T. V., Bitton G., Farrah S. R. and Morel J. L. (1988). Novel approach for modifying microporous filters for virus concentration from water. *Applied and Environmental Microbiology*, **54**(6), 1325–1329. https://doi.org/10.1128/aem.54.6.1325-1329.1988

Qureshi K., Bhatti I. and Shaikh M. S. (2011). Development of bio-coagulant from mango pit for the purification of turbid water. *Sindh University Research Journal-SURJ (Science Series)*, **43**(1), 105–110.

Rajoriya S., Saharan V. K., Pundir A. S., Nigam M. and Roy K. (2021). Adsorption of methyl red dye from aqueous solution onto eggshell waste material: kinetics, isotherms and thermodynamic studies. *Current Research in Green and Sustainable Chemistry*, **4**, 100180, https://doi.org/10.1016/j.crgsc.2021.100180

Ramadan B. S. and Purwono (2017). Challenges and opportunities of microbial fuel cells (MFCs) technology development in Indonesia. MATEC Web of Conferences. EDP Sciences, Vol. **101**, p. 02018, https://doi.org/10.1051/matecconf/201710102018

Rashed A. O., Esawi A. M. and Ramadan A. R. (2020). Novel polysulfone/carbon nanotube-polyamide thin film nanocomposite membranes with improved water flux for forward osmosis desalination. *ACS Omega*, **5**(24), 14427–14436, https://doi.org/10.1021/acsomega.0c00973

Rasool M.A., Tavakoli B., Chaibakhsh N., Pendashteh A.R. and Mirroshandel A.S. (2016). Use of a plant-based coagulant in coagulation-ozonation combined treatment of Leachate from a waste dumping site. *Ecological Engineering* **90**, 431–437.

Ravindra K., Mor S. and Pinnaka V. L. (2019). Water uses, treatment, and sanitation practices in rural areas of Chandigarh and its relation with waterborne diseases. *Environmental Science and Pollution Research*, **26**, 19512–19522, https://link.springer.com/article/10.1007/s11356-019-04964-y https://doi.org/10.1007/s11356-019-04964-y

Rehman K., Fatima F., Waheed I. and Akash M. S. H. (2018). Prevalence of exposure of heavy metals and their impact on health consequences. *Journal of Cellular Biochemistry*, **119**(1), 157–184, https://doi.org/10.1002/jcb.26234

Renuka V., Ravishankar C. N., Elavarasan K., Zynudheen A. A. and Joseph T. C. (2019). Production and characterization of chitosan from shrimp shell waste of parapeneopsisstylifera. *International Journal of Current Microbiology and Applied Sciences*, **8**(11), 2077, https://doi.org/10.20546/ijcmas.2019.811.240

Reynolds T. D. and Richards P. A. (1996). Unit Operations and Process in Environmental Engineering. 2nd edn. PWS Publishing Company, Boston.

Rupiasih N. N., Purnomo R. R. and Sumadiyasa M. (2016). Preparation and application of chitosan membranes to filter silver from X-ray film processing wastes. *Journal of Physics: Conference Series*, **710**(1), 012009, https://doi.org/10.1088/1742-6596/710/1/012009

Sachit E. (2018). Biological Treatment Unit, In *Wastewater Treatment Plant* (pp. 60–87). Environmental Engineering Department, Mustansiriyah University. https://uomustansiriyah.edu.iq/media/lectures/5/5_2018_10_11!10_33_33_PM.pdf

Saleem M. and Bachmann, R. T. (2019). A contemporary review on plant-based coagulants for applications in water treatment. *Journal of Industrial and Engineering Chemistry*, **72**, 281–297.

Saravanan J., Priyadharshini D., Soundammal A., Sudha G. and Suriyakala K. (2017). Wastewater treatment using natural coagulants. *SSRG International Journal of Civil Engineering* **4**(3), 40–42.

Sasáková N., Pavolová H., Nowakowicz-Debek B., Vargová M., Laktičová K. V., Kachnič J. and Hromada R. (2021). Ozonization of water, retention of ozone and devitalization of in water by ozone. *Folia Veterinaria*, **65**(3), 40–50, https://doi.org/10.2478/fv-2021-0026

Sayre I. M. (1988). International standards for drinking water. *Journal-American Water Works Association*, **80**(1), 53–60, https://doi.org/10.1002/j.1551-8833.1988.tb02980.x

Šćiban M. B., Antov M. G. and Klašnja M. T. (2006). Extraction and partial purification of coagulation active components from common bean seed. *Acta Periodica Technologica*, 37–43, https://doi.org/10.2298/APT0637037S

Šćiban M., Klašnja M., Antov M., and Škrbić, B. (2009). Removal of water turbidity by natural coagulants obtained from chestnut and acorn. *Bioresource technology*, **100**(24), 6639–6643.

Sekhar C., Venkatesan N. and Vidhyavathi A. (2018). Factors influencing moringa cultivation in Tamil, Nadu: an economic analysis. *Horticulture International Journal*, **2**(5), 223–229.

Sepúlveda E. S. C. A. E., Sáenz C., Aliaga E. and Aceituno C. (2007). Extraction and characterization of mucilage in *Opuntia* spp. *Journal of Arid Environments*, **68**(4), 534–545, https://doi.org/10.1016/j.jaridenv.2006.08.001

Seyyednejad S., Motamedi H., Vafei M. and Bakhtiari A. (2004). The antibacterial activity of *Cassia fistula* organic extracts. *Jundishapur Journal of Microbiology*, **7**(1), 1–4.

Shabiimam M. A., Kazi T., Anas P. and Shifa S. (2018). Treatment of water using various filtration techniques: review study. Conference Paper.

Sharma D. K. (2017). Numerations on phytochemical, pharmacological and ethnobotanical properties of *Cassia fistula* Linn: yellow shower. *The Journal of Phytopharmacology*, **6**(5), 300–306, https://doi.org/10.31254/phyto.2017.6509

Sharma S. and Bhattacharya A. J. A. W. S. (2017). Drinking water contamination and treatment techniques. *Applied Water Science*, **7**(3), 1043–1067, https://doi.org/10.1007/s13201-016-0455-7

Shmeis R. M. A. (2018). Water chemistry and microbiology. In: Comprehensive Analytical Chemistry, D. Barcelo (ed.) Elsevier, Vol. **81**, pp. 1–56.

Silva R. C., Gomes F. T., Andrade C. R. M., Monterio H. S., Dias A. C. R., Zagatto A. G. E. and Tornisileo L. V. (2013). Banana peel as an adsorbent for removing atrazine and ametryene from waters. *Journal of Agricultural Food Chemistry*, **61**(10), 2358–2363, https://doi.org/10.1021/jf304742h

Siong Y. K., Idris J. and Atabaki M. (2013). Performance of activated carbon in water filters. *Water Resources*, **32**(3), 1–19. Retrieved from https://www.researchgate.net/publication/234060484_Performance_of_activated_carbon_in_water_filters

Soto D., Urdaneta J., Pernía K., León O., Muñoz-Bonilla A. and Fernández-García M. (2015). Heavy metal (Cd2 + , Ni2 + , Pb2 + and Ni2 +) adsorption in aqueous solutions by oxidized starches. *Polymers for Advanced Technologies*, **26**(2), 147–152, https://doi.org/10.1002/pat.3439

Soto D., Urdaneta J., Pernía K., León O., Muñoz-Bonilla A. and Fernandez-García M. (2016). Removal of heavy metal ions in water by starch esters. *Starch-Stärke*, **68**(1–2), 37–46, https://doi.org/10.1002/star.201500155

Sowmeyan R., Santhosh J. and Latha R. (2011). Effectiveness of herbs in community water treatment. *International Research Journal of Biochemistry and Bioinformatics*, **1**(11), 297–303.

Subashree R., Suryapraba N. and Anusha G. (2017). Treatment of wastewater using banana and lemon peels as adsorbents. *International Journal of Engineering Sciences & Research*, **4**, 1289–1291.

Subramonian W., Wu T. Y. and Chai S. P. (2014). A comprehensive study on coagulant performance and floc characterization of natural *Cassia obtusifolia* seed gum in treatment of raw pulp and paper mill effluent. *Industrial Crops and Products* **61**, 317–324.

Swan A. P. (2008). U.S. Patent Application No. 11/807,645. Inlet Filter.

Tay J. H., Show K. Y. and Hung Y. T. (2004). Seafood processing wastewater treatment. In: Handbook of Industrial and Hazardous Wastes Treatment. 2nd edn. L. Wang, Y.-T. Hung, H. Lo and C. Yapijakis (eds.), Marcel Dekker, Inc., New York, pp. 647–684.

Trung D. M. and Luu T. L. (2021). Preparing modified activated carbon from *Cassia fistula* seed with HNO_3 for zinc removal in wastewater. In: Key Engineering Materials, Trans Tech Publications Ltd, Vol. **885**, pp. 85–93, https://doi.org/10.4028/www.scientific.net/KEM.885.85

UNICEF. (2008). UNICEF Handbook on Water Quality. United Nations Children's Fund, New York. Retrieved from https://www.ncbi.nlm.nih.gov/books/NBK234590/

UNICEF. (2019). Clean Drinking Water. Retrieved on 6th February 2023. Clean drinking water. https://www.unicef.org/india/what-we-do/clean-drinking-water

USEPA (US Environmental Protection Agency) (1999). Wastewater technology fact sheet: ozone disinfection. Retrieved on 15th September 2023 from https://www3.epa.gov/npdes/pubs/ozon.pdf

USEPA (US Environmental Protection Agency) (2022). Health Effects of exposures to mercury. Retrieved on 17th September 2023 from https://www.epa.gov/mercury/health-effects-exposures-mercury

Van Der Bruggen B., Vandecasteele C., Van Gestel T., Doyen W. and Leysen, R. (2003). A review of pressure-driven membrane processes in wastewater treatment and drinking water production. *Environmental progress*, **22**(1), 46–56.

Verma R. (2020). Use of Banana Peels in Water Purification. Biyani Girls College, Jaipur, India. Retrieved on 1st March 2023 from https://www.biyanicolleges.org/use-of-banana-peels-in-water-purification/

Vidya Vijay M., Sudarsan J. S. and Nithiyanantham S. (2019). Sustainability of constructed wetlands using biochar as effective absorbent for treating wastewaters. *International Journal of Energy and Water Resources*, **3**, 153–164, https://doi.org/10.1007/s42108-019-00025-9

Vijayaraghavan G., Sivakumar T. and Kumar A. V. (2011). Application of plant based coagulants for waste water treatment. *International Journal of Advanced Engineering Research and Studies*, **1**(1), 88–92.

Wang H., Park J. D. and Ren Z. J. (2015). Practical energy harvesting for microbial fuel cells: a review. *Environmental Science & Technology*, **49**(6), 3267–3277, https://doi.org/10.1021/es5047765

Water Conditioning and Purification. (2014). A brief history of ion exchange water treatment. Retrieved on 15th October 2023 from https://shorturl.at/rBXZ0

Whitehouse J. D. (1967). 'A study of the removal of pesticides from water'. *KWRRI Research Reports*, **185**, 1–191.

Wu Z., Chen X., Yuan B. and Fu M. L. (2020). A facile foaming-polymerization strategy to prepare 3D MnO_2 modified biochar-based porous hydrogels for efficient removal of Cd (II) and Pb (II). *Chemosphere*, **239**, 124745, https://doi.org/10.1016/j.chemosphere.2019.124745

Yee M. S. L., Hii L. W., Looi C. K., Lim W. M., Wong S. F., Kok Y. Y., Tan B-K., Wong C-Y. and Leong, C. O. (2021). Impact of microplastics and nanoplastics on human health. *Nanomaterials*, **11**(2), 496.

Yimer A. and Dame B. (2021). Papaya seed extract as coagulant for potable water treatment in the case of Tulte river for the community of Yekuset district, Ethiopia. *Environmental Challenges*, **4**, 100198, https://doi.org/10.1016/j.envc.2021.100198

Zainol N. A. and Fadli N. N. M. (2020). Surface water treatment using tamarind seed as coagulants via coagulation process. IOP Conference Series: Materials Science and Engineering. IOP Publishing, Vol. **864**, p. 012172, https://doi.org/10.1088/1757-899X/864/1/012172

Zhang R. (2001). Biology and Engineering of Animal Wastewater Lagoons. University of California, Davis.

Zhang J., Zhang F., Luo Y. and Yang H. (2006). A preliminary study on cactus as coagulant in water treatment. *Process Biochemistry*, **41**(3), 730–733, https://doi.org/10.1016/j.procbio.2005.08.016

Zhou H., Zhu H., Xue F., He H. and Wang S. (2020). Cellulose-based amphoteric adsorbent for the complete removal of low-level heavy metal ions via a specialization and cooperation mechanism. *Chemical Engineering Journal*, **385**, 123879, https://doi.org/10.1016/j.cej.2019.123879

doi: 10.2166/9781789064100_0043

Chapter 3

A comprehensive review of wetlands in India and their conservation

L. K. Bhardwaj[1], P. Rath[2], H. Jain[2] and M. Choudhury

[1]Amity Institute of Environmental Toxicology, Safety & Management, Amity University, Noida, Uttar Pradesh, India
[2]Amity Institute of Environmental Sciences, Amity University, Noida, Uttar Pradesh, India
[3]Environmental Research and Management Division, Voice of Environment (VoE), Guwahati, Assam, India
Corresponding author's email id-bhardwaj.laxmikant@gmail.com

ABSTRACT

Wetlands are highly valuable and productive environments for various organisms. This distinctive ecosystem sustains many habitats and biodiversity. In recent decades, they have faced significant endangerment because of urbanization or human intervention. The condition of wetlands in India is marked by a substantial decrease in both size and quality, with about half of the total wetlands having disappeared and the remaining ones degrading. Over time, wetlands have been transformed for many purposes due to urban development, industrialization, climate change, and intensified agriculture. The water quality in numerous wetlands has significantly deteriorated due to the presence of untreated sewage, industrial effluents, and agricultural runoff. This degradation has resulted in the loss of aquatic life, posed health hazards to nearby residents, and impacted pisciculture, and rice farming. In India, conservation efforts are being undertaken to tackle these difficulties by utilizing a combination of legal frameworks, legislative measures, and community engagement. Since the Ramsar Convention in 1971, numerous international initiatives have been endorsed to safeguard wetlands. Some wetlands near the endangered area have conservation management plans, whereas others do not. The wetlands in India cover around 58.2 million hectares of land and serve as important reservoirs of aquatic life. Research communities are prioritizing the preservation of these environments. This chapter presents a comprehensive analysis of the present condition of wetlands in India and highlights the imperative for their preservation. It emphasizes the primary obstacles encountered, and the endeavors being undertaken for its safeguarding and revival.

Keywords: Conservation, Ecosystem, India, Ramsar Convention, Wetlands

3.1 INTRODUCTION

Wetlands are defined as areas that lie between aquatic and terrestrial ecosystems (Mukherjee *et al.*, 2023) (Figure 3.1). They are commonly referred to as areas of high biodiversity (Reid *et al.*, 2005). Wetlands are highly valued ecosystems that provide a wide range of environmental services to enhance human civilization (Costanza *et al.*, 2014). The wetlands are protected by shallow water and retain water even in dry conditions, resulting in a consistently high and stable water table. They help

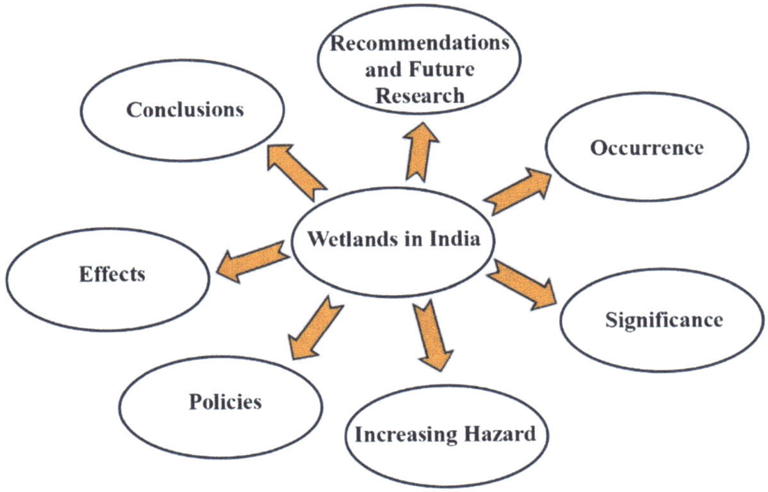

Figure 3.1 Graphical abstract.

reduce the impact of floods and trap suspended particles during flooding. Consequently, streams that pass-through wetland areas before reaching ponds have the ability to transport suspended particles and nutrients to the lakes, rather than streams that run directly into the lakes. These systems are very dynamic and seasonal, and they rival terrestrial ecosystems in terms of seasonal competition. They contribute to the preservation of species diversity on a global scale and play a crucial role in maintaining a thriving environment through many means. The wetlands have a crucial role in providing habitats for various plant and animal species, as well as performing vital ecological tasks including water purification, flood control, and carbon sequestration.

Wetlands can be classified into three categories: (i) areas of land that provide support for hydrophytes, (ii) areas of land that include undrained hydric soil, and (iii) areas of land that are saturated with water. They are categorized as coastal wetlands (estuarine and marine) and inland wetlands (riverine, lacustrine, and palustrine) based on their geological, hydrological, and ecological characteristics (Table 3.1). The wetlands serve various functions, including waste treatment, pollution interception, and toxic residue processing, which are chemicals in nature. Additionally, they provide

Table 3.1 Classification of wetlands.

S. No.	Major Categories		Location	Wetland Types
1	Coastal wetlands	Marine (undiluted salt water)	Open coast	Salt marsh, shrub wetland, mangrove swamp
		Estuarine (salt/ freshwater mix)	Estuaries (deltas, lagoons)	Shrub wetland, brackish marsh, mangrove swamp, salt marsh
2	Inland wetlands	Riverine (associated with rivers and streams)	River channels and floodplains	Delta marsh, freshwater marsh, bottomlands
		Lacustrine (associated with lakes)	Lakes and deltas	Shrub and forest wetlands, Freshwater marsh
		Palustrine (shallow ponds, freshwater wetlands)	Ponds, peatlands, uplands, groundwater seeps	Groundwater spring oasis, tundra peatland, ephemeral ponds

flood control, coastal protection, groundwater recharge, and sediment traps, and contribute to atmospheric equilibrium, which are physical/hydrological functions.

The wetland could be further sub-categorized into the following types, based on their physical characteristics, hydrology, vegetation, and ecological functions (Sarkar and Das, 2016):

(i) **Marshes**
- *Freshwater marshes:* They are found in non-tidal areas and are dominated by herbaceous vegetation. They often occur near rivers, lakes, or in floodplains.
- *Saltwater or coastal marshes*: They are located in tidal zones along coastlines and are influenced by the ebb and flow of tides. Vegetation is adapted to saline conditions.

(ii) **Swamps**
- *Freshwater swamps*: They are typically found in forested areas with slow or stagnant water flow. They are dominated by woody vegetation, such as trees and shrubs.
- *Saltwater or coastal swamps*: They occur in tidal areas and are characterized by salt-tolerant trees and shrubs. Examples include mangrove swamps.

(iii) **Bogs**
- *Ombrotrophic bogs*: They receive water and nutrients primarily from precipitation. They are acidic and support specialized vegetation such as sphagnum moss and carnivorous plants.
- *Minerotrophic bogs*: They receive water from both precipitation and mineral-rich groundwater. Vegetation is less acidic and may include sedges and shrubs.

(iv) **Fens**
- They are similar to bogs but receive nutrients from groundwater with higher mineral content. Fens are less acidic and support different plant species, including sedges and grasses.

(v) **Prairie potholes**
- They are found in the northern parts of North America, particularly in the Prairie Pothole Region. These shallow depressions collect water during wet periods, providing crucial habitat for waterfowl.

(vi) **Playa lakes**
- They occur in arid and semi-arid regions, particularly in the Great Plains of North America. Playa lakes are temporary, shallow depressions that fill with water during rainy events.

(vii) **Tidal flats and salt pans**
- Coastal areas influenced by tides, where flats are periodically submerged and exposed. Salt pans may form in arid regions due to the evaporation of saline water.

(viii) **Mangrove forests**
- Coastal wetlands characterized by salt-tolerant trees and shrubs. Mangroves provide essential habitat for many marine species and act as a buffer against storm surges.

(ix) **Riparian wetlands**
- These wetlands are located along the banks of rivers and streams. They are influenced by periodic flooding and support a variety of aquatic and terrestrial species.

(x) **Constructed wetlands**
- They are man-made wetlands designed to replicate natural wetland functions. They are often used for wastewater treatment, stormwater management, and habitat creation.

In recent decades, there has been a growing level of human activity in wetland areas. Consequently, these lands have become contaminated with multiple contaminants. Moreover, the invaluable significance of India's wetlands is endangered by a variety of human-induced stresses. The health and viability of these ecosystems face significant difficulties due to rapid urbanization, agricultural development, pollution, climate change, and unsustainable resource usage. These areas are subject to continuous discharge of domestic garbage, industrial sewage, and agricultural run-offs (Gupta *et al.*, 2023). Multiple researchers have documented various pollutants in these regions, as well as in different lake systems globally (Alam *et al.*, 2023; Bhardwaj & Jindal, 2019, 2020, 2022; Bhardwaj *et al.*, 2021, 2023; Bhardwaj & Sharma, 2021; Bhardwaj & Yadav, 2023). In light of the pressing need to tackle these dangers, efforts to preserve and responsibly manage wetlands have gained significant traction in recent times. This chapter focuses on the current state of wetland sites in India, their significance, the risks they face, and approaches to wetland management. The chapter also provides a comprehensive analysis of their ecological importance, distribution, and the numerous obstacles they encounter. This chapter will explore the importance of wetlands in preserving ecological balance. It will also examine the current legislative frameworks, conservation laws, and various tactics used to protect these unique habitats.

3.2 OCCURRENCE OF WETLANDS IN INDIA

India possesses a wide range of distinct wetland ecosystems, characterized by changes in climate, support systems, and topography. The habitats include Himalayan wetlands such as Tso Morari, Pangong Tso, Noorichan, Hanlay marshes, Malgam, Kranchu lakes, Barak valley, Dal, Chantau, Chushul, Haigam, Anchar, Beels, Wular, Nainital, Haukersar, Bhimtal, and Naukuchital. They also include Indo-Gangetic wetlands that extend from the Indus River to the Brahmaputra River. (iii) Coastal wetlands, including lagoons and mangroves, are found in the states of Andhra Pradesh, Orissa, West Bengal, Maharashtra, Karnataka, Gujarat, Kerala, Tamil Nadu, and Goa. The offshore coral reefs of Lakshadweep, Gulf of Mannar, Andaman and Nicobar Islands, and Gulf of Kutch, as well as the wetlands in the Deccan region, including the water tanks in every town, are notable ecological areas.

Wetlands are found in the dry sections of certain states such as Rajasthan and Gujarat, as well as in the deltaic areas of both the western and eastern coasts. They also exist in the wet regions of southern peninsular India, the highlands of central India, and the Lakshadweep and Andaman & Nicobar Islands. The Indo-Gangetic floodplain, renowned as the largest wetland in India, is of significant size. Wetlands rely on main rivers such as Narmada, Krishna, Ganga, Tapti, Brahmaputra, Kaveri, and Godavari, either directly or indirectly. Multiple ponds, tanks, and reservoirs contribute to the diversity of wetlands and enhance the country's wetland resources. India requires the presence of beneficial wetlands in order to achieve sustained food production for human consumption. Furthermore, they are essential for the sustenance of diverse species of flora and fauna. The criteria for wetlands of national importance include uniqueness, presence of water birds, diverse flora and fauna, consistent water flow, presence of fish and migratory birds, a variety of species, and a well-established biological community. The primary issue affecting the wetlands of India is the human population.

As of February 2022, India possesses the most extensive collection of wetlands, encompassing 49 locations and spanning an area of 10,93,636 hectares (Mohan, 2022). In their study, Kumari *et al.* (2020) found that India possesses 63% of its land area as natural wetlands, covering a total of 66,23,067 hectares. Additionally, they observed that 37% of the country's wetlands are human-made inland wetlands, spanning an area of 39,41,832 hectares. India has a total of 27,403 wetlands, consisting of 3,959 coastal wetlands and 23,444 interior wetlands (Rajakumar & Sivasubraminium, 2012). Figure 3.2 displays the percentage distribution of wetland areas in India. Wetlands encompass around 4.63% of India's total land area, with paddy farming accounting for over 70% (Prasad *et al.*, 2002). India has fishponds covering an area of 1.6 million hectares (Kolli, 2020). According to Gopal (2014), fishponds and paddy fields were identified as components of wetlands. In 1993, 40.9 million hectares of land out of a total area of 58.2 million hectares were allocated for rice cultivation (Kolli, 2020). According to Prasad *et al.* (2002), freshwater wetlands constitute 20% of India's biodiversity.

The government recognizes 648,507 hectares as wetlands in India (Paul *et al.*, 2011). According to the study by the Ministry of Environment, Forest and Climate Change (MoEFCC) (2012), 4.63% of the wetlands are protected by organizations such as the National Wetland Conservation Programme (NWCP), as well as the National Lake Conservation Plan (NLCP). According to the Worldwide Fund for Nature India (WWF-India), it has been reported that wetlands are considered to be one of the

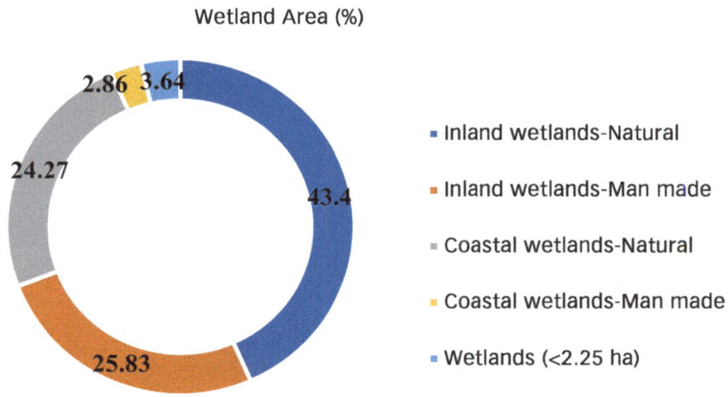

Figure 3.2 Distribution of wetland areas in India.

greatest threats to all ecological units in India (Anand *et al.*, 2012). After the convention of Ramsar in 1971, several laws have been directed in India to preserve the ecosystem of wetlands (Prasad *et al.*, 2002). India signed the international treaty of the Ramsar Convention on 1st February 1982. MoEFCC has proclaimed 75 wetland sites that are covered under the Ramsar Convention (Table 3.2). In 2022, 26 new Ramsar sites were included, and some examples are Pichavaram Mangrove & Pallikaranai Marsh Reserve Forest from Tamil Nadu, Karikili Bird Sanctuary, Sakhya Sagar from Madhya Pradesh, Pala wetland from Mizoram, etc. Tamil Nadu has the maximum number of lakes (4369) in India followed by Uttar Pradesh (3684) and West Bengal (1327) (Chandra *et al.*, 2021). Kol Wetland is the second-largest wetland in India and is situated in Kerala (Nisari & Sujatha, 2021).

3.3 SIGNIFICANCE OF WETLANDS

Wetlands are ecosystems that have significant ecological, economic, and social value because they possess a remarkable capacity to filter and purify water, earning them the nickname 'kidneys of the Earth' (Mitsch *et al.*, 2015). These distinct ecosystems, distinguished by the existence of water, play a vital role in the overall environment, providing a wide range of advantages that reach well beyond their immediate confines. They offer a multitude of services to the human race. Wetlands have multiple roles, including storing water, filtering water, recycling nutrients, reducing the impact of floods, maintaining the flow of streams, replenishing groundwater, and providing fuel, fodder, fish, drinking water, wildlife habitat, and controlling runoff in urban areas (Hammer & Bastian, 2020). The functions and values of the wetland are contingent upon its geographical placement. They are significant due to their role in facilitating food production and serving as resting grounds for migrating birds. Additionally, they offer designated spaces for the reproduction and nourishment of species. They serve as the natural habitat for fish, frogs, snakes, birds, and various other aquatic species. India is home to around 2400 subspecies and species of birds, as reported by Prasad *et al.* in 2002. They play a crucial role in ensuring the well-being, physical condition, and housing of the individuals residing in or around them. Additionally, they serve as crucial repositories for various plant genetic resources. According to Stefanakis (2019), they can enhance water quality by eliminating different types of contaminants (both inorganic and organic) using a combination of biological, chemical, and physical methods. Aside from the naturally occurring wetlands, there are also various wetlands that have been created by humans, which have an impact on the diversity of plant and animal species. The wetlands created by human activity are numerous and serve various purposes such as generating energy, supporting fisheries, facilitating irrigation, controlling floods, and providing water supplies.

Table 3.2 List of Ramsar sites in India with their area.

S. No.	Ramsar Wetland Sites	Location	Area (hectare)
1	Kolleru Lake	Andhra Pradesh	901
2	Deepor Beel	Assam	40
3	Kabartal Wetland	Bihar	26.2
4	Nanda Lake	Goa	0.42
5	Khijadia Wildlife Sanctuary	Gujarat	5.12
6	Nalsarovar Bird Sanctuary		120
7	Thol Lake Wildlife Sanctuary		6.99
8	Wadhvana Wetland		6.3
9	Bhindawas Wildlife Sanctuary	Haryana	4.12
10	Sultanpur National Park		1.43
11	Chandertal Wetland	Himachal Pradesh	0.49
12	Pong Dam Lake		156.62
13	Renuka Wetland		0.2
14	Hokera Wetland	Jammu and Kashmir	13.75
15	Hygam Wetland Conservation Reserve		8.02
16	Shallbugh Wetland Conservation Reserve		16.75
17	Surinsar Mansar Lakes		3.5
18	Wular Lake		189
19	Ranganathittu Bird Sanctuary	Karnataka	5.18
20	Asthamudi Wetland	Kerala	61.4
21	Sasthamkotta Lake		3.73
22	Vembanad Kol Wetland		1512.5
23	Tso Kar Wetland	Ladakh	95.77
24	Tsomoriri Lake		120
25	Bhoj Wetlands	Madhya Pradesh	32.01
26	Sakhya Sagar		2.48
27	Sirpur Wetland		1.61
28	Yashwant Sagar		8.23
29	Lonar Lake	Maharashtra	4.27
30	Nandur Madhameshwar		14.37
31	Thane Creek		65.21
32	Loktak Lake	Manipur	266
33	Pala Wetland	Mizoram	18.5
34	Ansupa Lake	Odisha	2.31
35	Bhitarkanika Mangroves		650
36	Chilka Lake		1165
37	Hirakud Reservoir		654
38	Satkosia Gorge		981.97
39	Tampara Lake		3
40	Beas Conservation Reserve	Punjab	64.29
41	Harike Lake		41

(Continued)

Table 3.2 List of Ramsar sites in India with their area (*Continued*).

S. No.	Ramsar Wetland Sites	Location	Area (hectare)
42	Kanjli Lake		1.83
43	Keshopur Miani Community Reserve		3.44
44	Nangal Wildlife Sanctuary		1.16
45	Ropar Lake		13.65
46	Keoladeo Ghana NP	Rajasthan	28.73
47	Sambhar Lake		240
48	Chitrangudi Bird Sanctuary	Tamil Nadu	2.6
49	Gulf of Mannar Marine Biosphere Reserve		526.72
50	Kanjirankulam Bird Sanctuary		0.97
51	Karikili Bird Sanctuary		0.58
52	Koonthankulam Bird Sanctuary		0.72
53	Pallikaranai Marsh Reserve Forest		12.48
54	Pichavaram Mangrove		14.79
55	Point Calimere Wildlife and Bird Sanctuary		385
56	Suchindram Theroor Wetland		0.94
57	Udhayamarthandapuram Bird Sanctuary		0.44
58	Vaduvur Bird Sanctuary		1.13
59	Vedanthangal Bird Sanctuary		0.4
60	Vellode Bird Sanctuary		0.77
61	Vembannur Wetland		0.2
62	Rudrasagar Lake	Tripura	2.4
63	Bakhira Wildlife Sanctuary	Uttar Pradesh	28.94
64	Haiderpur Wetland		69.08
65	Nawabganj Bird Sanctuary		2.25
66	Parvati Agra Bird Sanctuary		7.22
67	Saman Bird Sanctuary		5.26
68	Samaspur Bird Sanctuary		7.99
69	Sandi Bird Sanctuary		3.09
70	Sarsai Nawar Jheel		1.61
71	Sur Sarovar		4.31
72	Upper Ganga River		265.9
73	Asan Conservation Reserve		4.44
74	East Kolkata Wetlands	West Bengal	125
75	Sundarbans Wetland		4230

Hydrologic circumstances have the power to alter the characteristics of wetlands, including the salinity of the soil, pH levels, sediment qualities, and the availability of nutrients. The chemical and physical changes directly impact the biota, leading to significant alterations in species composition and ecosystem productivity (Pižl *et al.*, 2023). Phytoplankton, dyes, suspended inorganic debris, and organic detritus serve as indicators for assessing these changes, which are detectable by the naked eye. Mangrove vegetation exhibits flood tolerance. This coastal flora is highly prized for its use as firewood, local medicinal purposes, cattle feed, as well as for supporting fish and shellfish populations.

The significance of wetlands can be succinctly summarized within the below framework (Aslam *et al.*, 2021):

- *Biodiversity hotspots*: Wetlands are among the most biologically diverse ecosystems globally, providing habitats for a staggering array of plant and animal species. From migratory birds that rely on wetlands as crucial stopover points to amphibians, fish, and diverse plant communities, wetlands support a rich tapestry of life.
- *Water purification*: Wetlands act as natural water purifiers, trapping and filtering sediments, nutrients, and pollutants from the water. They play a pivotal role in maintaining water quality, ensuring the health of aquatic ecosystems, and providing clean water for various human uses.
- *Flood regulation*: Wetlands act as natural buffers against floods by absorbing and slowing down floodwaters. Their capacity to store and release water helps mitigate the impact of extreme weather events, protecting downstream areas from devastating floods.
- *Climate change mitigation*: Wetlands are significant carbon sinks, sequestering large amounts of carbon dioxide and helping mitigate the impacts of climate change. Their ability to store carbon in the form of organic matter contributes to global efforts to reduce greenhouse gas emissions (Bhardwaj, 2023; Kumar *et al.*, 2020; Choudhury *et al.*, 2021a).
- *Shoreline stabilization*: Coastal wetlands, including mangroves and salt marshes, play a crucial role in stabilizing shorelines. They act as natural barriers, protecting coastal areas from erosion caused by storms and tidal action.
- *Livelihoods and recreation*: Wetlands provide essential resources for human livelihoods, including fisheries, agriculture, and tourism. They offer recreational opportunities, attracting birdwatchers, nature enthusiasts, and ecotourists, contributing to local economies.
- *Cultural and spiritual significance*: Many communities around the world have cultural and spiritual connections to wetlands. These ecosystems often hold sacred value and are integral to the identity and traditions of indigenous peoples.

3.4 INCREASING DANGER TO WETLANDS

Wetlands are highly vulnerable ecosystems on a global scale. These areas, which are environmentally fragile, are damaged by the introduction of sewage waste, agricultural run-off, and industrial wastewater (Prasad *et al.*, 2009). According to Upadhyay *et al.* (2019), the release of wastewater into the wetland is a significant contributor to the degradation of the wetland's ecology. The primary factor contributing to the decline of wetland regions is the growth in population. Hence, the exponential growth of the human population and the inappropriate utilization of land and watersheds have resulted in a significant depletion of the wetland resources in the country. These have demonstrated hydrological stress, contamination, and associated impacts.

An increase in the human population may lead to a higher demand for wetland products such as fish, fiber, water, wood, and medicinal plants. Rapid industrialization and urbanization exert a profound impact on wetlands. The urban wetlands are at risk of extinction due to their utilization as dumping grounds for municipal garbage. According to the research conducted by Zutshi and Prasad (2008), the majority of freshwater wetlands are at risk, with only a small number being destroyed as a result of human activity. An increase in population and urbanization can exacerbate the quality of lake water. The degradation of water quality in wetlands is attributed to undesirable environmental, social, and economic outcomes (Bassi *et al.*, 2014). Excessive utilization of water resources can disturb the normal functioning of wetlands and lead to the depletion of wetland areas (Donnelly *et al.*, 2020). The wetlands of the nation have been adversely affected by water pollution, excessive development, vegetation loss, and road construction.

Overfishing and overhunting result in the depletion of several functions of wetlands, such as the support of food chains, human recreational activities, and the provision of habitats for wildlife. Destruction of wetland habitats diminishes the richness of both wetland-dependent wildlife and vegetation. Cattle grazing can have a substantial impact on wetlands by eradicating indigenous plant species. Activities such as fishing and grazing have also led to the loss of wetlands. Wetland habitats are harmed not just by external factors but also by the exploitation of wetlands for agriculture and fish production. However, wetlands have been threatened by a multitude of factors leading to habitat reductions. Currently, the significance of wetlands is being disregarded, leading to a threat to the origin of these benefits.

Ostad-Ali-Askari's (2022) analysis reveals that human-induced blockage is the primary factor responsible for the loss of wetlands. Approximately 74% of India's population resides in rural areas and relies on natural resources. The depletion of wetlands can have severe effects on these individuals (Prasad *et al.*, 2002). In India, densely populated areas near wetlands will see more significant repercussions compared to sparsely populated places. According to Chandra *et al.* (2021), over 50,000 lakes are polluted and classified as 'lifeless'. Restoring modified wetlands to their original state has proven to be extremely difficult once they are taken over for purposes other than wetland use.

3.4.1 Threats to wetlands in the Indian context

Wetlands in India face a number of growing threats that have disturbed their health, biodiversity, and the invaluable services they provide to both the environment and human communities (Shan *et al.*, 2021; Paonam and Chatterjee, 2022). The complex interplay of anthropogenic activities, climate change, and inadequate conservation measures contribute to the escalating challenges confronting these critical ecosystems.

- *Urbanization and infrastructure development*: One of the primary threats to wetlands in India is the rapid pace of urbanization and infrastructure development. As cities expand and populations increase, wetlands often bear the brunt of land conversion for residential, commercial, and industrial purposes. Encroachment and reclamation lead to habitat loss, disrupting the delicate balance of these ecosystems and undermining their ability to perform vital functions such as water purification and flood regulation.
- *Agriculture and Intensive Land Use*: Agricultural expansion and intensive land use practices pose significant threats to wetlands. Excessive use of fertilizers and pesticides in agricultural areas contributes to nutrient runoff, causing water pollution in wetland ecosystems. Drainage for agriculture also alters natural hydrological patterns, negatively impacting the health of wetland habitats.
- *Pollution and contamination*: Wetlands in India face contamination from various sources, including industrial discharge, untreated sewage, and agricultural runoff. Pollutants such as heavy metals, pesticides, and nutrients degrade water quality and harm the flora and fauna within wetland ecosystems. The cumulative impact of pollution poses a severe threat to the ecological integrity of these areas.
- *Invasive species*: The introduction and spread of invasive species in wetlands disrupt native ecosystems and outcompete indigenous flora and fauna. Invasive plants and animals can alter habitat structures, nutrient cycling, and food webs, leading to the decline of native biodiversity.
- *Climate change impacts*: Climate change poses a growing threat to wetlands in India. Altered precipitation patterns, rising temperatures, and extreme weather events contribute to changes in water levels, salinity, and the overall hydrological balance of wetland ecosystems. These changes can lead to shifts in vegetation, loss of habitat, and increased vulnerability to other stressors.
- *Over-extraction of water*: Excessive withdrawal of water for agriculture, industry, and domestic use disrupts the natural water balance of wetlands. Reduced water levels can lead to the drying up of wetland habitats, affecting the flora and fauna dependent on these ecosystems.
- *Poor management and conservation practices*: Inadequate management and conservation practices exacerbate the threats facing wetlands. Weak enforcement of existing regulations, insufficient monitoring, and a lack of awareness contribute to the degradation of these ecosystems. Effective wetland conservation requires integrated and sustainable management approaches.

- *Sand mining*: Unregulated sand mining in and around wetlands poses a significant threat. Extraction of sand alters the morphology and hydrology of wetland areas, leading to habitat loss and increased vulnerability to erosion.
- *Land reclamation for agriculture and development*: Conversion of wetlands for agricultural activities and development projects, including the creation of reservoirs and dams, results in the loss of crucial wetland habitats. Such alterations disrupt natural hydrological processes and impact the overall ecological balance of these ecosystems.
- *Insufficient legal protections and enforcement*: While India has laws and regulations aimed at conserving wetlands, their implementation and enforcement often fall short. Inconsistent application of legal frameworks and a lack of awareness among stakeholders contribute to the ongoing threats faced by wetlands.

Addressing these growing threats to wetlands in India requires a comprehensive and collaborative approach involving policymakers, local communities, environmental organizations, and researchers. Efforts to raise awareness, strengthen legal protections, and implement sustainable land-use practices are essential to ensuring the resilience and long-term survival of these critical ecosystems.

3.5 POLICIES FOR THE CONSERVATION AND MANAGEMENT OF WETLANDS IN INDIA

Wetlands do not get delineated by any specific jurisdictions related to administration. They are a universal resource, and the principal stakeholders can play a major role in the conservation. The conservation of wetlands was aimed to protect and preserve the regions where sources of water occur at or near the earth's surface such as bogs, marshes, and swamps. Although few wetlands have been successfully protected after the implementation of the Wildlife Protection Act whereas numerous wetlands are becoming simple targets for anthropogenic exploitation and are under grave threat of extinction. The reason is that selected wetlands have received attention under the NWCP, while the remaining are in a neglected state. In India, the conservation of these lands is indirectly affected by a number of legislative as well as policy measures.

Some of the important legislations as per subsequent years are as follows: Indian Fisheries Act of 1857; Indian Forest Act of 1927; Wildlife Protection Act of 1972; Water Prevention and Control of Pollution Act of 1974; Territorial Water and Marine Zone Act of 1976; Forest Conservation Act of 1980; Indian Maritime Zone Act of 1980; Environmental Protection Act of 1986; Wildlife Protection Amendment Act of 1991; Coastal Zone Regulation Act of 1991; National Conservation Strategy and Policy on Environment and Development Act of 1992; National Policy and Biodiversity Act of 1999; Kolkata Wetlands Conservation and Management Act of 2006.

The wetland conservations as well as the management are the responsibility of the MoEFCC. And these ecosystems are conserved under the NWCP which was launched in 1985–1986 by MoEFCC. The purpose of this programme was to prevent the nation's wetlands from further deterioration and to ensure their judicious usage for the assistance of local people and general biodiversity conservation. Under this programme, MoEFCC has labeled 115 wetlands situated in 24 states and two union territories (Shan *et al.*, 2021). The coordination of this programme is the responsibility of the central government while in the management and conservation of wetlands, state government/UTs play an important role. Under the establishment of the Environment Protection Act, MoEFFCC formed the Wetland Management and Conservation Act in 2017, which gives more power to the state government.

In spite of the efforts by MoEFCC, the Asian Wetland Bureau, as well as the WWF, there is no complete inventory of all the wetlands in India. So, the National Wetland Committee of the MoEFCC recommended a nationwide inventory of wetlands to be undertaken. It also recommended collaborative and sustainable management, regular monitoring of irrigation and cropping patterns, coordinated approach, water quality analysis, water resource management, and more research on the conservation of the wetlands. It is notable that the UK has specified 161 wetlands as Ramsar sites, but India has delineated a mere six sites till date while India is a country of mega diversity (Singha *et al.*, 2022). The committee made by MoEFCC suggests that efficient coordination between the several ministries such as industry, energy, water resources, fisheries, transport, and agriculture, is essential for the safeguarding of these ecosystems. There is definitely much ground to be included in the preservation attempts of wetlands.

3.5.1 Wetland conservation take in India (2020–2023)

Wetland conservation research in India is a multidisciplinary field that encompasses various parameters and aspects to understand, protect, and manage wetland ecosystems effectively. While conducting wetland conservation research, studies are focused on considering some following parameters (as shown in Figure 3.3), such as (Biswas Roy *et al.*, 2022).

(i) *Biodiversity assessments*: To study the diversity of plant and animal species in the wetland, including resident and migratory species, identification and documentation of rare or endangered species present.
(ii) *Hydrology and water quality*: Monitoring water flow/water levels/hydrological cycles within the wetland, analyzing water quality parameters (pH, dissolved oxygen, nutrient levels, and pollutant concentrations).
(iii) *Ecosystem services*: Identifying and quantifying the ecosystem services provided by the wetland, such as water purification, flood control, carbon sequestration, and recreation opportunities.
(iv) *Socioeconomic factors*: Such as eco-tourisms.

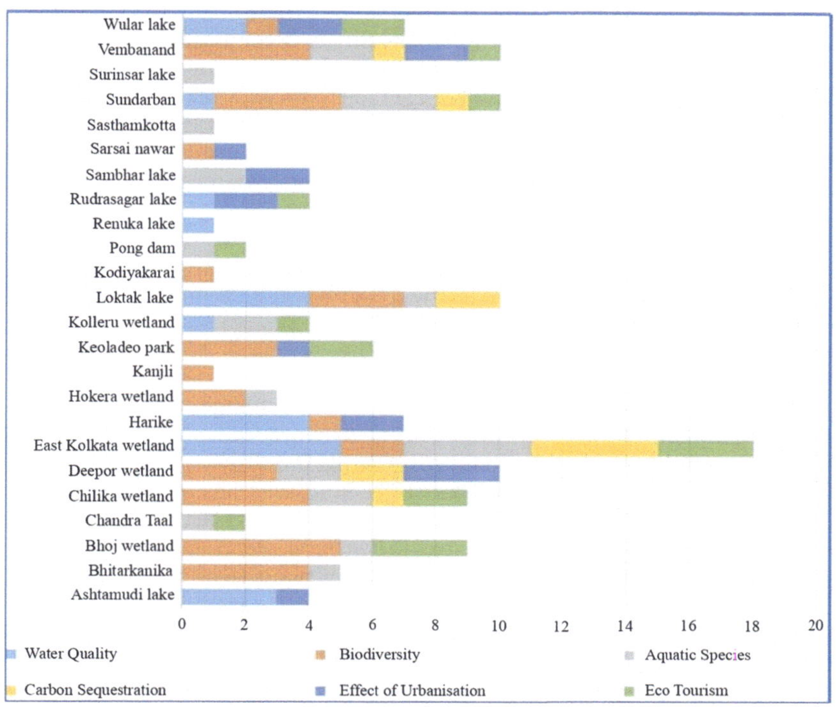

Figure 3.3 Some major research areas on wetlands/Ramsar sites of India (Biswas Roy *et al.*, 2022).

(v) *Geospatial analysis and remote sensing*: Such as the effects of urbanization, human activities, changes in land-use patterns over time

- *Chilika Lake Conservation, Odisha*: Explore the efforts to conserve and restore Chilika Lake, one of India's largest coastal lagoons, and the ecological and socioeconomic impacts of those efforts from 2020 to 2023. Conservation of wetlands along with increasing eco-tourism, and improving the local livelihood is being worked upon (Ojha & Rout, 2022; Sethy & Senapati, 2023).
- *Restoration of Vembanad Wetland, Kerala*: A recent study has reported that one of the problems was the water weed proliferation that has the potential to disrupt their current distribution pattern. Investigate the restoration and conservation initiatives carried out in Vembanad, the largest lake in the state of Kerala and India's second-largest Ramsar wetland, during the specified time period (Sinclair *et al.*, 2020).
- *Bhoj Wetlands, Madhya Pradesh*: Studies have shown that the impact of urban expansion on wetlands has led to an increase in built-up, barren land, and marshy land by 16.97%, 4.79%, and 0.7%, respectively, while simultaneously decrease in agricultural land, vegetation, and water bodies by 11.84%, 10.5%, and 0.13%, respectively (Das & Mehrotra, 2023). Such studies have helped to focus on the conservation strategies and challenges faced in protecting the Bhoj Wetlands, a Ramsar site in Madhya Pradesh, and assess the changes between 2020 and 2023.
- *Chambal River Basin conservation*: Studies have been conducted on wetlands and have revealed the degradation of wetland health, declined ecosystem services, and land cover changes. Therefore, there has been a need to examine the conservation programmes and achievements in the Chambal River Basin, which is home to a diverse range of wetland ecosystems, including the National Chambal Sanctuary (Imdad *et al.*, 2022).
- *Sundarbans mangrove conservation*: To safeguard the Sundarbans, which is the largest mangrove forest in the world, and its wetland habitats from dangers like climate change and habitat destruction, efforts are being made to promote eco-engineering and restoration techniques. These include implementing layers of plantation with true mangrove family members such as *Bruguiera* spp. and *Ceriops* spp., utilizing species like *Rhizophora mucronata* that possess stilt roots, and reducing grazing pressure after the plantation phases (Naz & Chowdhury, 2022; Rahman, 2019).
- *Pong Dam Wetland, Himachal Pradesh*: Examine the strategies implemented to preserve the Pong Dam Wetland and its significance as a crucial resting place for migrating avifauna. According to Malik and Rai (2019), the agricultural area has had a 159% increase, while the amount of barren land has dropped by 53%.
- *Tso Kar Lake conservation, Ladakh*: The research is centered on analyzing the conservation efforts in the high-altitude wetlands of Ladakh, specifically Tso Kar Lake, and its biological importance in relation to climate change (Chandan *et al.*, 2019).
- *Dal Lake revival, Jammu and Kashmir*: Efforts are underway to investigate the current initiatives aimed at revitalizing and preserving Dal Lake, a prominent urban wetland in the area. Around 15.6% of the state's entire land area has been designated as part of the protected area network. Efforts are being made to protect the flagship species, including *Cervus elaphus hanglu*, *Capra falconeri*, *Panthera uncia*, and *Ursus thibetanus*, through coordinated research and detailed management plans (Khuroo *et al.*, 2020).
- *Wular Lake Restoration, Jammu and Kashmir*: Examine the conservation initiatives carried out in Wular Lake, the largest body of freshwater in India, between 2020 and 2023. Research is currently underway to analyze and evaluate several factors of water quality, such as physicochemical concentration values, land use change systems, and water quality index (Jamal *et al.*, 2022).

- *Keoladeo National Park, Rajasthan*: Prior research has demonstrated that there has been a significant impact caused by increasing human intervention, urbanization, traditional farming practices, industrial growth, and excessive utilization of ecological resources. For instance, a study has determined that significant alterations in land use/land cover and the widespread proliferation of *P. juliflora* have negatively impacted the habitat quality of bird species. Research has mostly examined the evaluation of conservation efforts undertaken to protect Keoladeo National Park and its interconnected wetlands, which play a crucial role in supporting migratory birds and biodiversity (Choudhary *et al.*, 2021b; Uchoi & Sen, 2023).

3.6 EFFECT OF CLIMATE CHANGE ON WETLAND

The wetlands are directly affected by several climate change causes, including rising temperatures, alterations in precipitation patterns, and elevated sea levels.

3.6.1 Coastal wetlands

Coastal wetlands may see a reduction in their area as a result of the increase in sea level, rendering them more susceptible to the effects of climate change. The primary impacts of climate change on coastal regions include elevated salinity levels in freshwater and estuaries, increased flooding, the transportation of nutrients and sediments, altered patterns of pollution, and the loss of ecosystems. The biodiversity of coastal wetlands adapts to both temporary and permanent fluctuations in water levels. Rising sea levels will lead to alterations in species compositions and a decline in the production and functionality of wetlands. As a result of these alterations, species have the ability to relocate to different regions. Rising sea levels will compel coastal wetland systems to relocate further inland. Nevertheless, the movement of people could be impeded by the utilization of land in inland areas. For instance, numerous coastal wetlands will be incapable of relocating into the interior due to the existence of particular human land activities in close proximity to the coastal region.

According to Boorman (1990), the coastal region is particularly susceptible to climate change as a result of flooding and rising sea levels. The IPCC (1996) conducted a study on the consequences of climate change and concluded that the global rise in sea levels, ranging from 1 to 5 m, is mostly attributed to the thermal expansion of seawater and the melting of ice caps and glaciers. According to the Intergovernmental Panel on Climate Change (IPCC) in 1996, sea levels are projected to increase by 1–9 m in the next few decades. India possesses a greater number of coastal wetlands, accounting for around 36% of the overall wetland area. The wetlands in Lakshadweep, Gulf of Mannar, Gujarat, Andaman and Nicobar, and Orissa coast are similar to coral reefs and have different concentrations of mangroves. These wetlands also serve as nesting sites for turtles.

3.6.2 Mangrove ecosystem

India possesses the most extensive and varied mangrove habitats globally. The mangrove ecosystem in the vicinity of the Bay of Bengal has developed as a result of the natural accumulation of sediments, together with the separation of intertidal areas. The physical characteristics of this habitat are mostly shaped by deltaic formations. The system comprises of drainage channels connected to both underwater and above-water embankments, as well as areas of land affected by tides. Above the average height of high tide, there exist marshes that are located at the edges and are of little significance, as well as sandbars and islands that are influenced by the tides. The equilibrium between seawater and freshwater sustains the biodiversity of these ecosystems. The alteration in freshwater flow and the elevation of sea levels caused by shifts in precipitation patterns will directly influence the diversity and composition of species.

Table 3.3 Effect of climate change on wetlands of India and the world.

	India			Worldwide	
	Impact	**Adaptation Measures**		**Impact**	**Adaptation Measures**
Altered Precipitation Patterns:	Changes in precipitation patterns can affect the water levels of wetlands. Irregular rainfall may lead to drought conditions, impacting the availability of water in some wetland ecosystems.	Improved water management practices and conservation efforts are crucial to mitigate the effects of altered precipitation.	Loss of Biodiversity	Climate change contributes to shifts in species distribution, affecting the biodiversity of wetland ecosystems worldwide.	Conservation initiatives, protected area management, and restoration efforts are crucial for preserving global wetland biodiversity.
Sea-Level Rise	Coastal wetlands in regions like the Sundarbans are vulnerable to sea-level rise, resulting in saltwater intrusion and habitat loss.	Coastal protection measures, mangrove restoration, and community-based adaptation strategies are essential for mitigating sea-level rise impacts.	Permafrost Thaw	Wetlands in northern regions with permafrost are susceptible to thawing, leading to changes in hydrology and vegetation.	Monitoring permafrost conditions, implementing conservation measures, and promoting sustainable land use are vital for adaptation.
Temperature Increase	Rising temperatures can alter the hydrological cycle, leading to increased evaporation and potential changes in wetland vegetation.	Sustainable land-use planning, habitat restoration, and monitoring of temperature-sensitive species are essential for adaptation.	Methane Emissions	Climate change can amplify methane emissions from wetlands, contributing to greenhouse gas concentrations.	Sustainable wetland management, restoration, and global efforts to reduce greenhouse gas emissions are critical for mitigating methane release.
Extreme Weather Events	Increased frequency and intensity of extreme weather events, such as cyclones and floods, pose a threat to wetland ecosystems.	Implementing early warning systems, restoring natural buffers, and developing resilient infrastructure is critical for adapting to extreme weather events.	Shifts in Wetland Distribution	Changes in temperature and precipitation patterns can lead to shifts in the distribution of wetland types globally.	Protected area planning, habitat restoration, and international collaboration are essential for addressing shifts in wetland distribution.
Glacial Retreat	Himalayan wetlands may be affected by glacial retreat, altering water availability and flow patterns.	Monitoring glacial changes, sustainable water management, and conservation of high-altitude wetlands are essential for adaptation.	Rising Sea Levels	Low-lying coastal wetlands worldwide face threats from rising sea levels, leading to saltwater intrusion and habitat loss.	Coastal protection, restoration of natural buffers, and global efforts to reduce greenhouse gas emissions are essential for adaptation.

3.6.3 High-altitude wetlands

India possesses a substantial quantity of elevated wetland regions that are commonly referred to as alpine wetlands. These wetland areas possess exceptional biodiversity. The rising temperature of the lake water is attributed to climate change. Animals and plants in high-altitude wetland areas are particularly vulnerable to little temperature fluctuations due to their lack of alternative habitats.

3.6.4 Coral reefs

Coral reefs have a high degree of sensitivity to fluctuations in lake temperature. These maritime environments have the highest level of biological diversity. These organisms can only flourish in the warm waters of tropical and subtropical regions of oceans and seas. Temporary rises in water temperatures (1–2°C) can lead to coral 'bleaching', whereas increases in water temperature over the normal range (~3–4°C) can result in substantial coral death. Corals may be adversely affected by the escalating storm surges and elevating sea levels. Coral reefs have received significant global attention in the context of climate change scenarios (IUCN/UNEP, 1993). India possesses a larger quantity of corals that are categorized as natural wetlands. Elevated levels of carbon dioxide (CO_2) in the atmosphere have a detrimental impact on the formation of coral reefs as they cause alterations in the pH of seawater. It is concluded that climate change will affect the growth of coral reef systems (Bijlsma, 1996).

3.6.5 Inland natural wetlands

Wetlands encompass rivers, lakes, and waterlogged areas, making up about 37% of the whole wetland area. The change in the hydrological cycle will impact them through alterations in transpiration, runoff, precipitation, and evaporation. Semi-arid and arid regions are susceptible to variations in precipitation, and these areas are most impacted when there is a decrease in rainfall.

3.6.6 Carbon cycle

Wetlands have a significant impact on the global carbon cycle. Worldwide, the carbon reservoir in wetlands is estimated to be approximately 230 gigatons out of a total of 1943 gigatons. Wetlands have a constant increase in carbon stocks because of their poor nutrient availability and anaerobic properties. These wetlands are also the source of methane emissions and production. Conversion of wetlands to agricultural land leads to the significant release of NO_2 and CO_2 while reducing methane emissions. This form of evaluation has not yet been conducted in India. Nevertheless, the act of preserving, protecting, or restoring wetland ecosystems can serve as a feasible component of a comprehensive approach to mitigating climate change.

Both in India and globally, successful adaptation and mitigation methods entail a blend of policy interventions, community involvement, scientific inquiry, and international cooperation. Table 3.3 provides a concise overview of the impact of climate change on wetlands, specifically focusing on India and the global scale.

3.7 CONCLUSIONS

Wetlands in their natural state are highly advantageous ecosystems that offer numerous advantages to people. They provide assistance to a wide range of distinct and exceptional environments. They undertake a strenuous endeavor to uphold the economic, cultural, and ecological vitality of plant and animal life. Given that public assets offer numerous advantages, it is imperative that their security and administration be a shared responsibility. Currently, these habitats are being polluted as a result of a rise in population and human activities such as tourism, excessive use, overfishing, and the discharge of wastewater from residential and industrial sources. The rapid rate at which they are vanishing from the land is a matter of significant worry to researchers and scientists, who are now prioritizing the preservation of these habitats.

While there are numerous regulations in place, there is currently no specific legislation dedicated to the preservation and safeguarding of wetlands. Wetland conservation has been insufficiently prioritized, and the management is currently organizing awareness campaigns on wetland conservation in collaboration with local institutions and NGOs. India possesses a vast array of plant and animal life and should be actively working to protect the ecological diversity of these habitats through the governing authority. Investigation is crucial in the development of national policy to determine the evolving characteristics of these ecosystems. In order to bolster the legal frameworks, it is necessary to enhance and enforce current laws and regulations pertaining to wetland conservation. This should be done while also taking into account, updating, and enhancing these legal frameworks to tackle growing difficulties and integrate new scientific discoveries. These investigations will assist management in formulating policies to effectively mitigate wetland contamination. A systematic understanding will assist management in comprehending the lucrative benefits of wetlands, so aiding in the establishment of their significance and the successful implementation of planning procedures. Integrated management necessitates the promotion and consideration of the ecological, social, and economic dimensions of wetlands. It also involves the implementation of landscape-level planning to safeguard entire watersheds and their corresponding wetlands. Collaboration across diverse stakeholders, regional groups, and business sectors is crucial for the development of an effective management strategy. Regular and systematic monitoring of the wetlands is essential.

3.8 RECOMMENDATION, AND FUTURE RESEARCH

For the successful preservation of wetlands, it is crucial to adopt a comprehensive approach that includes the participation of many stakeholders and the implementation of diverse strategies. Wetlands are vital ecosystems that offer a wide range of ecological, economic, and social advantages. These include providing habitat for a variety of species, filtering water, regulating floods, and offering recreational activities. First and foremost, it is crucial to incorporate corporate and educational institutions into conservation initiatives. These organizations have the ability to have a substantial impact on achieving ecological balance by advocating for sustainable practices and increasing public knowledge about the significance of wetlands. Corporate enterprises have the ability to implement environmentally conscious practices that reduce their negative effects on wetlands, while educational institutions can provide knowledge to students and the general public regarding wetland ecosystems and measures for conservation. Partnerships between these institutions and conservation groups can result in creative solutions and increased public involvement. A comprehensive assessment of wetlands is crucial for efficient protection. This inventory should comprehensively record the variety of plant and animal species found in these environments, along with their ecological significance. Gaining knowledge about the many species and their functions within the wetland can provide valuable insights for making informed decisions about its management and identifying areas that require conservation efforts. In addition, the inventory should encompass an evaluation of the advantages that wetlands offer to nearby towns, such as the purification of water, management of floods, and provision of recreational activities. This comprehensive approach emphasizes the importance of wetlands and provides a rationale for investing in their preservation. Environment impact assessments (EIAs) are essential for evaluating potential risks to wetlands resulting from significant development projects. Prior to embarking on any significant projects or alterations that may have an effect on wetland areas, it is essential to carry out an EIA to assess potential hazards and suggest strategies to minimize them. This proactive strategy guarantees the protection of wetlands from detrimental activity and minimizes any negative consequences.

An important stage is to assess the socio-economic and conservation advantages of wetlands. Wetlands provide a variety of advantages, including the support of local economies and the improvement of biodiversity. An exhaustive assessment of these advantages serves to illustrate the wider significance of wetlands and reinforces the argument for their preservation. This assessment should take into account not just the immediate financial gains but also the secondary benefits offered by robust wetland ecosystems. Utilizing cutting-edge technologies to minimize the amount of water needed for agriculture can greatly aid in the preservation of wetlands. Agricultural practices frequently place significant strain on wetland resources due to the extraction of water and the introduction of pollutants. By allocating resources to water-efficient technology and implementing sustainable practices, the strain on wetlands can be decreased, resulting in improved land and water management. These aims can be accomplished by advancements in irrigation systems, water recycling, and precision agriculture. Adequate funding is essential for the effective preservation of wetlands. It is imperative for national and regional governments to give sufficient resources to bolster conservation efforts. In addition, soliciting financial assistance from foreign organizations and philanthropists helps get the required funding for extensive projects and restoration endeavors. Public–private partnerships can also play a crucial role in combining resources and specialized knowledge.

Restoring impaired wetlands is equally crucial to conserving pristine ones. Restoration initiatives have the potential to revive deteriorated wetland habitats, thereby reinstating their biological roles and enhancing their ability to sustain biodiversity. Concurrently, it is imperative to undertake measures to safeguard and oversee current wetlands in order to avert more deterioration. An effective national wetland conservation strategy should prioritize cooperative and enduring management practices. This strategy should encompass a wide range of stakeholders, including government agencies, local communities, conservation organizations, and the commercial sector. Implementing collaborative management approaches can optimize the efficacy of conservation efforts and guarantee the inclusion of a wide range of opinions. Policymakers must possess comprehensive knowledge of the significance of wetland preservation. Presenting them with pertinent data, study discoveries, and instances of success can aid in garnering backing for conservation programs and catalyzing legislative alterations. Increased bilateral collaboration in resource management can also bolster conservation endeavors, promoting shared obligations and cooperative approaches to tackle wetland conservation difficulties. Ultimately, a thorough and synchronized strategy for wetland conservation is crucial for safeguarding these vital ecosystems. To guarantee the preservation and responsible utilization of wetlands for future generations, it is imperative to incorporate several stakeholders, utilize technology, obtain finance, and encourage collaborative management.

ACKNOWLEDGMENT

The authors are very grateful to Amity University, Noida for providing the platform to conduct this study.

REFERENCES

Alam S., Bhardwaj L. K., Mallick R. and Rai S. (2023). Estimation of heavy metals and fluoride ion in vegetables grown nearby the stretch of River Yamuna, Delhi (NCR), India. *Indian Journal of Environmental Protection*, **43**(1), 64–73.

Anand A., Chandan P. and Singh R. B. (2012). Homestays at Korzok: supplementing rural livelihoods and supporting green tourism in the Indian Himalayas. *Mountain Research and Development*, **32**(2), 126–36, https://doi.org/10.1659/MRD-JOURNAL-D-11-00109.1

Aslam A., Parthasarathy P. and Ranjan R. K. (2021). Ecological and societal importance of wetlands: a case study of North Bihar (India). In: Wetlands conservation: current challenges and future strategies, 55–86.

Bassi N., Kumar M. D., Sharma A. and Pardha-Saradhi P. (2014). Status of wetlands in India: A review of extent, ecosystem benefits, threats and management strategies. *Journal of Hydrology: Regional Studies*, **2**, 1–19, https://doi.org/10.1016/j.ejrh.2014.07.001

Bhardwaj L. K. (2023). A Comprehensive Review on the Climate Change and Its Impact on Health. https://doi.org/10.20944/preprints202305.0159.v1

Bhardwaj L. K. and Jindal T. (2019). Contamination of lakes in Broknes peninsula, East Antarctica through the pesticides and PAHs. *Asian-Journal of Chemistry*, **31**(7), 1574–1580. https://doi.org/10.14233/ajchem.2019.22022

Bhardwaj L. K. and Jindal T. (2020). Persistent organic pollutants in lakes of Grovnes Peninsula at Larsemann Hill area, East Antarctica. *Earth Systems and Environment*, **4**, 349–58, https://doi.org/10.1007/s41748-020-00154-w

Bhardwaj L. K. and Jindal T. (2022). Evaluation of coliform and faecal coliform bacteria in the lakes of Broknes and Grovnes Peninsula, East Antarctica. *Nature Environment & Pollution Technology*, **2**, 21.

Bhardwaj L. K. and Sharma A. (2021). (Microplastics (MPs) in Drinking Water: Uses, Sources & Transport. https://doi.org/10.20944/preprints202104.0498.v1

Bhardwaj L. K. and Yadav P. (2023). Microplastic Contamination, an Emerging Threat to the Freshwater Environment and Human Health: A Systematic Review. https://doi.org/10.20944/preprints202312.0990.v1

Bhardwaj L. K., Sharma S. and Jindal T. (2021). Occurrence of polycyclic aromatic hydrocarbons (PAHs) in the lake water at Grovnes Peninsula over East Antarctica. *Chemistry Africa*, **4**, 965–80, https://doi.org/10.1007/s42250-021-00278-0

Bhardwaj L. K., Sharma S. and Jindal T. (2023). Estimation of physico-chemical and heavy metals in the lakes of Grovnes & Broknes Peninsula, Larsemann Hill, East Antarctica. *Chemistry Africa*, **4**, 1–8.

Bijlsma J. (1996). Effects of climate change on coral reefs. Contribution to IPCC - Intergovernmental Panel on Climate Change, 1996. Climate Change 1995 – Impacts, Adaptations and mitigation of climate change: scientific technical analysis. Contribution of Working Group II to the second Assessment Report of the IPCC, Cambridge University Press, Cambridge.

Biswas Roy M., Nag S., Halder S. and Kumar Roy P. (2022). Assessment of wetland potential and bibliometric review: a critical analysis of the Ramsar sites of India. *Bulletin of the National Research Centre*, **46**(1), 1, https://doi.org/10.1186/s42269-022-00740-0

Boorman L. A. (1990). Impact of sea level changes on coastal areas. In: Landscape ecological impacts of climate change, M. M. Boer and R. S. de Groot (eds.), IOS Press, Amsterdam, pp. 379–391.

Chandan P., Singh A. and Rajashekariah K. (2019). Biodiversity and regional significance of Himalayan high altitude wetlands. *The State of Ecology of the Tibetan Plateau (Preprint).*

Chandra K., Bharti D., Kumar S., Raghunathan C., Gupta D., Alfred J.R.B. and Chowdhury B.R. (2021). Faunal Diversity in Ramsar Wetlands of India, 1–292 (Jointly Published by the Director, Zoological Survey of India and Wetland Division, Ministry of Environment, Forest and Climate Change, Government of India). file:///C:/Users/Admin/Downloads/Faunal_Diversity_of_Ramsar_Wetlands_of_India_E_verion_02.02.2021.pdf

Choudhury M., Sharma A., Singh P. and Kumar D. (2021a). Impact of climate change on wetlands, concerning Son Beel, the largest wetland of North East, India. In: Global climate change, Elsevier, pp. 393–414.

Choudhary A., Deval K. and Joshi P. K. (2021b). Study of habitat quality assessment using geospatial techniques in Keoladeo National Park, India. *Environmental Science and Pollution Research*, **28**, 14105–14114, https://doi.org/10.1007/s11356-020-11666-3

Costanza R., De Groot R., Sutton P., Van der Ploeg S., Anderson S. J., Kubiszewski I., Farber S. and Turner R. K. (2014). Changes in the global value of ecosystem services. *Global Environmental Change*, **26**, 152–158, https://doi.org/10.1016/j.gloenvcha.2014.04.002

Das N. and Mehrotra S. (2023). Impact of urban expansion on wetlands: A case study of Bhoj Wetland, India. *Journal of the Indian Society of Remote Sensing*, **51**(8), 1697–1714, https://doi.org/10.1007/s12524-023-01728-7

Donnelly J. P., King S. L., Silverman N. L., Collins D. P., Carrera-Gonzalez E. M., Lafón-Terrazas A. and Moore J. N. (2020). Climate and human water use diminish wetland networks supporting continental waterbird migration. *Global Change Biology*, **26**(4), 2042–59, https://doi.org/10.1111/gcb.15010

Gopal B. (2014). Mangroves are wetlands, not forests: some implications for their management. In: Mangrove ecosystems of Asia: status, challenges and management strategies, pp. 439–53.

Gupta P., Sharma A. and Bhardwaj L. K. (2023). Solid Waste Management (SWM) and Its Effect on Environment & Human Health. Preprints 2023, 2023090384. https://doi.org/10.20944/preprints202309.0384.v1

Hammer D. A. and Bastian R. K. (2020). Wetlands ecosystems: natural water purifiers?. In: Constructed wetlands for wastewater treatment, pp. 5–19.

Imdad K., Rihan M., Sahana M., Parween S., Ahmed R., Costache R. and … Tripathi R. (2022). Wetland health, water quality, and resident perceptions of declining ecosystem services: a case study of Mount Abu, Rajasthan, India. *Environmental Science and Pollution Research*, 1–27.

Intergovernmental Panel on Climate Change (IPCC) (1996). Climate Change 1995 – Impacts, adaptations and mitigation of climate change: scientific technical analysis, Contribution of Working Group II to the second Assessment Report of the IPCC, Cambridge University Press, Cambridge.

IUCN/UNEP (1993). Reefs at risk: a programme of action. IUCN, Gland, Switzerland, p. 24.

Jamal S., Ahmad W. S., Ajmal U., Aaquib M., Ashif Ali M., Babor Ali M. and Ahmed S. (2022). An integrated approach for determining the anthropogenic stress responsible for degradation of a Ramsar site–Wular lake in Kashmir, India. *Marine Geodesy*, **45**(4), 407–434, https://doi.org/10.1080/01490419.2022.2034686

Khuroo A. A., Mehraj G., Muzafar I., Rashid I. and Dar G. H. (2020). Biodiversity conservation in Jammu and Kashmir state: current status and future challenges. *Biodiversity of the Himalaya: Jammu and Kashmir State*, **18**, 1049–1076, https://doi.org/10.1007/978-981-32-9174-4_41

Kolli M. K. (2020). Lake Catchment Interaction Analysis by Using Remote Sensing and GIS Techniques – the case study of Kolleru Lake, South India.

Kumar D., Choudhury M. and Rathore A. (2020). Valuation of ecosystem services benefits of Son Beel wetland in Assam, India: a case study of natural solutions to climate change water. Climate Change Conference, Hamburg University, Germany.

Kumari R., Shukla S. K., Parmar K., Bordoloi N., Kumar A. and Saikia P. (2020). Wetlands conservation and restoration for ecosystem services and halt biodiversity loss: An Indian perspective. In: Restoration of wetland ecosystem: a trajectory towards a sustainable environment, pp. 75–85.

Malik M. and Rai S. C. (2019). Drivers of land use/cover change and its impact on Pong Dam wetland. *Environmental Monitoring and Assessment*, **191**, 1, https://doi.org/10.1007/s10661-019-7347-x

Ministry of Environment, Forests and Climate Change Annual report. (2012). MoEFCC, Government of India, New Delhi.

Mitsch W. J., Bernal B. and Hernandez M. E. (2015). Ecosystem services of wetlands. *International Journal of Biodiversity Science, Ecosystem Services & Management*, **11**(1), 1–4, https://doi.org/10.1080/21513732.2015.1006250

Mohan M. S. (2022). The Socio-legal Consequences of Nature Rights in lake Restoration and Governance.

Mukherjee S., Rizvi S. S., Biswas G., Paswan A. K., Vaiphei S. P., Warsi T. and Mitran T. (2023). Aquatic Ecosystems Under Influence of Climate Change and Anthropogenic Activities: Potential Threats and Its Mitigation Strategies. *Hydrogeochemistry of Aquatic Ecosystems*, 307–331.

Naz A. and Chowdhury A. (2022). Eco-engineering and mangrove restoration methods to stabilize earthen embankments and establishing bio-shield against natural disasters: A case study from Sundarban Ramsar wetland, India. In: Assessing, mapping and modelling of mangrove ecosystem services in the Asia-Pacific region, pp. 183–198.

Nisari A. R. and Sujatha C. H. (2021). Assessment of trace metal contamination in the Kol wetland, a Ramsar site, Southwest coast of India. *Regional Studies in Marine Science*, **47**, 101953, https://doi.org/10.1016/j.rsma.2021.101953

Ojha A. and Rout J. (2022). Restoration and conservation of wetlands: a geospatial approach. In: Anthropogeomorphology: a geospatial technology based approach, Springer International Publishing, Cham, pp. 617–634.

Ostad-Ali-Askari K. (2022). Review of the effects of the anthropogenic on the wetland environment. *Applied Water Science*, **12**(12), 260, https://doi.org/10.1007/s13201-022-01767-4

Paonam J. and Chatterjee S. (2022). Threat perception of stakeholders: case study of Loktak Lake, a Ramsar site under Montreux record in North East India. *Wetlands*, **42**(8), 114, https://doi.org/10.1007/s13157-022-01630-x

Paul M., Chanda M. and Gupta S. (2011). Strategy and scenario for wetland conservation in India. *Chronicles of Young Scientists*, **2**(2), 79, https://doi.org/10.4103/2229-5186.82976

Pižl V., Sterzyńska M., Tajovský K., Starý J., Nicia P., Zadrożny P. and Bejger R. (2023). Effects of hydrologic regime changes on a taxonomic and functional trait structure of earthworm communities in mountain wetlands. *Biology*, **12**(3), 482, https://doi.org/10.3390/biology12030482

Prasad A. D., Venkataramana G. V. and Thomas M. (2009). Fish diversity and its conservation in major wetlands of Mysore. *Journal of Environmental Biology*, **30**(5), 713.

Prasad S. N., Ramachandra T. V., Ahalya N., Sengupta T., Kumar A., Tiwari A. K., Vijayan V. S. and Vijayan L. (2002). Conservation of wetlands of India – a review. *Tropical Ecology*, **43**(1), 173–86.

Rahman F. (2019). Wildlife conservation and mangrove interpretation centre, Karamjal, Sundarban: A case study of a site-specific architectural project in a mangrove forest. *Nakhara: Journal of Environmental Design and Planning*, **17**, 55–70, https://doi.org/10.54028/NJ2019175570

Rajakumar R. and Sivasubraminium C. (2012). Seasonal water quality status in Udhayamarthandapuram bird sanctuaries (Tamil Nadu). *Asian Journal of Environmental Science*, **7**(1), 78–86.

Reid W. V., Mooney H. A., Cropper A., Capistrano D., Carpenter S. R., Chopra K., Dasgupta P., Dietz T., Duraiappah A. K., Hassan R. and Kasperson R. (2005). Ecosystems and human well-being-synthesis: a report of the millennium ecosystem assessment. Island Press.

Sarkar P. and Das T. (2016). Wetland ecosystem services and its valuation with special reference to India – A review. In: Biodiversity and environmental conservation K. Upadhaya (Ed), Discovery Publishing House, New Delhi, India, pp. 59–76.

Sethy M. K. and Senapati A. K. (2023). Perceptions towards ecotourism practice and the willingness to pay: evidence from Chilika coastal wetland ecosystem, Odisha. *International Journal of Geoheritage and Parks*, **11**(3), 497–513, https://doi.org/10.1016/j.ijgeop.2023.08.001

Shan V., Singh S. K. and Haritash A. K. (2021). Present status, conservation, and management of wetlands in India. In: Advances in energy and environment: select proceedings of TRACE 2020, Springer, Singapore, pp. 235–256.

Sinclair M., Ghermandi A., Moses S. A. and Joseph S. (2020). Ecosystem service assessment and mapping for sustainable management of wetlands in Kerala, India. In: Environmental assessments: scenarios, modelling and policy, p. 200.

Singha K. B., Singh N. B. and Singh K. G. (2022). Factors determining the willingness to pay for the conservation of biodiversity of Sone Beel, Assam, India. *Cogito*, **14**(3), 2066–7094.

Stefanakis A. I. (2019). The role of constructed wetlands as green infrastructure for sustainable urban water management. *Sustainability*, **11**(24), 6981, https://doi.org/10.3390/su11246981

Uchoi A. and Sen M. (2023). Keoladeo national park: a managerial approach to a unique protected area cum world heritage site of India. *Journal of Heritage Management*, 24559296221147032.

Upadhyay P., Pruski L. O., Kaleita A. L. and Soupir M. L. (2019). Effects of land management on inundation of prairie pothole wetlands in the Des Moines Lobe using AnnAGNPS. *Agricultural Water Management*, **213**, 947–56, https://doi.org/10.1016/j.agwat.2018.12.016

Zutshi B. and Prasad S. R. (2008). Impact of pollution on fresh-and marine water resources – A review. *Pollution Research*, **27**(3), 461–6.

doi: 10.2166/9781789064100_0063

Chapter 4

Constructed wetland for wastewater treatment: design, operation, and performance

Suparna Deb[1], Bhavesh Choudhary[1], Shubham Kashyap[1] and Utkarsh Choudhary[2]*

[1]College of Fisheries, Central Agricultural University (Imphal), Lembucherra, Agartala, Tripura (West), 799210, India
[2]Yadav Measurements Private Limited, Udaipur, Rajasthan, 313003, India
**Corresponding author email: bhaveshchoudhary@yahoo.com*

ABSTRACT

Constructed wetlands have emerged as a sustainable and efficient approach for treating wastewater, offering numerous environmental and economic benefits. This chapter examines the design, operation, and performance of contracted wetlands as a viable solution for wastewater treatment especially in fishery sector. The design aspect highlights the critical factors considered during the planning phase of constructed wetlands. It discusses the selection approaches of appropriate wetland types, such as free water surface (FWS) and subsurface flow (SSF), based on specific pollutant removal requirements and available land space. Factors such as hydraulic retention time (HRT), vegetation type and layout configuration are examined in the context of maximizing treatment efficiency and ensuring successful integration into the fishery environment. The operation section delves into the practical implementation and management of constructed wetlands for wastewater treatment in the fishery sector. It explores the importance of regular monitoring and maintenance to sustain optimum performance. The significance of understanding wetland ecology and the roles of microorganism and aquatic plants in pollutant removal are also discussed, highlighting the importance of ecological balance within the system. The performance assessment of constructed wetlands focuses on their removing various pollutants commonly found in water, organic matter, and potentially harmful substances. Moreover, this chapter discusses the environmental effect or benefits of constructed wetlands, such as habitat creation and biodiversity enhancement, which can contribute to the overall structure of promoting a healthy fishery ecosystem. The potential economic advantages are also discussed in this chapter. Overall, this provides valuable insights into design, operation, and performance of constructed wetlands for wastewater treatment in the fishery sector. Understanding the critical factors involved in their implementation and their capacity to improve water quality can contribute significantly to sustainable development in the fishery industry, ensuring both environmental protection and economic viability.

Keywords: Wetland, waste-water treatment, design, environmental impact, ecological balance, fishery.

4.1 INTRODUCTION

4.1.1 Background and significance of constructed wetlands in wastewater treatment

Wetlands are known as the 'kidneys of the Earth' because they are among the most vital and valuable ecosystems on the planet are among the most significant ecosystems on Earth, which perform an important function in flood mitigation and contaminated water filtration, and also providing habitat for a variety of plants and animals. They also respond to climate change but it has been degraded and reduced substantially over the last 50 years as land cover has changed in some areas (Guo *et al.*, 2017). Wetlands are areas with low water levels, frequently near the ground surface, that are occupied by active plants during the growing season and water saturation period (Barducci *et al.*, 2009). Article 1 of the Ramsar Convention defines wetlands as 'areas of marsh, fen, peatland, or water, whether natural or artificial, permanent or temporary, with water that is static or flowing, fresh, brackish, or salt, including areas of marine water the depth of which at low tide does not exceed six metres'.

Wetlands are divided into two types: natural and manmade (Zhang *et al.*, 2010). Wetlands can be found anywhere, including the tundra and the tropics. We do not know exactly how much of the earth's surface is made up of wetlands, but according to the UNEP-World Conservation Monitoring Centre, wetlands cover around 570 million hectares (5.7 million km^2), or roughly 6% of the Earth's land area. Mitsch and Gosselink (2015) estimate that wetland covers approximately 4–6% of the Earth's land surface. Mangroves span around 240,000 km^2 of coastal area, with an estimated 600,000 km^2 of coral reefs remaining worldwide.

Wetlands may improve water quality, protect shorelines, recharge groundwater, reduce flood and drought severity, and provide unique habitats for a variety of plants and animals. Wetlands have received increased attention from scientists and government officials in order to preserve the aquatic system's biodiversity (Fickas *et al.*, 2016; Zhang *et al.*, 2010). A wetland is a broad term that encompasses coastal wetlands, peatlands, mangrove forests, estuary wetlands, marshes, and other forms of wetlands. The coastal wetland serves as an essential source of nutrient cycling capacity for maintaining water quality (Schmidt & Skidmore, 2003). Peatlands play a significant part in the carbon cycle because they store carbon dioxide (CO_2), the primary greenhouse gas, from the atmosphere as peat; yet, they also emit enormous amounts of CO_2 and methane (CH_4), the second most important greenhouse gas (Gorham, 1991). Many fish and crustacean species rely on mangrove forests for habitat, and they also play a vital role in coastal protection and water quality. Despite their importance, around 50% of wetlands have disappeared since the 1900s (Nicholls, 2004). Wetland studies are therefore critical to protecting wetlands and mitigating climate change (Guo *et al.*, 2017).

It is important to note that wetlands cover 5–8% of the earth's surface (Mitsch & Mander, 2018; Xu *et al.*, 2019), with 29.83 million km^2 distributed across Asia (9.2 million), South America (7.95 million), and North America (5.65 million), accounting for 78% of the global percentage. Africa has 19% (5.6 million) (Boone & Bhomia, 2017), primarily vegetated coastal wetlands that store organic and inorganic carbon in greater quantities. One of the primary environmental benefits given by natural wetlands is carbon sequestration, which occurs both in the soil and in biomass. In sediments, it has a 2–3 ratio with biomass storage (Zhao *et al.*, 2019). Marín-Muñiz *et al.* (2014) found that swampy soils stored more carbon than marshes (0.92 ± 0.12 kgC/m^2 yr vs. 0.31 ± 0.08 kgC/m^2 yr) in a 3:1 ratio. Riparian and coastal wetlands can store 50 times more carbon in soil than other land forest types (Liu *et al.*, 2020). Other more conservative writers, such as Donato *et al.* (2011), demonstrate that the first 30 cm of soil in wetlands such as mangroves can store twice as much carbon per hectare as boreal forests. This thickness or layer is especially vulnerable to land-use changes. Adame *et al.* (2013), citing the Intergovernmental Panel on Climate Change (IPCC), claim that mangroves can store twice or thrice as much carbon as tropical and temperate forests.

Tropical wetlands have the largest soil organic carbon stocks. Köchy *et al.* (2015) and Villa and Bernal (2018) confirm that they can store up to one-third of global carbon. Mangroves store 218 tonnes of carbon per year from the atmosphere, making them an important component of 'blue carbon' (Cui *et al.*, 2018), which refers to the carbon exchanged between coastal habitats with vegetation (Choudhary *et al.*, 2024). As a result, determining natural wetland conditions and their evolution over time is critical for the global context because it provides information about the value of ecosystem services per unit area, including places where vegetation loss occurs. They are desirable to restore and maintain, as well as to improve through priority policies (Sanderman *et al.*, 2018), because of their ability to act as natural carbon sinks.

Vegetation loss means a loss of functions not just in ecosystem services for humans, but also in the breakdown of the global carbon cycle, as well as the water and nutrient cycles (Bernal & Mitsch, 2013). Furthermore, global warming caused by climate change resulting from the industrial revolution jeopardizes future human life and has made it difficult to reduce petrol emissions while conserving ecosystems such as wetlands that absorb CO_2 (Zhang *et al.*, 2017). Despite the benefits of wetlands, dense populations in coastal areas, combined with increased poverty, account for 1–3% of annual deforestation (Alongi, 2012). As a result of the damage caused to their vegetation, it is anticipated that the population surrounding wetlands, which benefits from their presence, will approach half of the world's population within a radius of 100 km from the coast (Sun *et al.*, 2017). Davidson (Liu *et al.*, 2020) calculated the area losses by continents since 1900 and changed the deforestation rate from 64% to 71%, with coastal wetlands seeing greater rates. Wetlands confront several hazards, such as drainage, pollution, and development. Nonetheless, their significance is apparent. Understanding and enjoying these underwater wonderlands allow us to work towards their protection and assure their continued viability for future generations.

4.1.2 According to Ramsar convention types of wetlands

Five major wetland types are generally recognized:

(i) *Marine/coastal wetlands*: including coastal lagoons, rocky shores, seagrass beds and coral reefs.
(ii) *Estuarine wetlands*: including deltas, tidal marshes and mudflats, and mangrove swamps.
(iii) *Lacustrine wetlands*: wetlands associated with lakes.
(iv) *Riverine wetlands*: wetlands along rivers and streams.
(v) *Palustrine wetlands*: meaning 'marshy' – marshes, swamps and bogs.
Human-made wetlands include fish and shrimp ponds, farm ponds, irrigated agricultural area (such as rice paddies), salt pans, dams, reservoirs, gravel pits, wastewater treatment ponds and canals. The Ramsar Convention established a Ramsar Classification of Wetland varieties, which contains 42 varieties divided into three categories: marine and coastal wetlands, inland wetlands, and human-made wetlands.

Marshes

A marsh is a wetland dominated by herbaceous plants like grasses, rushes, and sedges. Small bushes frequently develop along the boundary, marking the transition to drier land. Marshes typically form on the shallow borders of lakes and rivers. They provide home for a wide range of aquatic invertebrates, many of which live and feed on rotting plants. Invertebrates support a diverse range of fish, amphibians (frogs), reptiles (snakes and turtles), waterfowl, water birds, and wetland mammals such as muskrats.

Bog

Bogs are acidic, low-oxygen wetlands formed when organic waste accumulates faster than it decays. Bog soils are wet and acidic peats composed of sphagnum moss and other vegetation that decomposes slowly. The majority of northern bogs originate in contained glacial depressions known as kettle lakes, where there is limited inflow and outflow, and precipitation is the primary water source rather than streams or groundwater. A mat of moss or other plant forms on the lake's edge and gradually spreads across its surface. The bog may eventually fill in completely, leaving no open water visible. Bogs are deficient in nitrogen and other nutrients that plants require due to a lack of oxygen and their acidic pH. A distinct population of carnivorous plants has adapted to these conditions. Pitcher plants, sundews, and bladderworts get their nutrition from the meat (insects) they collect and consume. Bogs along the lakeshore are typically filled-in lakebeds with sphagnum bases and shrubs that can endure acidic soils, such as leatherleaf, bog rosemary, bog laurel, and cranberries. Bog communities also include several orchid species. The lakeshore has four large bog sections, including a shrubby one near

Sand Point, a filled-in bog northeast of Beaver Lake, multiple bog pockets around Legion Lake, and a classic bog lake east of Twelvemile Beach campground.

Swamp

Swamps are wooded wetlands with unique tree and soil types. The Upper Peninsula and Pictured Rocks swamps are mostly conifer swamps, with northern white cedar, black spruce, and tamarack as the dominant species, while balsam fir, eastern hemlock, and white pine may also be present. Swamps and lowland woods are quite similar and often used interchangeably. Swamps, on the other hand, are often wetter for longer periods of the year and have deeper standing water than lowland forests. Acidic (black spruce/tamarack) and alkaline conifer swamps are found across the national lakeshore, but alkaline swamps supporting northern white cedar/tamarack ecosystems are more abundant. Many have grown in the park's river basins, especially along Miners River and Sevenmile Creek. Old growth northern white cedars from the Beaver Basin Wilderness marshes are among the park's oldest trees. Cedar swamps also serve as habitat for a variety of wildlife species, including deer and snowshoe hare, which require important winter habitat. More than 80 species of wildlife in the Upper Peninsula are known to spend portions of their lives in cedar swamps.

Fens

Fens are a significant and distinctive wetland form. Fens are peat-forming wetlands that rely on groundwater input, take thousands of years to grow, and cannot be quickly regenerated once destroyed. Fens are also hotspots for biodiversity. They are frequently home to unusual flora, insects, and small mammals. Larger animals such as deer and livestock graze in this sort of wetland. Fens are also beneficial to people. They are important groundwater discharge locations and can be used to identify shallow aquifers. Vegetation in all wetlands helps to recycle nutrients, trap eroding soil, and filter out polluting pollutants like nitrates. Furthermore, fens feature strongly in nearly all models of CO_2-induced global change because they are a significant sink for atmospheric carbon (Weixelman & Cooper 2009).

Peatlands

Bogs and fens are two types of wetlands that can accumulate peat. They are both examples of peatlands. Peat is organic materials (dead plant remains) deposited in water-soaked circumstances as a result of incomplete decomposition. Due to the saturated soil, the rate of organic matter creation exceeds the rate of breakdown, resulting in peat formation. Peatlands form and exist only when the hydrologic regime results in perennial soil saturation. In the United States and Canada, a bog is defined as any peat-covered terrain with at least 20–40 cm of peat accumulation within the top 80 cm of the soil profile. In Europe, 30 cm of peat is enough to identify a site as bog. The thickness of the peat is crucial to the ecosystem's operation. The peat must be thick enough to hold a considerable amount of water, and the plants' roots must receive all or almost all of their water and nutrients from the peat body.

Peatlands make up between 50 and 70% of all wetlands worldwide. Peatlands are the most extensive and prolific wetland type. They encompass about four million square kilometres, or 3% of the land surface, and contain 10% of all freshwater and one-third of the world's soil carbon, making them critical in regulating atmospheric carbon dioxide levels. Peatlands are particularly abundant in boreal areas. The presence of a well-developed moss ground layer, as well as abundant shrubs or sedges with limited or no tree cover distinguishes the ecology of these environments.

4.1.3 Advantages and limitations of constructed wetland system

Wetlands have played a vital role in human history and cultures for centuries. They provide a variety of benefits, including:

- *Flood control*: Wetlands act like sponges, soaking up excess water and preventing flooding.
- *Water purification*: Wetlands filter pollutants from water, improving water quality.

- *Habitat for wildlife*: Wetlands provide important habitat for a wide variety of plants and animals, including many endangered species.
- *Recreation*: Wetlands offer opportunities for fishing, hunting, birding, and other forms of recreation.

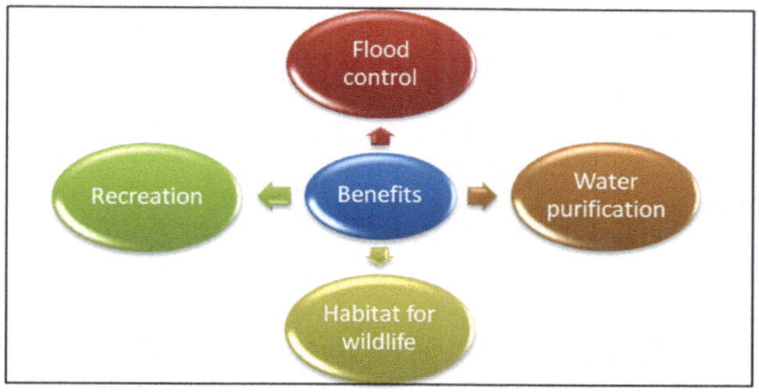

Figure 4.1 Advantages of constructed wetland system.

Threats to wetlands

Unfortunately, wetlands are under threat from a variety of human activities, such as:

- *Drainage*: Wetlands are often drained for agriculture, development, or other purposes.
- *Pollution*: Wetlands can be polluted by agricultural runoff, industrial waste, and other sources.
- *Climate change*: Rising sea levels and changing weather patterns are also threatening wetlands.

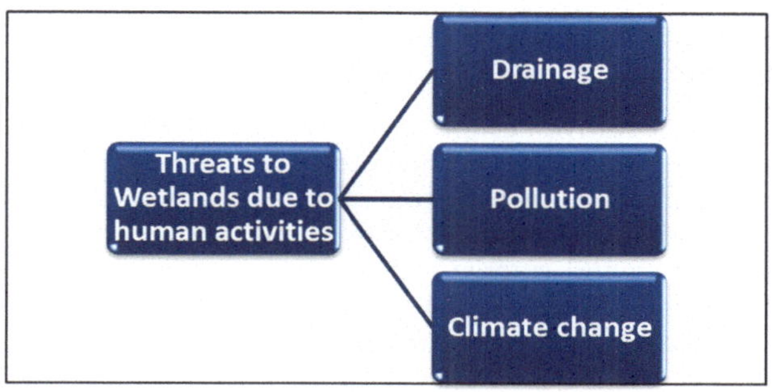

Figure 4.2 Anthropogenic threats to wetland system.

4.1.4 Role of constructed wetlands in sustainable wastewater management

Constructed wetlands (CWs) play an important role in sustainable wastewater treatment because they replicate the natural processes of natural wetlands that filter water. They provide a number of advantages that make them an appealing alternative to standard, energy-intensive treatments. Here is a deeper look at their contributions:

(1) **Effective pollution removal:** CWs effectively remove several contaminants, including:
  - *Nutrients*: They remove nitrogen and phosphorus from receiving water bodies, reducing oxygen depletion (eutrophication).
  - *Organic debris*: Microorganisms in the wetland degrade organic debris, lessening its negative impact.
  - *Pathogens*: Plant roots and filters effectively reduce pathogen levels.
  - *Heavy metals*: CWs can sequester or immobilize heavy metals, minimizing their negative impact on aquatic life.

(2) **Sustainable advantages:**
- *Low energy consumption*: CWs operate passively with minimal energy input, significantly reducing their environmental footprint.
- *Cost-effectiveness*: Compared to traditional treatment plants, CWs often have lower construction and operation costs, especially in rural areas.
- *Biodiversity enhancement*: CWs attract diverse plant and animal life, contributing to local biodiversity and ecosystem health.
- *Visual appeal*: CWs can be aesthetically pleasing, integrating seamlessly into landscapes and providing recreational opportunities.

(3) **Versatile applications:**
- CWs can treat various types of wastewater, including:
- *Municipal wastewater*: From small communities to large cities.
- *Agricultural wastewater*: From livestock farms to aquaculture facilities.
- *Industrial wastewater*: From various industries depending on the type and pre-treatment needs.
- *Stormwater runoff*: Improving water quality before entering waterways.

4.2 DESIGN CONSIDERATIONS

Constructed wetlands have been used for a variety of purposes

- According to Rodgers and Dunn (1994), constructed wetlands can act as a buffer zone between valuable aquatic systems and dangerous point or non-point source contaminants that can harm downstream ecosystems.
- Constructed wetlands can effectively remove metals from wastewater and acid mine drainage through lithic geochemical cycles (Hammer, 1992; Hupp *et al.*, 1993).
- Treat industrial effluents and recalcitrant materials (Rogozinski *et al.*, 1992).

The primary goal of this integrative design was to transfer these selected metals from the water column (aqueous forms) to organic surfaces and into ligand-associated forms such as particulates, which can settle relatively quickly and reduce metal bioavailability in the water column. By choosing the right hydroperiod, hydrosoil, and vegetation for this artificial wetland, properties that aid in metal transfers and transformations should emerge (e.g., the hydrosoil should establish a negative redox potential and basic pH value upon flooding). The physical and chemical properties of Cu, Pb, and Zn, along with the intrinsic function of the wetland, make the necessary transfers and transformations of these aquatic metals thermodynamically conceivable or likely (Hawkins *et al.*, 1997).

4.2.1 Selection of suitable site and site-specific factors

4.2.1.1 Hydroperiod selection

Water carries elements like metals in wetlands, and the hydroperiod has a direct impact on wetland chemical and physical features (e.g., hydrosoil redox and pH). The hydro period chosen for this built wetland corresponded to the maximum hydraulic retention time (HRT) attainable at many refineries or other industrial facilities. A maximal HRT was desirable to allow any kinetically constrained transformations or transfers to proceed as much as possible. As a result, a 24-hour HRT was chosen for this wetland, with the option to enhance or reduce the HRT by adjusting flow or water depth within the created wetlands. A water depth of 30 cm was chosen to produce a hydraulic regime compatible with the chosen vegetation while also maintaining a decreasing hydrosoil environment, which is critical for the removal of the selected metals from the water column. As a result, the selected water depth was a compromise between the depth required to maintain negative redox in the hydrosoil and the depth accepted by the chosen flora.

4.2.1.2 Hydrosoil selection criteria

In addition to pH and redox requirements, hydrosoil selection criteria included appropriate chemical and physical characteristics (e.g., organic carbon), ability to sequester metals while reducing toxicity, minimal background metal and organic contaminant concentrations, compatibility with wetland plant species, and nutrient availability for plant growth.

CW has been used to treat a wide range of wastewaters, including pre-treated municipal wastewater, livestock waste, industrial effluent, biomedical and pharmaceutical waste, agriculture and aquaculture runoff, landfill leachate, greywater, acid mine drainage, food processing waste, lake and river water, stormwater, and others. The specific chemicals to be treated vary depending on the source of the wastewater (Kadlec & Knight, 1996). Nutrients, particularly nitrogen and phosphorus, are the principal pollutants in municipal and agricultural waste streams. Oxygen-demanding wastes are also found in urban and agricultural wastewater, as well as some industrial effluents such paper mill sludge (Vymazal, 2014). Industrial waste streams may contain halogens, metals, metalloids, and synthetic organic compounds. Biomedical waste, along with waste from municipal, greywater, and animal sources, is likely to contain harmful bacteria and medications. Acid mine drainage may cause concern since it acidifies receiving waters. Furthermore, many of these wastewater streams contain significant amounts of dissolved and suspended particles (Vymazal, 2014).

To reduce these numerous pollutants, CW takes several fundamental forms (Vymazal, 2010). Surface waters enter free water surface (FWS) created wetlands via a stream, swale, ditch, or effluent pipe. Water flows through the sealed wetland basin or series of basins, contacting the substrate surface, vegetation, and atmosphere. FWS systems rely on algal mats and macrophytes for their ability to absorb and metabolize pollutants, as well as the binding sites they give for microbial biofilms. In FWS wetlands, the key pollutant treatment methods include biomass absorption, microbial metabolism, substrate and organic detritus adsorption, ultraviolet irradiation, volatilization, and sedimentation (Vymazal, 2014). Contaminants are taken up and sequestered by biomass absorption, which is predominantly accomplished by plants. Adsorption is the process by which pollutants cling to the surface of detritus, vegetation, or substrate particles; these surfaces also support biofilm growth and microbial metabolism. FWS can be especially successful at removing organic pollutants, suspended particles, and nitrogen from influent wastewater (Vymazal, 2014). Floating treatment wetlands (FTW) are innovative systems that suspend macrophytes on buoyant mats, allowing the roots to dangle into the water column (Colares *et al.*, 2020). FTW has showed promise in removing nutrients, metals, and organic pollutants from surface waters (Shahid *et al.*, 2018).

Subsurface CW systems contain both horizontal flow (HF) and vertical flow (VF). In HF designs, wastewater is fed into a bed of porous substrate and emergent plant rhizosphere, where it rises over an impervious liner before exiting the system (Vymazal & Kropfelov, 2008). This design maximizes substrate and rhizosphere exposure while minimizing ambient exposure, making it suitable for organic and nitrogen microbial metabolism, suspended particles filtration, and ion sorption (Vymazal, 2010). Vertical flow (VF) subsurface systems are similar, but they occasionally send wastewater to the surface of a substrate bed, where it percolates into drainage structures below (Stefanakis *et al.*, 2014).

VF systems allow for increased oxygenation, which can be beneficial for the aerobic treatment of organic pollutants and nitrification. A version is the 'tidal flow' (TF) CW, which allows wastewater to entirely flood a subterranean bed, hold it for a period of time, and then drain it completely (Cooper, 2005). This periodic fluctuation allows for alternate periods of oxidation and reduction, which improves the treatment efficiency of certain pollutants. Many CW are hybrid designs, which use different types of CW in succession to produce a complex network of aerobic and anaerobic zones, settling ponds, filtration, sorption sites, and microbial communities to encourage numerous removal pathways (Patyal *et al.*, 2021; Sayadi *et al.*, 2012).

4.2.2 CW performance and management

Physical, chemical, and biological treatment methods in CW vary by design, hence CW conformations are tailored to specific contaminant treatment goals. CW effectiveness is essentially determined by the interactions of waste products with the CW's vegetation, substrate, microorganisms, and water column (Patyal *et al.*, 2021). Physical waste removal mechanisms include flocculation, precipitation, sedimentation, and filtration rely on interactions between the water column and the substrate (Kadlec & Wallace, 2008). Chemical removal activities, including as interactions with ions and adsorbent surfaces, ion exchange, and redox reactions, take place primarily on the substrate. For this reason, CW research has focused on substrate, examining the relative performance of various adsorbent substrate materials such as alum sludge, limestone, coal slag, sand, rice husks, charcoal, and many more (Vymazal *et al.*, 2021). The different physicochemical properties of substrate materials can result in a range of removal efficiencies for different contaminants; according to Patyal *et al.* (2021) substrate materials varied in removal efficiency for oxygen-demanding wastes from 71.8 to 82%, total phosphorus from 77 to 80%, and total nitrogen from 52 to 82%.

Biological processes such as microbial metabolism, phytoremediation, biosorption, and predation occur in both the substrate and the water column (Kadlec & Wallace, 2008). Macrophyte stems, roots, and leaves restrict flow rate and increase surface area for microbial biofilms. Macrophytes also emit carbon compounds and oxygen, especially in the rhizosphere, which promotes chelation and aerobic microbial metabolism (Kataki *et al.*, 2021). Macrophyte metabolism helps with waste retention and elimination by phytoaccumulation, phytodegradation, volatilization, and sequestration (Varma *et al.*, 2021).

The physicochemical features of CW are carefully managed. For example, hydraulic parameters are key components in CW design (Rahman *et al.*, 2020). The estimated hydraulic loading rate (HLR) and pollutant concentration determine the CW's required surface area and volume (Trang *et al.*, 2010). The frequency and duration of flooding and drawdown episodes are also important because they control the amount of oxygen in the water column and substrate. HRT, or average time that influent wastewater remains in the CW, is determined by the CW design, HLR, and hydrologic regime (White *et al.*, 2004). A longer HRT typically leads to a larger percentage of pollutant elimination for a particular system (Ghosh & Gopal, 2010).

Scientists discovered a significant increase in retention of N and coliform bacteria as HRT was raised from 0.3 to 9.3 days in an FWS municipal wastewater treatment wetland. In an HF CW system, Ghosh and Gopal (2010) demonstrate a higher nitrogen retention (83–100%) using a 4-day HRT than a 1-day HRT (21–77%). Many CW designs use long, sinuous pathways, baffles, or many cells to extend the HRT and increase system efficiency (Kadlec, 2019).

Other elements that can affect CW performance include season, temperature, pH, oxygen availability, changes in biomass productivity, bioturbation, and meteorological events (Varma *et al.*, 2021). These, too, can be handled to some degree. In cold-weather locations, greenhouses (Varma *et al.*, 2021) and bed heating can help raise temperatures. Artificial aeration, pH buffering, insulation, and bio-augmentation are all employed to maintain key parameters for efficient waste disposal (Dong *et al.*, 2012). Vymazal *et al.* (2021) and Ingrao *et al.* (2020) examine difficulties in new wastewater streams and novel pollutants in the context of tighter effluent contaminant concentration regulations.

The optimum macrophyte species for CW systems differs according on region, design, and waste stream, but the plants must be strong, with early establishment, rapid growth, huge biomass, and resistance to the severe conditions that wastewater offers (Brisson & Chazarenc, 2009). While plant species selection is regarded as a key aspect of CW design, the literature is conflicting regarding the advantages or drawbacks of one macrophyte species over another (Brisson & Chazarenc, 2009; Vymazal, 2013). The most commonly employed species include *Typha latifolia*, *Phragmites australis*, *Typha angustifolia*, *Juncus effusus*, *Scirpus lacustris*, *Scirpus californicus*, and *Phalaris arundinacea*. Subsurface CW frequently employ emergent macrophytes to take advantage of the increased treatment provided by a rhizosphere. Vymazal (2011) discovered that macrophytes from the genera *Phragmites*, *Typha*, and *Scirpus* are more common in subsurface systems.

Many CW (especially FWS) systems remove or eliminate nuisance species such as burrowing mammals (e.g., muskrat, beaver), mosquitos, bioturbators, and aggressive grazers (Li *et al.*, 2021). However, wetland creatures can have a positive impact on CW performance. Li *et al.* (2021) go over a long list of invertebrates, fish, birds, reptiles, and mammals that can improve pollutant retention through adsorption and bioaccumulation by diversifying the microbial population, consuming pathogens, and promoting plant development.

4.3 CONSTRUCTED WETLANDS APPLICATIONS

Constructed wetlands (CWs) are manmade ecosystems that resemble natural wetlands to treat wastewater, manage stormwater, and boost biodiversity. They provide a sustainable and cost-effective alternative to standard treatment procedures, making them an important tool for a wide range of applications. Let us look at some of the primary ways CWs are employed today:

(1) **Wastewater treatment**
- *Municipal wastewater*: CWs can effectively treat domestic wastewater from small communities to large cities. They remove organic matter, nutrients, and pathogens through a combination of natural processes like sedimentation, filtration, and microbial activity (Hammer, 2020).
- *Industrial wastewater*: CWs can be adapted to treat a variety of industrial wastewaters, depending on the specific contaminants present. Pretreatment may be necessary for some types of wastewater (Stefanakis, 2018).
- *Agricultural wastewater*: CWs can treat runoff from farms and animal feedlots, reducing pollution from fertilizers, pesticides, and manure (Patil *et al.*, 2021).

(2) **Stormwater management**
- *Stormwater runoff*: CWs can capture and filter stormwater runoff, preventing pollutants from entering rivers and streams. They also help to reduce flooding and erosion (Pier *et al.*, 2015).
- *Flood control*: CWs can act as natural sponges, absorbing excess water during heavy rains and releasing it slowly, reducing the risk of flooding downstream (Barras *et al.*, 2002).

(3) **Habitat enhancement and biodiversity:**
- *Habitat creation*: CWs provide vital habitat for a variety of plants, animals, and microorganisms. They can be designed to attract specific species, such as birds, fish, and amphibians (Ware & Callaway, 2019).
- *Wetland restoration*: CWs can be used to restore degraded natural wetlands, improving water quality and biodiversity.

(4) **Other applications**
- *Groundwater remediation*: CWs can be used to clean up contaminated groundwater by removing pollutants such as nitrates and heavy metals.
- *Aquaculture*: CWs can be used to create ponds for fish farming, providing a natural and sustainable alternative to traditional aquaculture methods.
- *Educational and recreational opportunities*: CWs can be used as educational tools to teach people about the importance of wetlands and provide recreational opportunities such as birdwatching and hiking.

4.4 PLANT SPECIES SELECTION AND ROLE IN TREATMENT PROCESSES

Vegetation criteria

Selection criteria for the wetland vegetation included:

(1) Effects upon hydrosoil pH and redox,
(2) Compatibility with selected hydrosoil,
(3) Compatibility with the local climate,

(4) Ability to withstand hydroperiod fluctuations and changes in HRT
(5) Resistance to herbivory
(6) Cost and availability
(7) Nonexotic species.

Species used
Plants

- Large aquatic plants, including cattails, reeds, and bulrushes, play an important role in CWs. They deliver oxygen, filter contaminants, and keep the soil stable (Stefanakis *et al.*, 2021).
- Emergent plants, such as rushes and sedges, have roots in the water and stems that rise to the surface. They provide shelter and food for a variety of species while also aiding in erosion prevention (Shelef *et al.*, 2013).

Animals

- Amphibians, including frogs, toads, and salamanders, thrive in CWs due to abundant food supplies such as insects and worms (Hayslett, 2003).
- Birds such as ducks, geese, herons, and kingfishers frequent manmade wetlands for nesting, feeding, and resting.
- CWs may support diverse fish species based on water quality and design. Minnows, carp, and catfish are some of the more prevalent fish.
- Mulamoottil *et al.* (1998) found that invertebrates, such as insects, worms, molluscs, and crustaceans, play an important role in decomposition and nutrient cycling in CWs.

4.5 CHALLENGES OF CONSTRUCTED WETLANDS (CWS)

While built wetlands (CWs) present a potential approach to wastewater treatment, stormwater management, and habitat creation, they also face a number of obstacles that must be addressed for their successful implementation and long-term viability. Here are some important hurdles to consider.

(1) *Land availability*: CWs require a large amount of land, which can be difficult in highly populated places or when there is competition for property (Avellán & Gremillion, 2019).
(2) *Treatment efficiency*: CWs are vulnerable to temperature, flow rate, and pollution load changes, which can have an impact on treatment efficiency. They may be ineffective for treating highly contaminated wastewater or specific types of contaminants, necessitating extra pretreatment or alternate solutions (Wang *et al.*, 2017).
(3) *Maintenance and monitoring*: CWs require regular maintenance, which includes vegetation management, silt removal, and monitoring of water quality and contaminant levels. Inadequate maintenance can cause blockage, reduced treatment efficiency, and environmental imbalances (Kumar & Dutta, 2019).
(4) *Public perception*: Some communities may have negative attitudes towards CWs due to concerns about odours, mosquitoes, or potential effects on property values. Public education and involvement are critical for overcoming these fears and gaining community support.
(5) *Climate change*: Climate change might harm CWs by altering precipitation patterns, temperature extremes, and water availability, potentially compromising their performance and resilience (Stefanakis, 2019).
(6) *Cost considerations*: While CWs may be more cost-effective than typical treatment facilities in the long term, the initial building and land acquisition expenditures can be prohibitively expensive. Finding funding and establishing the economic benefits of CWs are critical to wider adoption (Chen *et al.*, 2023).

7 *Design and engineering challenges*: Wetland ecology, hydrology, and technical knowledge are required for effective CW design and construction (Jiang & Chui, 2022). Improper design can cause operational issues and lower treatment performance (Liu *et al.*, 2024).
8 *Invasive species*: CWs are susceptible to invasion by non-native plant species, which can disturb ecological balance and harm native species (Matthews *et al.*, 2009). Effective monitoring and management strategies are required to prevent and control invasive species.
9 *Limited understanding*: Despite increased research and development, there is still much to learn about the long-term performance and environmental consequences of CW. More research and monitoring are required to improve the design, management, and integration of CWs into landscapes (Jiang & Chui, 2022).

4.6 CONCLUSION

The chapter highlights the effectiveness of constructed wetlands as a viable and sustainable option of treating wastewater. These systems are capable of not only reducing pollution levels but also enhancing the biodiversity and health of the ecosystem. However, even though individual wetlands have very high nitrogen attenuation rates, large-scale multiplicity will need sophisticated and intelligent designing and planning to position it for maximum effectiveness. One of the important aspects of such systems is the long-term monitoring situation that will help understand how ecological succession occurs, what species become available and their preferences, thus helping in designing better wetland systems in the future. In combining remote sensing and landscape theory, it has been shown how a range of dimensional indicators could be used to provide an assessment of the state of wetlands, which for practical management purposes is more effective.

REFERENCES

Adame M. F., Kauffman J. B., Medina I., Gamboa J. N., Torres O., Caamal J. P., Reza M. and Herrera-Silveira J. A. (2013). Carbon stocks of tropical coastal wetlands within the karstic landscape of the Mexican Caribbean. *PLoS ONE*, **8**, e56569, https://doi.org/10.1371/journal.pone.0056569

Alongi D. M. (2012). Carbon sequestration in mangrove forests. *Carbon Management*, **3**, 313–322, https://doi.org/10.4155/cmt.12.20

Avellán T. and Gremillion P. (2019). Constructed wetlands for resource recovery in developing countries. *Renewable and Sustainable Energy Reviews*, **99**, 42–57, https://doi.org/10.1016/j.rser.2018.09.024

Barducci A., Guzzi D., Marcoionni P. and Pippi I. (2009). Aerospace wetland monitoring by hyperspectral imaging sensors: a case study in the coastal zone of San Rossore Natural Park. *Journal of Environmental Management*, **90**, 2278–2286, https://doi.org/10.1016/j.jenvman.2007.06.033

Barras C., Mazik Sr R. A., Brucker J. C., Mores J. M., Carl A. B., Teacher P. T., ... and Moyer R. Turtle Creek Watershed Association Board of Directors (2002).

Bernal B. and Mitsch W. J. (2013). Carbon sequestration in freshwater wetlands in Costa Rica and Botswana. *Biogeochemistry*, **115**, 77–93, https://doi.org/10.1007/s10533-012-9819-8

Boone J. K. and Bhomia R. K. (2017). Ecosystem carbon stocks of mangroves across broad environmental gradients in west-central Africa: global and regional comparisons. *PLoS ONE*, **12**, e0185943, https://doi.org/10.1371/journal.pone.0185943

Brisson J. and Chazarenc F. (2009). Maximizing pollutant removal in constructed wetlands: should we pay more attention to macrophyte species selection? *Science of The Total Environment*, **407**, 3923–3930, https://doi.org/10.1016/j.scitotenv.2008.05.047

Chen J., Guo F., Wu F. and Bryan B. A. (2023). Costs and benefits of constructed wetlands for meeting new water quality standards from China's wastewater treatment plants. *Resources, Conservation and Recycling*, **199**, 107248, https://doi.org/10.1016/j.resconrec.2023.107248

Choudhary, B., Dhar, V. and Pawase, A.S., (2024). Blue carbon and the role of mangroves in carbon sequestration: Its mechanisms, estimation, human impacts and conservation strategies for economic incentives. *Journal of Sea Research*, **199**, p.102504. https://doi.org/10.1016/j.seares.2024.102504

Colares G. S., Dell'Osbel N., Wiesel P. G., Oliveira G. A., Lemos P. H. Z., da Silva F. P., Lutterbeck C. A., Kist L. T. and Machado Ê. L. (2020). Floating treatment wetlands: a review and bibliometric analysis. *Science of The Total Environment*, **714**, 136776, https://doi.org/10.1016/j.scitotenv.2020.136776

Cooper P. F. (2005). The performance of vertical flow constructed wetland systems with special reference to the significance of oxygen transfer and hydraulic loading rates. *Water Science & Technology*, **51**, 81–90.

Cui X., Liang J., Lu W., Chen H., Liu F., Lin G., Xu F., Luo Y. and Lin G. (2018). Stronger ecosystem carbon sequestration potential of mangrove wetlands with respect to terrestrial forests in subtropical China. *Agricultural and Forest Meteorology*, **249**, 71–80, https://doi.org/10.1016/j.agrformet.2017.11.019

Donato D. C., Kauffman J. B., Murdiyarso D., Kurnianto S., Stidham M. and Kanninen M. (2011). Mangroves among the most carbon-rich forests in the tropics. *Nature Geoscience*, **4**, 293–297, https://doi.org/10.1038/ngeo1123

Dong H., Qiang Z., Li T., Jin H. and Chen W. (2012). Effect of artificial aeration on the performance of vertical-flow constructed wetland treating heavily polluted river water. *Journal of Environmental Sciences*, **24**, 596–601, https://doi.org/10.1016/S1001-0742(11)60804-8

Fickas K. C., Cohen W. B. and Yang Z. (2016). Landsat-based monitoring of annual wetland change in the Willamette Valley of Oregon, USA from 1972 to 2012. *Wetlands Ecology and Management*, **24**, 73–92, https://doi.org/10.1007/s11273-015-9452-0

Ghosh D. and Gopal B. (2010). Effect of hydraulic retention time on the treatment of secondary effluent in a subsurface flow constructed wetland. *Ecological Engineering*, **36**, 1044–1051, https://doi.org/10.1016/j.ecoleng.2010.04.017

Gorham E. (1991). Northern peatlands: role in the carbon cycle and probable responses to climatic warming. *Ecological Applications*, **1**, 182–195, https://doi.org/10.2307/1941811

Guo M., Li J., Sheng C., Xu J. and Wu L. (2017). A review of wetland remote sensing. *Sensors*, **17**(4), 777, https://doi.org/10.3390/s17040777

Hammer, D. A. (1992). Designing constructed wetlands systems to treat agricultural nonpoint source pollution. *Ecological Engineering*, **1**(1–2), 49–82.

Hammer D. A. (ed.). (2020). Constructed Wetlands for Wastewater Treatment: Municipal, Industrial and Agricultural. CRC Press, Boca Raton, London, New York, Washington D.C.

Hawkins W. B., Rodgers Jr. J. H., Gillespie Jr. W. B., Dunn A. W., Dorn P. B. and Cano M. L. (1997). Design and construction of wetlands for aqueous transfers and transformations of selected metals. *Ecotoxicology and Environmental Safety*, **36**(3), 238–248, https://doi.org/10.1006/eesa.1996.1505

Hayslett M. S. (2003). Natural history of the mole salamander *Ambystoma talpoideum* in Virginia.

Hupp, C. R., Woodside, M. D. and Yanosky, T. M. (1993). Sediment and trace element trapping in a forested wetland, Chickahominy River, Virginia. Wetlands, **13**, 95–104.

Ingrao C., Failla S. and Arcidiacono C. (2020). A comprehensive review of environmental and operational issues of constructed wetland systems. *Current Opinion in Environmental Science & Health*, **13**, 35–45, https://doi.org/10.1016/j.coesh.2019.10.007

Jiang L. and Chui T. F. M. (2022). A review of the application of constructed wetlands (CWs) and their hydraulic, water quality and biological responses to changing hydrological conditions. *Ecological Engineering*, **174**, 106459, https://doi.org/10.1016/j.ecoleng.2021.106459

Kadlec R. H. (2019). Treatment Marshes for Runoff and Polishing. CRC Press, Boca Raton, FL, USA.

Kadlec R. H. and Knight R. L. (1996). Treatment Wetlands. CRC Press, Boca Raton, FL, USA.

Kadlec R. H. and Wallace S. (2008). Treatment Wetlands. CRC Press, Boca Raton, FL, USA.

Kataki S., Chatterjee S., Vairale M. G., Dwivedi S. K. and Gupta D. K. (2021). Constructed wetland, an eco-technology for wastewater treatment: a review on types of wastewater treated and components of the technology (macrophyte, biofilm and substrate). *Journal of Environmental Management*, **283**, 111986, https://doi.org/10.1016/j.jenvman.2021.111986

Köchy M., Hiederer R. and Freibauer A. (2015). Global distribution of soil organic carbon–part 1: masses and frequency distributions of SOC stocks for the tropics, permafrost regions, wetlands, and the world. *Soil*, **1**, 351–365, https://doi.org/10.5194/soil-1-351-2015

Kumar S. and Dutta V. (2019). Efficiency of constructed wetland microcosms (CWMs) for the treatment of domestic wastewater using aquatic macrophytes. In: Sobti, R., Arora, N., Kothari, R. (eds) Environmental Biotechnology: For Sustainable Future. Springer, Singapore, 287–307, https://doi.org/10.1007/978-981-10-7284-0_11

Li Q., Long Z., Wang H. and Zhang G. (2021). Functions of constructed wetland animals in water environment protection–a critical review. *Science of the Total Environment*, **760**, 144038, https://doi.org/10.1016/j.scitotenv.2020.144038

Liu H., Yi Y., Yue Y. and Cui B. (2020). Reducing the likelihood of carbon loss from wetlands by improving the spatial connections between high carbon patches. *Journal of Cleaner Production*, **267**, 121819, https://doi.org/10.1016/j.jclepro.2020.121819

Liu S., Zhang Y., Feng X. and Pyo S. H. (2024). Current problems and countermeasures of constructed wetland for wastewater treatment: a review. *Journal of Water Process Engineering*, **57**, 104569, https://doi.org/10.1016/j.jwpe.2023.104569

Marín-Muñiz J. L., Hernández M. E. and Moreno-Casasola P. (2014). Comparing soil carbon sequestration in coastal freshwater wetlands with various geomorphic features and plant communities in Veracruz, Mexico. *Plant and Soil*, **378**, 189–203, https://doi.org/10.1007/s11104-013-2011-7

Matthews J. W., Peralta A. L., Soni A., Baldwin P., Kent A. D. and Endress A. G. (2009). Local and landscape correlates of non-native species invasion in restored wetlands. *Ecography*, **32**(6), 1031–1039, https://doi.org/10.1111/j.1600-0587.2009.05863.x

Mitsch W. J. and Gosselink J. G. (2015). Wetlands, Wiley, Hoboken, New Jersey, 5th ed.

Mitsch W. J. and Mander Ü. (2018). Wetlands and carbon revisited. *Ecological Engineering*, **114**, 1–6, https://doi.org/10.1016/j.ecoleng.2017.12.027

Mulamoottil G., McBean E. A. and Rovers F. (1998). Constructed Wetlands for the Treatment of Landfill Leachates. Routledge, Lewis Publishers, CRC Press, Boca Raton, London, New York, Washington D.C.

Nicholls R. J. (2004). Coastal flooding and wetland loss in the 21st century: changes under the SRES climate and socio-economic scenarios. *Global Environmental Change*, **14**, 69–86, https://doi.org/10.1016/j.gloenvcha.2003.10.007

Patil R. R., Pohekar K. N. and Rani N. (2021). Constructed wetland: a sustainable approach for wastewater treatment. In: Environment, Development and Sustainability in India: Perspectives, Issues and Alternatives, Published by Springer Nature Singapore Pte Ltd, Singapore, pp. 227–242.

Patyal V., Jaspal D. and Khare K. (2021). Materials in constructed wetlands for wastewater remediation: a review. *Water Environment Research*, **93**, 2853–2872, https://doi.org/10.1002/wer.1648

Pier B. M., Dresser B. R., Lee J. J., Boylen C. W. and Nierzwicki-Bauer S. A. (2015). Ecological analysis before and after planting in a constructed wetland in the Adirondacks. *Wetlands*, **35**, 611–624, https://doi.org/10.1007/s13157-015-0652-4

Rahman M. E., Bin Halmi M. I. E., Bin Abd Samad M. Y., Uddin M. K., Mahmud K., Abd Shukor M. Y., Sheikh Abdullah S. R. and Shamsuzzaman S. M. (2020). Design, operation and optimization of constructed wetland for removal of pollutant. *International Journal of Environmental Research and Public Health*, **17**, 8339, https://doi.org/10.3390/ijerph17228339

Rodgers Jr, J. H., & Dunn, A. (1992). Developing design guidelines for constructed wetlands to remove pesticides from agricultural runoff. *Ecological Engineering*, **1**(1–2), 83–95.

Rogozinski, L. R., Laubaucher, R. C. and Farmer, J. M. (1992) Groundwater treatment via constructed wetlands. Presented at the petroleum hydrocarbons conference, November 4–6, 1992, Houston.

Sanderman J., Hengl T., Fiske G., Solvik K., Adame M. F., Benson L., Bukoski J. J., Carnell P., Cifuentes-Jara M., Donato D., Duncan C., Eid E. M., zu Ermgassen P., Lewis C.J. E., Macreadie P. I., Glass L., Gress S., Jardine S.L., Jones T. G., Nsombo E. N., Rahman M. M., Sanders C. J., Spalding M. and Landis E. (2018). A global map of mangrove forest soil carbon at 30 m spatial resolution. *Environmental Research Letters*, **13**, 055002, https://doi.org/10.1088/1748-9326/aabe1c

Sayadi M. H., Kargar R., Doosti M. R. and Salehi H. (2012). Hybrid constructed wetlands for wastewater treatment: a worldwide review. *Proceedings of the International Academy of Ecology and Environmental Sciences*, **2**, 204.

Schmidt K. and Skidmore A. (2003). Spectral discrimination of vegetation types in a coastal wetland. *Remote Sensing of Environment*, **85**, 92–108, https://doi.org/10.1016/S0034-4257(02)00196-7

Shahid M. J., Arslan M., Ali S., Siddique M. and Afzal M. (2018). Floating wetlands: a sustainable tool for wastewater treatment. *CLEAN – Soil, Air, Water*, **46**, 1800120, https://doi.org/10.1002/clen.201800120

Shelef O., Gross A. and Rachmilevitch S. (2013). Role of plants in a constructed wetland: current and new perspectives. *Water*, **5**(2), 405–419, https://doi.org/10.3390/w5020405

Stefanakis A. I. (ed.). (2018). Constructed wetlands for industrial wastewater treatment.

Stefanakis, A. I. (2019). The role of constructed wetlands as green infrastructure for sustainable urban water management. *Sustainability*, **11**(24), 6981.

Stefanakis A. I., Calheiros C. S. and Nikolaou I. (2021). Nature-based solutions as a tool in the new circular economic model for climate change adaptation. *Circular Economy and Sustainability*, **1**, 303–318, https://doi.org/10.1007/s43615-021-00022-3

Stefanakis A., Akratos C. S. and Tsihrintzis V. A. (2014). Vertical Flow Constructed Wetlands: Eco-Engineering Systems for Wastewater and Sludge Treatment. Elsevier, Amsterdam, the Netherlands.

Sun X., Li Y., Zhu X., Cao K. and Feng L. (2017). Integrative assessment and management implications on ecosystem services loss of coastal wetlands due to reclamation. *Journal of Cleaner Production*, **163**, S101–S112, https://doi.org/10.1016/j.jclepro.2015.10.048

Trang N. T. D., Konnerup D., Schierup H. H., Chiem N. H. and Brix H. (2010). Kinetics of pollutant removal from domestic wastewater in a tropical horizontal subsurface flow constructed wetland system: effects of hydraulic loading rate. *Ecological Engineering*, **36**, 527–535, https://doi.org/10.1016/j.ecoleng.2009.11.022

Varma M., Gupta A. K., Ghosal P. S. and Majumder A. (2021). A review on performance of constructed wetlands in tropical and cold climate: insights of mechanism, role of influencing factors, and system modification in low temperature. *Science of the Total Environment*, **755**, 142540, https://doi.org/10.1016/j.scitotenv.2020.142540

Villa J. A. and Bernal B. (2018). Carbon sequestration in wetlands, from science to practice: an overview of the biogeochemical process, measurement methods, and policy framework. *Ecological Engineering*, **114**, 115–128, https://doi.org/10.1016/j.ecoleng.2017.06.037

Vymazal J. (2010). Constructed wetlands for wastewater treatment. *Water*, **2**, 530–549, https://doi.org/10.3390/w2030530

Vymazal J. (2011). Constructed wetlands for wastewater treatment: five decades of experience. *Environmental Science & Technology*, **45**, 61–69, https://doi.org/10.1021/es101403q

Vymazal J. (2013). Emergent plants used in free water surface constructed wetlands: a review. *Ecological Engineering*, **61**, 582–592, https://doi.org/10.1016/j.ecoleng.2013.06.023

Vymazal J. (2014). Constructed wetlands for treatment of industrial wastewaters: a review. *Ecological Engineering*, **73**, 724–751, https://doi.org/10.1016/j.ecoleng.2014.09.034

Vymazal J. and Kropfelov L. (2008). Wastewater Treatment in Constructed Wetlands with Horizontal Subsurface Flow. Springer, Dordrecht, The Netherlands.

Vymazal J., Zhao Y. and Mander Ü. (2021). Recent research challenges in constructed wetlands for wastewater treatment: a review. *Ecological Engineering*, **169**, 106318, https://doi.org/10.1016/j.ecoleng.2021.106318

Wang M., Zhang D. Q., Dong J. W. and Tan S. K. (2017). Constructed wetlands for wastewater treatment in cold climate–a review. *Journal of Environmental Sciences*, **57**, 293–311, https://doi.org/10.1016/j.jes.2016.12.019

Ware J. and Callaway R. (2019). Public perception of coastal habitat loss and habitat creation using artificial floating islands in the UK. *Plos One*, **14**(10), e0224424, https://doi.org/10.1371/journal.pone.0224424

Weixelman, D., & Cooper, D. J. (2009). Assessing proper functioning condition for fen areas in the Sierra Nevada and Southern Cascade Ranges in California: A user guide. United States Department of Agriculture, US Forest Service. R5-TP–028.

White J. R., Ramesh Reddy K. and Moustafa M. Z. (2004). Influence of hydrologic regime and vegetation on phosphorus retention in everglades stormwater treatment area wetlands. *Hydrological Processes*, **18**, 343–355, https://doi.org/10.1002/hyp.1379

Xu S., Liu X., Li X. and Tian C. (2019). Soil organic carbon changes following wetland restoration: a global meta-analysis. *Geoderma*, **353**, 89–96, https://doi.org/10.1016/j.geoderma.2019.06.027

Zhang L., Wang M.-H., Hu J. and Ho Y.-S. (2010). A review of published wetland research, 1991–2008: ecological engineering and ecosystem restoration. *Ecological Engineering*, **36**, 973–980, https://doi.org/10.1016/j.ecoleng.2010.04.029

Zhang T., Cao G., Cao S., Zhang X., Zhang J. and Han G. (2017). Dynamic assessment of the value of vegetation carbon fixation and oxygen release services in Qinghai lake basin. *Acta Ecologica Sinica*, **37**, 79–84, https://doi.org/10.1016/j.chnaes.2016.12.005

Zhao C., Liu S., Jiang Z., Wu Y., Cui L., Huang X. and Macreadie P. I. (2019). Nitrogen purification potential limited by nitrite reduction process in coastal eutrophic wetlands. *Science of The Total Environment*, **694**, 133702, https://doi.org/10.1016/j.scitotenv.2019.133702

doi: 10.2166/9781789064100_0077

Chapter 5

Integrated multi-trophic aquaculture (IMTA) for wastewater treatment and resource recovery: a sustainable approach

Bhavesh Choudhary[1,2], Suparna Deb[1], Nayan Chouhan[1], Sangeetha Azmeera[2] and Vinayaka Choudhary[3]*

[1]*College of Fisheries, Central Agricultural University (Imphal), Lembucherra, Agartala, Tripura (West), 799210, India.*
[2]*College of Fisheries, Dr. Balasaheb Sawant Konkan Krishi Vidyapeeth, Shirgaon, Ratnagiri 415629, Maharashtra, India.*
[3]*Regional Institute of Education (NCERT, New Delhi), Ajmer, Rajasthan, 305004, India.*
**Corresponding author email: bhaveshchoudhary@yahoo.com*

ABSTRACT

IMTA stands for integrated multi-trophic aquaculture, which is an innovative approach to aquaculture that aims to optimize and minimize resource utilization and also minimize environmental impacts. Although IMTA is not specifically designed for wastewater treatment and resource recovery, it can be applied in such contexts to enhance the overall efficiency and sustainability of these processes. In traditional wastewater treatment systems, the focus is primarily on removing pollutants and treating wastewater to meet regulatory standards of effluent. However, IMTA has more holistic approaches to create more sustainable and efficient ecosystems. In IMTA wastewater treatment systems, the principle involves utilizing different organisms in a cascading trophic structure to maximize resource utilization and minimize waste. Wastewater treatment is carried out via primary treatment, secondary treatment, and tertiary systems. In secondary treatment systems, filter-feeder organisms are mostly cultured, whereas in tertiary treatment systems macro algae, shellfishes, and seaweeds can be cultivated as water contains excess nutrients such as nitrogen and phosphorous and thus reduces the chances of eutrophication and water pollution. The harvested shellfish and macro algae can be used as valuable resources, shellfish sold for human consumption, whereas macro algae can be processed into products such as biofuels, fertilizers, or animal feed. Thus, the system becomes more efficient and reduces environmental impacts. It allows for conversion of waste streams into valuable products and promotes a circular economy approach.

Keywords: Wastewater treatment, resource recovery, eutrophication, circular economy, IMTA, sustainable aquaculture approach.

5.1 INTRODUCTION

A growing global population will need healthy food, and the aquaculture sector is recognized as a vital enterprise in this regard (FAO, 2020). Compared to wild-catch fisheries, aquaculture offers a more dependable and sustainable supply of food. As a result, it has been predicted that fish farms would generate nearly two-thirds of all seafood by 2030 (Kobayashi *et al.*, 2015).

Aquaculture has a number of issues; the most significant of which is the environmental one, despite its quick expansion. According to Skriptsova and Miroshnikova (2011), metabolites from domesticated animals and external inputs such as fertilizers and feeds used in aquaculture systems are the primary contributors to the accumulation of nutrients in aquatic ecosystems. The main pollutants found in aquaculture effluents are phosphorus and nitrogenous (nitrites, nitrates, and ammonia) compounds. In intensive aquaculture, large quantities of inputs, such as feeds and fertilizers (50–60% of operational costs), are employed to achieve high production objectives in a short period of time. According to several investigations, fish excretion, respiration, and uneaten feed all contribute to the loss of around 85% of phosphorus, 80–88% of carbon, and 52–95% of nitrogen to the aquaculture environment.

Aquaculture systems might provide in-situ removal and utilization of inorganic and organic nutrients through biological processes to offer environmental sustainability, according to numerous studies (Buschmann *et al.*, 2001; Huo *et al.*, 2012; Samocha *et al.*, 2015; Skriptsova & Miroshnikova, 2011), whereas organic and particulate wastes could efficiently be utilized by filter-feeding animals, namely bivalve molluscs (Irisarri *et al.*, 2015; Shpigel & Blaylock, 1991). The introduction of macroalgae into aquaculture systems is one of the most efficient methods to use nutrients such as nitrogen and phosphorus in the particular system. Integrated multi-trophic aquaculture (IMTA) refers to the integration of these inorganic (macroalgae/aquatic weed) and organic extractive (bivalves) species known as un-fed species with cultivated or fed species in the same system.

The goal of IMTA is to broaden the scope of fed aquaculture, such as finfish or shrimp, by incorporating extractive aquaculture. This involves capturing both inorganic nutrients, such as those found in seaweeds, and organic nutrients, through the use of suspension-feeders and deposit-feeders. These nutrients are then utilized to promote the growth of co-cultured species that have a commercial value (Chopin *et al.*, 2012). Pond culture has been a traditional practice in Asia for many centuries. However, the effective use of organic particles produced through aquaculture as a food source for marine bivalves in open-water IMTA systems is hindered by the limited time available to capture solid wastes carried by the horizontal flow of particles, as well as the size and concentration of these particles (Cranford *et al.*, 2013). According to physiological model projections, in order for salmon aquaculture solids to have a net decrease in the organic loading from an IMTA site, they would need to make up at least 10–20% of a mussel's total diet (Reid *et al.*, 2013a).

There is evidence to suggest that properly designed IMTA sites might help improve the situation. In fact, mussels that are produced in close proximity to fish farms have been found to consume at least 20% of their diet from fish-derived sources, according to studies conducted by Lander *et al.* (2012) and Reid *et al.* (2013b). However, research conducted so far on the potential combined growth benefits of IMTA systems on bivalves has shown inconsistent findings. Various research studies have shown proof of growth advantages for shellfish or proof of organic absorption from farms in open-water IMTA systems (Handå *et al.*, 2012; Jiang *et al.*, 2013). However, some studies have indicated that there is no substantial impact (Irisarri *et al.*, 2013). Due to the variation in conditions, the effectiveness of multi-trophic culture has primarily been demonstrated in inland ponds where the connection between the different trophic levels is more direct (Yuan *et al.*, 2013). However, it is difficult to determine the advantages of IMTA in open-water or coastal farms, except in cases of extremely high culture density commonly found in Southeast Asia (Ferreira *et al.*, 2012).

Deposit-feeding sea cucumbers have demonstrated potential in recycling the bigger particles of debris through their feeding activities when cultivated alongside other species. This not only helps in cleaning up the environment but also contributes to the profitability of farms (Yuan *et al.*, 2013). Sea cucumbers may consume waste from finfish, which leads to improved development and survival (Orr *et al.*, 2014). Additionally, they have been successfully produced along with shrimp in a co-culture system (Martínez-Porchas *et al.*, 2010). Co-culture experiments involving sea cucumbers, mussels, and fish revealed that detritus-feeders can consume up to 70% of the material that settles on the seafloor. These feeders were found to reduce the total organic carbon and total nitrogen contents of fish faeces

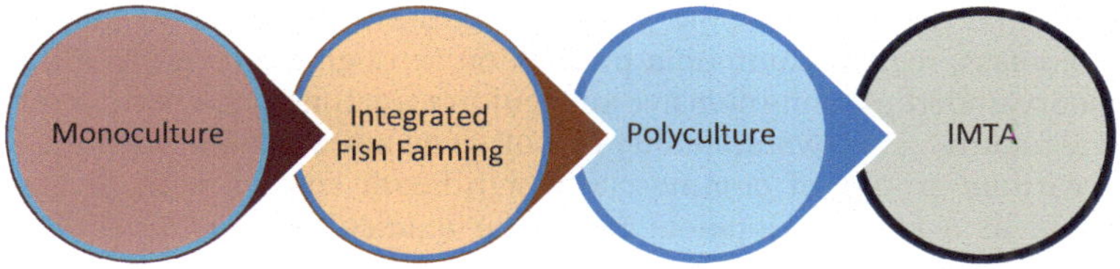

Figure 5.1 Evolution and origin of IMTA.

by an average of 60.3 and 62.3%, respectively (Hannah *et al.*, 2013). Therefore, this chapter describes the origin and evolution of IMTA, suitable species, basic setup, beneficial role, and future challenges in wastewater treatment.

5.2 ORIGIN AND EVOLUTION OF IMTA

Thousands of years ago, in ancient civilizations such as China and Egypt, when farmers engaged in a sort of agriculture combined with fish culture, is when IMTA first emerged. It looked like a logical step for people to consider using waste from fish culture to create secondary crops as well as secondary revenue. It took time for IMTA to develop; it started with integrated fish farming, progressed through polyculture, and finally included ecological engineering into aquaculture. This resulted in the creation of current IMTA, also known as Polyculture 2.0 (Nissar *et al.*, 2022).

According to Chopin and Robinson (2004), the term 'modern IMTA' was first used to describe a culture system that included species from various trophic levels or nutritional levels (Figure 5.1).

5.3 IMPORTANCE OF IMTA

According to Zhang *et al.* (2011), recirculating aquaculture systems (RASs) can help in reduce water use. In such systems, the water in fish farming is treated before being transported to a restoration system and utilized again. IMTA is a further advancement of RAS technology. According to Alexander *et al.* (2016), IMTA systems have been recognized as a crucial technology to sustainably enhance freshwater fish output. The extractive species use the waste products of the major cultivated species as a source of energy or nutrients for growth. Despite the fact that IMTA seems like a sort of polyculture that entails raising several species in the same system, the former is a precision aquaculture system with a greater nutrient utilization efficiency than the latter (Milstein, 2005). Generally speaking, polyculture with extractive species does not address environmental mitigation. Thus, by transforming wastes from the primary species into resources for aquatic weeds and filter-feeders to improve fertilizer usage efficiency and sustainability of the system, commercial aquaculture might reduce its negative environmental effect through IMTA (Kumar *et al.*, 2000). The circular economy, which treats trash as a resource and generates a second source of revenue, is the foundation of IMTA systems (Winans *et al.*, 2017). In IMTA, economically significant extractive species consume waste generated by other species, hence treating wastewater. One major benefit of IMTA is that it lessens the environmental effect of aquaculture by preventing the discharge of eutrophic waters into natural waters (Ridler *et al.*, 2007) while producing a useful byproduct (FAO, 2018).

5.4 BIOREMEDIATION AND BIOMITIGATION OF AQUACULTURE WASTEWATER

Generally speaking, intense, semi-intensive, and extensive aquaculture make up land-based aquaculture systems. In extensive aquaculture, the treatment and cultivation processes often occur in the same pond and need little to no energy input (feed, fertilizer, aeration, etc.). A mixture of autotrophic activities, fuelled by nitrifying bacteria and photoautotrophic processes in the water column, that absorb ammonium directly into algal biomass, regulate ammonia production through oxidation to nitrate (Ebeling *et al.*, 2006). The same theory underlies semi-intensive aquaculture treatment, but because of the larger animal numbers and energy requirements, water quality can often be more difficult to manage.

By integrating various groups of commercially significant species with diverse feeding habits, which has been cited as a key development factor for aquaculture sustainability, IMTA can help achieve the concept of bio-mitigation. Utilizing the mutualism between certain detritivorous fish and phytoplankton is a key component of IMTA. In IMTA, a farm site is integrated with a variety of complementary species, optimizing nutrient utilization and lowering solid waste. For waste treatment to enable more ecologically responsible aquaculture practices, a comprehensive integrated solution based on more effective reactor systems must be developed (Zhang & Kitazawa, 2016). The bioremediation of dissolved nutrient loads in waterbodies requires the selection of suitable algae species, aquatic weeds, and strains. Land-based aquaculture has been used to successfully show bioremediation activities, according to Lawton *et al.* (2013).

5.5 SUITABLE SPECIES FOR IMTA AND THEIR ROLE

A careful selection of species is necessary due to the integrated nature of IMTA systems (Rosa *et al.*, 2020). Typically, the fed component comprises carnivorous fish or shrimp that are fed either rough fish or pellets. The food that the environment provides sustains other biological components of an IMTA operation, such as bivalves and seaweeds (Strand *et al.*, 2019). Fish/shrimp, fish/seaweed/shellfish, shellfish/shrimp, and seaweed/shrimp are only a few of the species that have been discussed in the literature.

5.5.1 Factors determine the selection criteria of species

- From a trophic standpoint, the chosen species should complement one another; specifically, animals at lower trophic levels (low-trophic animals) should be able to consume waste produced by creatures at higher trophic levels (high-trophic animals). These configurations boost the yield of farmed aquatic plants while decreasing water pollution (Viji, 2015) (Figure 5.2).
- Ideally, local species that are well suited to their habitat should be used to choose the species. Having a rapid rate of development is a crucial aspect of a species' requirements in the IMTA (Largo *et al.*, 2016).
- Native species should be given priority because they will prevent invasion and reduce the acclimatization time (Nissar *et al.*, 2022).
- As environmental circumstances cannot be controlled, the location for an IMTA system construction should be carefully chosen. A thorough understanding of environmental factors, such as organic matter, dissolved inorganic matter, and particle size, is therefore required (Chopin, 2017; Sasikumar & Viji, 2016).
- Selected species ought to be able to accumulate a sizable amount of biomass (Chopin *et al.*, 2010).
- The grown specialties should be in high demand on the market, either in their raw or processed form, or their derivatives should be profitable (Carras *et al.*, 2019; Shi *et al.*, 2013).
- Before making investments, it is best to co-culture a species with a known market value and establish buyers in the market.
- A decision needs to be taken on the growth potential of biofiltering organisms. A biofiltering organism should either have a high growth potential and be able to be collected on a regular basis, or it should have a high value that would make up for lower biomass growth.

5.5.2 Finfish

One of the main revenue generators for IMTA systems is the culture of high-trophic finfish (Ghosh *et al.*, 2016; Hoang *et al.*, 2018). However, the local diet and market value have a major role in choosing

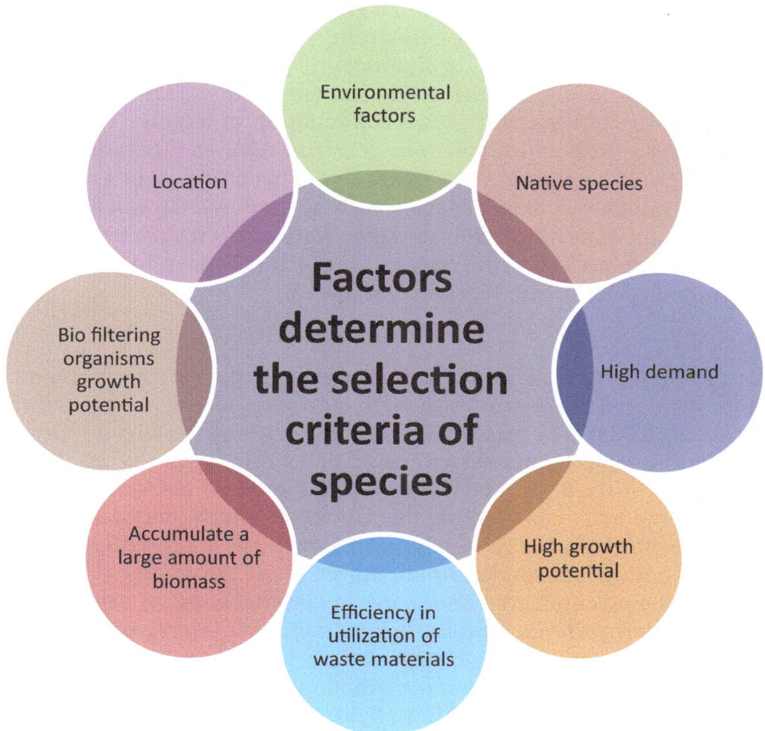

Figure 5.2 Factors that determine the selection criteria of species.

the best finfish species to employ in IMTA systems. Therefore, it is important to give local high-value species first priority (Holanda *et al.*, 2020). *Anoplopoma fimbria*, *Oncorhynchus tshawytscha*, *Oncorhynchus mykiss*, *Pagrus major*, *Salmo salar*, *Seriola quinqueradiata*, *Sparus aurata*, *Thunnus orientalis*, *Mugil cephalus*, *Sebastes schlegeli*, *Gadus* spp., and *Dicentrarchus labrax*.

5.5.3 Crustaceans

In the IMTA, crustaceans are thought to be one of the primary feeding species (Chopin, 2015; Rosa *et al.*, 2020). Crop degradation, threats to ecosystem's integrity, and competition with wild populations are some of the environmental issues associated with farming crustaceans (Nederlof *et al.*, 2022). The main crustaceans raised in the IMTA are shrimps, prawns, crabs, and lobsters (Chang *et al.*, 2020). Crustacean farming has long been practiced in conjunction with other agricultural practices; for example, Vietnamese rice farmers have grown species such as *Penaeus merguiensis*, *Penaeus indicus*, *Penaeus monodon*, or *Metapenaeus ensis* in their rice paddies (Rosa *et al.*, 2020). *Penaeus vannamei*, *Penaeus stylirostris*, *Pandalus platyceros*, and *Fenneropenaeus chinensis* are also cultured in IMTA.

5.5.4 Mussels

It is possible for filter-feeders to remove organic particulate matter from the water column. As they offer economic benefits and lessen the environmental effect of agriculture, mussels are among the species with excellent possibilities for usage as filter-feeders (Granada *et al.*, 2016). Prospective candidates for use as extractive species in IMTAs include mussels such as *Mytilus edulis*, *Mytilus trossulus*, *Patinopecten yessoensis*, and *Crassostrea gigas* (Chopin *et al.*, 2012; Hargrave *et al.*, 2022). In a tropical context, an integrated system combining the bivalves (*Crassostrea madrasensis*) and finfish (*Etroplus suratensis*) was successful in significantly limiting eutrophication (Sukhdhane *et al.*, 2018). In a study by Largo *et al.* (2016), the establishment of an IMTA system in the tropical open waters of southern Cebu, the Philippines, uses a combination of locally available species, including seaweeds (*Gracilaria heteroclada* and *Eucheuma denticulatum*) as inorganic extractive species and donkey's ear abalone (*Haliotis asinina*) as feed species. This study is an excellent example of how to successfully use very valuable endemic species such as donkey's ear abalone and *G. heteroclada*. Shellfish is typically thought of as a reservoir for some finfish pathogens; hence, adding shellfish as a filter-feeder to the IMTA might change the dynamics of fish pathogen infection and therefore lower the risk of illness (Granada *et al.*, 2016; Pietrak *et al.*, 2012).

5.5.5 Seaweeds

Seaweeds serve as primary producers in natural ecosystems, converting inorganic substances into biomass that is then devoured by creatures at higher trophic levels (Leston *et al.*, 2011; Nederlof *et al.*, 2022; Torres *et al.*, 2008). They are very good at removing dissolved inorganic nutrients and accumulating their considerable biomasses (Kang *et al.*, 2021; Leston *et al.*, 2011; Nederlof *et al.*, 2022). They are therefore utilized in IMTA systems as inorganic extractive species, which reduce eutrophication by absorbing nutrients entering the water column (Nederlof *et al.*, 2022); they also aid in bioremediation (Samocha *et al.*, 2015). In addition, seaweeds are presently employed as a source of medication and food for both people and animals (Abreu *et al.*, 2014).

Numerous studies are being conducted on the use of *Gracilaria bursa pastoris*, *Gracilaria gracilis*, *Gracilaria chilensis*, *Gracilaria birdiae*, *Gracilaria lemaneiformis*, *Alaria esculenta*, *Chondrus crispus*, *Laminaria japonica*, *Palmaria palmata*, *Porphyra dioica*, *Asparagopsis armata*, *Macrocystis pyrifera*, *Saccharina latissima*, *Gracilariopsis longissima* (Rhodophyta), *Ulva rotundata*, *Ulva lactuca*, *Ulva ohnoi*, *Ulva rigida*, and *Ulva intestinalis* (Chlorophyta), *Chaetomorpha linum* as biofilters for IMTA systems (Granada *et al.*, 2016; Nederlof *et al.*, 2022).

5.5.6 Echinodermata

The use of echinoderms, especially holothurians as detritus- and deposit-feeders, can facilitate further optimization of IMTA. Their potential as powerful extractive species in IMTA has attracted research due to their value and apparent capacity to devour bigger particles (Cubillo *et al.*, 2016; Nelson *et al.*, 2012). Incorporating holothurians to feed on the settling particles missed by the suspension-feeders might help reduce the total organic carbon content and total organic matter of water. For oligotrophic environments such as the Mediterranean, such additions can be highly successful. The completed product (Beche de Mer) of *Holothuria scabra* and *Holothuria lessoni* sold for around US$369–1898/kg in China in 2016 (Purcell *et al.*, 2018), suggesting that the holothurians may have economic potential for an IMTA setup.

Sea cucumbers may clean up oyster waste and can eat waste (fish excrement, extra fish food, and algae), turning a useless product into sea cucumber biomass. This is similar to the co-culture of sea cucumbers with mussels and fish. Compared to individuals growing apart (250 m) from the fish farm, *Parastichopus californicus* shown rapid growth and high survival (99.5%) under sablefish net pens in an experimental IMTA system (Hannah *et al.*, 2013). As evidenced by its high absorption efficiency (>80%) when exposed to high carbon particulate matter (salmon food and faeces) (Nelson *et al.*, 2012), *Cucumaria frondosa* offers as a possible IMTA candidate.

5.5.7 Polychaetes

With regards to nutrient recovery from particulate organic matter (POM), a number of marine polychaetes have been identified as being particularly useful in IMTA, including species such as *Alitta virens* (Brown *et al.*, 2011; Guerrero & Cremades, 2012), *Hediste diversicolor* (Wang *et al.*, 2019;

Yousefi-Garakouei *et al.*, 2019), *Abarenicola pusilla* (Gómez *et al.*, 2017, 2019), *Arenicola marina*, *Perinereis aibuhitensis* (Guerrero & Cremades, 2012), *Perinereis nuntia* (Palmer, 2010), *Perinereis helleri* (Palmer, 2010; Palmer *et al.*, 2014), *Capitella* spp., and *Ophryotrocha craigsmithi* (Nederlof *et al.*, 2019). The Mediterranean polychaete *Sabella spallanzanii* has been shown to have the potential to filter and eliminate bacterial groups, including pathogenic human pathogens and vibrios (Stabili *et al.*, 2010). According to research on *S. spallanzanii* ability to filter *Vibrio alginolyticus*, they are highly efficient in filtering these bacteria (Licciano *et al.*, 2005), making them suitable as biofilters. By serving as bio-diffusors (Mermillod-Blondin, 2011; Queirós *et al.*, 2013), facilitating bioturbation (Shull, 2009) by creating burrows, and performing bio irrigation (Shull, 2009), ragworms have the power to create ecosystems. *P. nuntia vallata*, a polychaete, successfully consumes and assimilates the faeces of Japanese flounder, turning about half of the nitrogen into body tissue. In addition to improving the fitness of cocultured organisms, investigations on the culture of *Nereis diversicolor* utilizing eel sludge suggest the former may help minimize environmental wastes (Garcia-Alonso *et al.*, 2008).

5.5.8 Sponges

The bioproducts (biosilica, biosintering, biomedical agents, and collagen) that sponges produce have increased interest in incorporating them into an IMTA setup pertains to their role in filter-feeding augmented with the bioproducts (Gokalp *et al.*, 2019). The study by Pronzato *et al.* (1998) provides the first indication that sponges are taken into account as an extractive species for IMTA. Through their quick and effective filter feeding capacity, sponges can accumulate a significant quantity of organic waste that is present in water (Weisz *et al.*, 2008). Sponges are sedentary suspension-feeders that can eat organic particles as small as 0.1–50 μm (Maldonado *et al.*, 2010). In urban and aquaculture settings, sponges can be used to clean up organic pollution because their carbon absorption rates are equivalent to those of bivalves (Stabili *et al.*, 2006).

Sponge populations have been shown to increase in urban run-off from waste streams (Ledda *et al.*, 2014), coculture with mussels (Page *et al.*, 2011), and mariculture farms (coculture with sea bass) in several in-situ investigations. A number of sponges, including *Dysidea avara*, *Chondrosia reniformis* (Osinga *et al.*, 2010), *Chondrilla nucula* (Milanese *et al.*, 2003), and *Spongia officinalis* var. adriatica (Stabili *et al.*, 2006), have proven to be quite effective in removing waste and integrating with mariculture.

5.5.9 Coelenterata

Aquaculture facilities are commonly biofouled by coelenterates, particularly hydrozoans. Shell smothering, competition, net occlusion, and other detrimental effects on finfish and shellfish are all linked to benthic hydroids (Durr & Watson, 2010; Fitridge *et al.*, 2012). *Apostichopus japonicus* (sea cucumber) and *Rhopilema esculenta* (jelly fish) were co-cultured, and the results showed that the jelly fish increased the settling velocity of the particulate matter. Additionally, compared to those in monoculture, the sea cucumbers from the co-culture pond of *A. japonicus* and *R. esculenta* displayed a higher specific growth rate (SGR). In addition to providing a supplemental cash crop, use of jellyfish and sea cucumber in a co-culture may reduce the benthic nutrient load.

5.6 BASIC SETUP OF AN IMTA SYSTEM

To ensure economic and environmental sustainability, IMTA includes rebuilding a natural trophic network. In the case of marine exploitation through catch fisheries, a paradigm change has been observed, with aquaculture now providing more than 50% of the fin and shellfish consumed globally. The increasing demand for aquaculture products will eventually close the growing gap between supply and demand due to the global population boom and the stagnation of global fisheries since the mid-1980s (Hughes *et al.*, 2016). Due to this, technological advancements in aquaculture are unavoidable if we are to close the gap and provide enough food (Njiru *et al.*, 2019). Aquatic animals are typically raised in cages (floating enclosures inside aquatic environments), which is a technical innovation that allows for unlimited material exchange between the cage and the surrounding water body (Chakraborty *et al.*, 2010). According to Ombwa *et al.* (2018), organic pollution from fish cage leachates, such as uneaten meals and excrement, can lower oxygen levels both within and outside the cage. The use of these organic wastes as a food for the growth of algae might be facilitated by the mineralization of these wastes into inorganic components (Dias *et al.*, 2011).

According to IMTA may be either land-based, using ponds and tanks for the culture, or it can be situated in open water with a variety of species combinations. The fundamental component of an IMTA arrangement is two structures positioned closely together. The cage for the fed aquaculture component and the longlines for the extractive component are two examples of these. After providing appropriate access to the fed cages, extractive species integration must be completed (Chopin, 2013). To promote water flow from the supplied component to the extractive species, the land-based IMTA setup is made up of ponds or tanks that are stacked sequentially (Sumbing *et al.*, 2016). The water module is developed according to a flow through system or a recirculating water system.

The extractive component (shellfish and seaweed) is cultivated using the longline culture consisting of four components:

(1) *Backbone:* This component holds the buoys or floats, with a typical length of 100 m with a 5 m inter-backbone gap. Buoys are kept at an appropriate distance to support the bivalves grown on the backbone.
(2) *Warp:* It is a synthetic rope attached to the edge of the backbone, measuring thrice the depth of water, moored to the seabed using a heavy-weight anchor.
(3) *Lantern nets:* These nets are hung on the backbone and contain scallops or oysters, with a 2.3 m spacing between consecutive nets.
(4) *Horizontal kelp rope:* A longline system is typically used to culture kelp, with 174 kelp ropes strung between two parallel backbones and 40 kelp individuals per rope. The hanging ropes are used to join the kelp ropes to the backbone. The hanging ropes are used to join the kelp ropes to the backbone.

The protection from wind and waves, suitable depth (5–20 m), excellent water quality, and a plentiful plankton supply in addition to the mud–sand sediment for the longline's deployment are some requirements for shellfish production. The offshore IMTA setup is at risk from the prevalent hydrodynamic forces, particularly the orbital motion brought on by marine forces (Gaylord *et al.*, 2008; Holthuijsen, 2007; Troell *et al.*, 2009). Breaking waves and swell place significant stress on the fed and extractive components, making them unsuitable for offshore IMTA.

5.7 BENEFICIAL ROLE OF IMTA

5.7.1 Nutrient uptake

Studies from both open-water and land-based cultures have demonstrated that the nutrients produced by fed aquaculture species are appropriate for the development of seaweed (Troell *et al.*, 2009). The types of fish that are raised, the layout of the farm, the type of feed used, and other site-specific factors all have an impact on the development and performance of the seaweed. The percentage of accessible nutrients bound by these seaweeds at any particular moment can be used to evaluate the capacity of seaweeds in open-water cultures to extract nutrients from the water column. However, because net pen culturing takes place in open water, it is challenging to assess changes in water-nutrient concentrations because they will vary greatly depending on the direction of the current, the depth, the season, the hour of the day, and a host of other factors. Due to this variation, many researchers have come to varying findings on the efficacy of IMTA systems in reducing nutrient concentrations in the water column (Troell *et al.*, 2009).

Under natural circumstances (Abreu *et al.*, 2009; Buschmann *et al.*, 2008; Harrison & Hurd, 2001) and in fish pond effluents (Cohen & Neori, 1991; Neori *et al.*, 1991), researchers have examined the connections between seaweed density, light levels, and nutrient availability (concentration or flow). Under non-limiting circumstances, seaweed growth is influenced by ambient nutrient concentrations, saturation kinetics by light intensity, temperature, and other environmental parameters, as well as other environmental factors (Buschmann *et al.*, 2008). In a culture system, as seaweed biomass density rises, so does the ambient availability of both nutrients and light, which causes growth rates to drop precipitously. Additionally, according to Buschmann *et al.* (2001), the performances of biofiltration and seaweed growth are functionally independent. As nutrient levels decrease, the significance of water turbulence around the seaweeds for the rate of uptake increases (Gonen *et al.*, 1993; Msuya & Neori, 2008). Usually, timely thinning of the seaweed biomass (e.g. frequent harvesting) solves these limitations. Of course, in an offshore farm, frequent harvests may have logistical and financial implications.

5.7.2 Waste to wealth by utilizing waste material

Filter-feeders may eat particulate waste materials, such as waste feed and fish cage excrement, and phytoplankton that thrives on fish-excreted nutrients. Mussels and oysters grew more quickly next to fish cages as revealed in multiple experiments (Jones & Iwama, 1991; Wallace, 1980). The ambient concentration of particulate organic matter is the single most crucial factor affecting mussel growth rate. Therefore, it has been suggested that the occurrence of temporal and geographical fluctuation in food supply in natural water bodies may account for the variable degrees of success of experimental IMTA locations (Troell *et al.*, 2009).

5.7.3 Increase biodiversity

If IMTA is to design biodiversity for better economic and environmental performance, the importance of appropriate habitat is equally clear. The current fish farming infrastructure, which is constructed using steel, plastic, ropes, nets, buoys, and so on, provides substrates that are already liveable but are not often present in open seas. Within a few weeks, a range of creatures, such as mussels, sea cucumbers, anenomes, tunicates, nudibranchs, hydrozoans, and algae, often begin to colonize the area. These species either feed by the absorption of dissolved nutrients and light or the capture of tiny particles. Because of this and other factors, including economic ones, species such as mussels and kelps are already significant biodiversity components that have been incorporated into the IMTA method.

Due to the increased surface area, the inclusion of three-dimensional structures at IMTA sites is believed to provide a substantial potential to boost biodiversity (Robinson *et al.*, 2011). A series of species, including sea stars, urchins, fish, and crustaceans, are brought into existence through the colonization of new substrates. These species are either drawn to the newly accessible prey or the more varied environments that the colonized substrates afford. A diverse community of organisms representing various trophic levels that occupy the niches provided by the aquaculture system develops throughout time. If we wish to build effective IMTA systems, it is crucial that we comprehend the natural processes in biodiversity that occur around aquaculture sites.

5.7.4 Some advantages of IMTA

- *Wastewater bio-mitigation*: managing wastewater by using biofilters that are appropriate for the operation's ecology and ecological niche. Some of the problems brought on by monoculture can be resolved by using this approach (Correia *et al.*, 2020; Granada *et al.*, 2018).
- *Higher earnings through diversification*: the sale of commercial by-products increases the aquaculture operation's overall economic worth. The expenses of intricate bio-filtration systems are also partially covered by these economic benefits (Fonseca *et al.*, 2015; Ridler *et al.*, 2007).

- Supporting local economic development by creating direct and indirect jobs (Fonseca *et al.*, 2015; Knowler *et al.*, 2020).
- *Natural crop insurance*: the variety of goods offers protection from price swings for any one commodity as well as against crop loss due to diseases or bad weather (O'Neill *et al.*, 2022; Sasikumar & Viji, 2016).
- *Controlling disease*: according to Chopin (2013) and Granada *et al.* (2016), several seaweed species have antibacterial properties that help prevent or treat illnesses in farmed fish.
- *Increasing revenue by charging higher prices*: according to Carras *et al.* (2019), IMTA goods might command a higher price by participating in organic certification programmes or eco-labelling initiatives.
- Reducing or eliminating reliance on fish meal and fish oil (Granada *et al.*, 2016)

However, there are several drawbacks for IMTA:

- The installed artificial system is vulnerable to disease outbreaks. The culture species are more vulnerable to infection from other culture species and from external sources because they are kept in a non-natural setting (Granada *et al.*, 2016; Stabili *et al.*, 2010).
- Because the system here depends on inputs from many trophic levels (plants, fish, and invertebrates) to thrive, the likelihood of system collapse is enormous. The entire system may fail if one of these layers is compromised (Chopin, 2017).
- The IMTA regulation is still not in place to tightly regulate the system. In actuality, managing this can be challenging, resulting in inefficient production and decreased sustainability (Buck *et al.*, 2018).
- This system is reliant on outside resources such as water and fish feed. Production can abruptly stop when these resources become scarce or expensive (Wilfart *et al.*, 2020).
- An IMTA system can result in the overcrowding and overpopulation of organisms in culture. Additionally, excessive animal population causes higher levels of stress, accelerates the spread of disease, and results in high mortality (Chopin, 2017; Granada *et al.*, 2016).

5.7.5 Valorization

The Copenhagen World Climate Conference marked the introduction of the low carbon concept, which entails low emissions, low energy consumption, and low pollution. This concept has had an impact on fishery ecosystems as well, which are thought to respond abnormally under the combined influence of anthropogenic activities and climate change. There is an urgent need for a scientific method to recover and magnify biological carbon sinks in order to establish sustainable aquaculture practices and actively combat climate change. In the context of fisheries, a carbon sink refers to the procedures that use aquatic species to fish in water to absorb carbon dioxide before removing it after those aquatic products are harvested. An effective and environmentally beneficial interpretation of a carbon sink in the context of fisheries is IMTA, which promotes cultural variety, boosts economic growth, and reduces strain on ecosystems from aquaculture while also being fairly efficient. In addition to providing an effective food supplement, IMTA also offers a wide range of ecosystem services, with a service value ratio of 18:1 when compared to monoculture (Fang, 2018; Nissar *et al.*, 2022). These services range from the production of materials and the improvement of water quality to the regulation of air quality and climate. Diversification and optimization might make it easier to value an IMTA system. Diversifying IMTA can have positive effects on the economy and the environment, as seen by the ability of filter-feeders to absorb harmful bacteria, which reduces disease outbreaks and keeps a lid on human diseases. To analyse the densities and culturing conditions for optimal production, more and more experimental models can be tested to address the concept's infancy (Granada *et al.*, 2016).

5.8 FUTURE CHALLENGES IN IMTA

The difficulties with IMTA systems become clear as they are more frequently used. According to Sasikumar and Viji (2016) and Rosa *et al.* (2020), the following are the current difficulties faced by IMTA systems:

- A limited number of capable species.
- The necessity for careful monitoring of inputs and nutrient fluxes.
- *Higher capital expenditure*: due to the requirement for more advanced technology and engineering in open-sea farming, IMTA operations demand a bigger up-front investment.
- *Increased area requirement*: despite the fact that IMTA systems reduce the stress on natural resources and help the environment, rivalry between aquaculture operations and other marine and coastal industries still exists. Conflicts between stakeholders are common, and they can stem from issues such as site allocation to worries about environmental effects such as pollution.

5.9 CASE STUDIES

5.9.1 Case study 1

Aquaculture that is environmentally sustainable is driven by IMTA systems, which include the cultivation of species from several trophic levels. Using IMTA, uneaten feed, solid wastes, and dissolved nutrients are converted into edible and nutritious food. Options to promote sustainability on the social, environmental, and economic levels lead to its larger sustainability potentials. IMTA is investigated in this chapter as a means of achieving economically, ecologically, and socially sustainable development. They gathered information demonstrating how IMTA may help local economies, reduce the detrimental environmental consequences of aquaculture, increase competitiveness, and ensure long-term economic viability. The socio-economic and cost-effectiveness analyses that are currently available indicate that IMTA systems have favourable prospects due to product diversification, quicker production cycles, and IMTA product prices. However, they also demonstrate a difference between financial returns at the macroeconomic level and returns at the entrepreneurial unit level. The main reason for the previously poor participation with IMTA, according to the authors, is the absence of governance analysis or the improper institutional development with regards to aquaculture governance and management laws and regulations. This is partly due to inappropriate laws, regulations, and public and private sector decision-making policies and execution, which are highlighted by the dearth of assessments of aquaculture governance institutions (Hossain *et al.*, 2022).

5.9.2 Case study 2

The industry of marine aquaculture is expanding quickly. Researchers, business leaders, and policymakers are paying more and more attention to offshore aquaculture installations as a possible avenue for the aquaculture sector to grow significantly. At the same time, there has been a rise in interest in both onshore and offshore aquaculture systems that integrate farmed species (such as finfish) with nearby inorganic extractive species (such as seaweeds) and organic extractive species (such as suspension- and deposit-feeders). Based on several potential economic, societal, and environmental benefits, such as the recycling of waste nutrients from higher trophic-level species into the production of lower trophic-level crops with commercial value, such systems, referred to as IMTA, should greatly increase the sustainability of aquaculture. The exposed condition of the open ocean adds a number of technical and financial difficulties to the nearshore environments that IMTA faces. To develop offshore IMTA, it is necessary to identify the societal, economic, and environmental costs of producing finfish monocultures by internalizing the bioremediative services of extractive species in IMTA offshore systems. This can be carried out by comparing the advantages and risks of these large-scale systems with similarly scaled monocultures of high trophic-level finfish in offshore systems. According to Troell *et al.* (2009), the outcomes of these studies will be useful in assessing the viability of using the IMTA technique as a development strategy for offshore aquaculture.

5.9.3 Case study 3

It is imperative to apply ecological engineering techniques in order to create an aquaculture that is more sustainable. In light of this, systems for IMTA based on seaweeds are being developed to lessen the environmental issues brought on by various types of feed aquaculture. A number of macroalgal species, particularly those belonging to the genus *Gracilaria*, have demonstrated their efficacy as biofilters and have been underutilized as a source for agar production. To assess this species' potential as a biofilter element of an IMTA system, a single and turbot land-based aquaculture plant established a seaweed production system with 1200 L tanks. To investigate the impact of stocking density with adequate water exchange, an annual experiment was conducted. Although *Gracilaria vermiculophylla* was able to sustain a decent overall performance, the data show that successful productivity requires year-round changes to the culture circumstances. Generally speaking, the system's cultivation densities had a negative relationship with biomass output and nutrient removal. *G. vermiculophylla* of 0.7 ± 0.05 kg dw/m^2/month was produced in the tanks seeded with 3 kg fw/m^2, and this biomass removed 221 ± 12.82 g/m^2/month of carbon and 40.54 ± 2.02 g/m^2/month of nitrogen ($\pm 0.03\%$ of the monthly fish nitrogen input). The primary environmental elements influencing the development and nutrient removal capabilities of the seaweed were also temperature and light. This pilot IMTA system can be deployed in fish aquaculture enterprises after the necessary upscaling. According to Abreu *et al.* (2011), *G. vermiculophylla* has demonstrated its effectiveness as a part of land-based IMTA systems, offering potential economic and environmental advantages for fish farms.

5.9.4 Case study 4

In a different study, the effects of feeding juvenile *Oreochromis niloticus* (Nile tilapia) with three seaweeds produced in IMTA – *G. vermiculophylla*, *P. dioica*, and *Ulva* spp. – were assessed with respect to the fish's growth, body composition, and gut histology. Each of the three seaweeds was designed to replace 10% of the entire diet in three different experimental diets: *G. vermiculophylla*, *P. dioica*, and *Ulva* spp. There was no seaweed added to the control diet. Throughout the course of an 84-day study, diets were provided to duplicate groups of 25 young Nile tilapia, with an average body weight of 12.1 g. When the trial came to a conclusion, fish fed with the *G. vermiculophylla* diet showed much lower growth performance, but their feed conversion ratio increased significantly as well. Regarding body composition, none of the therapies had any negative impacts. With a notable decrease in villi length on the *G. vermiculophylla* diet, the addition of the three seaweeds to the diet clearly altered the anatomy of the fish's digestive system. Given that no detrimental effects on growth performance or body composition were noted, the study's findings point to the potential utility of *P. dioica* and *Ulva* spp. in replacing fishmeal in tilapia juvenile diets up to 10%. But adding 10% *G. vermiculophylla* to the diet appears to have the opposite effect, decreasing fish feed consumption and growth performance (Silva *et al.*, 2015).

5.10 FUTURE PERSPECTIVE

The prospects for IMTA appear bright. In an effort to replicate natural ecosystems, IMTA is a sustainable aquaculture technique that involves the co-cultivation of several species, including seaweeds, finfish, and shellfish. The following are some important factors: sustainability, which lessens pollution and nutrient overloading in aquatic environments by using the waste of one species as nutrients for another. Diversification that lowers the risks brought on by disease outbreaks and market swings by encouraging the cultivation of a range of species. Economic viability by providing farmers with more consistent income through the creation of several revenue streams from various species. Maintain ecosystem health by enhancing water quality and encouraging biodiversity makes it an important instrument for preserving and reviving the wellbeing of coastal ecosystems. Technological developments such as automated control and monitoring systems will improve the efficacy and efficiency of IMTA operations. The expansion of IMTA is anticipated to occur beyond conventional fish farming locations, thereby augmenting global food security and alleviating pressure on wild fisheries. Finally, given its alignment with the increasing demand for ethical and sustainable aquaculture methods and has a promising future.

5.11 CONCLUSION

IMTA involves the co-cultivation of species from distinct trophic levels in close proximity to one another. The organic and inorganic waste products resulting from one cultivated species are recycled and used as nutritional inputs for other species. Through product diversification, quicker production cycles, and price premiums on IMTA products, IMTA can lessen the ecological effects of aquaculture operations nearby, enhance public attitudes of aquaculture, and benefit aquaculture operators monetarily. It has been shown that using IMTA increases the farm's assimilative capacity and significantly lowers the environmental cost of aquaculture. Furthermore, adding extractive species (such as seaweeds and/or invertebrates) to fed monoculture processes already in place might boost farm revenues. Adopting IMTA can be even more profitable if there are favourable public perceptions of the company, as demonstrated by consumers' willingness to pay more for its goods. To show the true worth of IMTA to society, comparative bioeconomic models of IMTA still need to be developed, as well as competitive production systems' capacity to internalize externalities need to be assessed. It is also necessary to look into marketing opportunities such as promoting the eco-certification of IMTA products and more research into financial incentives such as the tools required to encourage the adoption of IMTA.

REFERENCES

Abreu M. L., Varela D. A., Henríquez L., Villarroel A., Yarish C., Sousa-Pinto I. and Buschmann A. H. 2009. Traditional vs. integrated multi-trophic aquaculture of *Gracilaria chilensis* C.J. Bird J. McLachlan & E.C. Oliveira: productivity and physiological performance. *Aquaculture*, **293**, 211–220, https://doi.org/10.1016/j.aquaculture.2009.03.043

Abreu M. H., Pereira R., Yarish C., Buschmann A. H. and Sousa-Pinto I. (2011). IMTA with *Gracilaria vermiculophylla*: productivity and nutrient removal performance of the seaweed in a land-based pilot scale system. *Aquaculture*, **312**(1–4), 77–87, https://doi.org/10.1016/j.aquaculture.2010.12.036

Abreu M. H., Pereira R. and Sassi J. F. (2014). Marine algae and the global food industry. In: Marine Algae: Biodiversity, Taxonomy, Environmental Assessment, and Biotechnology, L. Pereira and J. Magalhaes Neto (eds.), CRC Press, Boca Raton, FL, pp. 300–319, https://doi.org/10.1201/b17540

Alexander, K. A., Angel, D., Freeman, S., Israel, D., Johansen, J., Kletou, D., Meland M., Pecorino D., Rebours C., Rousou M., Shorten M. and Potts, T. (2016). Improving sustainability of aquaculture in Europe: stakeholder dialogues on integrated multi-trophic aquaculture (IMTA). *Environmental Science & Policy*, **55**, 96–106.

Brown N., Eddy S. and Plaud S. (2011). Utilization of waste from a marine recirculating fish culture system as a feed source for the polychaete worm, *Nereis virens*. *Aquaculture*, **322**, 177–183, https://doi.org/10.1016/j.aquaculture.2011.09.017

Buck B. H., Troell M. F., Krause G., Angel D. L., Grote B. and Chopin T. (2018). State of the art and challenges for offshore integrated multi trophic aquaculture (IMTA). *Frontiers in Marine Science*, **5**, 165, https://doi.org/10.3389/fmars.2018.00165

Buschmann A. H., Troell M. and Kautsky N. (2001). Integrated algal farming: a review. *Cashiers de Biologie Marine*, **42**, 83–90.

Buschmann A. H., Varela D. A., Hernández-González M. C. and Huovinen P. (2008). Opportunities and challenges for the development of an integrated seaweed based aquaculture activity in Chile: determining the physiological capabilities of *Macrocystis* and *Gracilaria* as biofilters. *Journal of Applied Phycology*, **20**, 571–577, https://doi.org/10.1007/s10811-007-9297-x

Carras M. A., Knowler D., Pearce C. M., Hamer A., Chopin T. and Weaire T. (2019). A discounted cash-flow analysis of salmon monoculture and integrated multi-trophic aquaculture in eastern Canada. *Aquaculture Economics & Management*, **24**(1), 43–63, https://doi.org/10.1080/13657305.2019.1641572

Chakraborty S. B., Mazumdar D. and Banerjee S. (2010). Determination of ideal stocking density for cage culture of monosex Nile tilapia (*Oreochromis niloticus*) in India. *Proceedings of the Zoological Society*, **63**(1), 53–59, https://doi.org/10.1007/s12595-010-0007-3

Chang Z. Q., Neori A., He Y. Y., Li J. T., Qiao L., Preston S. I., Liu P. and Li J. (2020). Development and current state of seawater shrimp farming, with an emphasis on integrated multitrophic pond aquaculture farms, in China – a review. *Reviews in Aquaculture*, **12**(4), 2544–2558, https://doi.org/10.1111/raq.12457

Chopin T. (2013). Aquaculture, integrated multi-trophic (IMTA). In: Sustainable Food Production, P. Christou, R. Savin, B. A. Costa-Pierce, I. Misztal and C. B. A. Whitelaw (eds.), Springer, New York, NY, 184–205, https://doi.org/10.1007/978-1-4614-5797-8_173

Chopin T. (2015). Marine aquaculture in Canada: well-established mono cultures of finfish and shellfish and an emerging integrated multi 862 trophic aquaculture (IMTA) approach including seaweeds, other invertebrates, and microbial communities. *Fisheries*, **40**(1), 28–31, https://doi.org/10.1080/03632415.2014.986571

Chopin T. (2017). Challenges of moving integrated multitrophic aqua culture along the R&D and commercialization continuum in the western world. *The Journal of Ocean Technology*, **12**(2), 34–47.

Chopin T. and Robinson S. (2004). Defining the appropriate regulatory and policy framework for the development of integrated multi-trophic aquaculture practices: introduction to the workshop and positioning of the issues. *Bulletin of the Aquaculture Association of Canada*, **104**, 4.

Chopin T., Troell M., Reid G. K., Knowler D., Robinson S. M. C., Neori A., Buschmann A. H. and Pang S. (2010). Integrated multi-trophic aquaculture, part 2. Global Seafood Alliance's, https://www.globalseafood.org/advocate/integrated-multi-trophic-aquaculture-part-2/ (accessed 1 September 2010)

Chopin T., Cooper J. A., Reid G., Cross S. and Moore C. (2012). Open-water integrated multi-trophic aquaculture: environmental biomitigation and economic diversification of fed aquaculture by extractive aquaculture. *Reviews in Aquaculture*, **4**(4), 209–220, https://doi.org/10.1111/j.1753-5131.2012.01074.x

Cohen I. and Neori A. (1991). *Ulva lactuca* biofilters for marine fishpond effluents: I. Ammonia uptake kinetics and nitrogen content. *Botanica Marina*, **34**, 475–482, https://doi.org/10.1515/botm.1991.34.6.475

Correia M., Azevedo I. C., Peres H., Magalhães R., Oliva-Teles A., Almeida C. M. R. and Guimarães L. (2020). Integrated multitrophic aquaculture: a laboratory and hands-on experimental activity to promote environmental sustainability awareness and value of aquaculture products. *Frontiers in Marine Science*, **7**, 156, https://doi.org/10.3389/fmars.2020.00156

Cranford P. J., Reid G. K. and Robinson S. M. (2013). Open water integrated multi-trophic aquaculture: constraints on the effectiveness of mussels as an organic extractive component. *Aquaculture Environment Interactions*, **4**(2), 163–173, https://doi.org/10.3354/aei00081

Cubillo A. M., Ferreira J. G., Robinson S. M. C., Pearce C. M., Corner R. A. and Johansen J. (2016). Role of deposit feeders in integrated multi-trophic aquaculture – a model analysis. *Aquaculture*, **453**, 54–66, https://doi.org/10.1016/j.aquaculture.2015.11.031

Dias J. D., Takahashi E. M., Santana N. F. and Bonecker C. C. (2011). Impact of fish cage-culture on the community structure of zooplankton in a tropical reservoir. *Iheringia Zoology Series*, **101**(1–2), 75–84, https://doi.org/10.2307/25470713

Durr S. and Watson D. I. (2010). Biofouling and antifouling in aquaculture. In: Biofouling, S. Durr and J. C. Thomason (eds.), Blackwell Publishing Ltd., John Wiley & Sons Ltd, The Atrium, Southern Gate, Chichester, West Sussex, PO19 8SQ, United Kingdom, pp. 267–287.

Ebeling J. M., Timmons M. B. and Bisogni J. J. (2006). Engineering analysis of the stoichiometry of photoautotrophic, autotrophic, and heterotrophic removal of ammonia-nitrogen in aquaculture systems, *Aquaculture*, **257**(1–4), 346–358, https://doi.org/10.1016/j.aquaculture.2006.03.019

Fang J. (2018). Training Module for Integrated for Multitrophic Aquaculture in PR China. UNDP/GEF Yellow Sea Large Marine Ecosystem (YSLME) Phase II Project.

FAO. (2018). The state of fisheries and aquaculture in the world 2018. In: FAO.

FAO. (2020). The State of World Fisheries and Aquaculture (SOFIA).

Ferreira J. G., Saurel C. and Ferreira J. M. (2012). Cultivation of gilthead bream in monoculture and integrated multi-trophic aquaculture. Analysis of production and environmental effects by means of the FARM model. *Aquaculture*, **358**, 23–34, https://doi.org/10.1016/j.aquaculture.2012.06.015

Fitridge I., Dempster T., Guenther J. and de Nys R. (2012). The impact and control of biofouling in marine aquaculture: a review. *Biofouling*, **28**(7), 649–669, https://doi.org/10.1080/08927014.2012.700478 (PMID: 22775076).

Fonseca T., David F. S., Ribeiro F. A. S., Wainberg A. A. and Valenti W. C. (2015). Technical and economic feasibility of integrating sea horse culture in shrimp/oyster farms. *Aquaculture Research*, **2015**, 1–10.

Garcia-Alonso J., Muller C. T. and Hardege J. D. (2008). Influence of food regimes and seasonality on fatty acid composition in the ragworm. *Aquatic Biology*, **4**, 7–13, https://doi.org/10.3354/ab00090

Gaylord B., Denny M. W. and Koehl M. A. R. (2008). Flow forces on seaweeds: field evidence for roles of wave impingement and organism inertia. *Biology Bulletin*, **215**, 295–308, https://doi.org/10.2307/25470713

Ghosh S., Ranjan R., Megarajan S., Pattnaik P., Dash B. and Edward L. (2016). Mixed culture of Pacific white shrimp *Litopenaeus vannamei* (Boone, 1931) and flathead grey mullet *Mugil cephalus* (Linnaeus, 1758) in floating cages. *Indian Journal of Fisheries*, **63**, 63–69.

Gokalp M., Wijgerde T., Sara A., de Goeij J. M. and Osinga R. (2019). Development of an integrated mariculture for the collagen- rich sponge *Chondrosia reniformis*. *Marine Drugs*, **17**, 29–43, https://doi.org/10.3390/md17010029

Gómez S., Hurtado C. F., Orellana J., Valenzuela-Olea G. and Turner A. (2017). *Abarenicola pusilla* (Quatrefages, 1866): a novel species for fish waste bioremediation from marine recirculating aquaculture systems. *Aquaculture Research*, **49**, 1–5, https://doi.org/10.1111/are.13562

Gómez S., Hurtado C. F. and Orellana J. (2019). Bioremediation of organic sludge from a marine recirculating aquaculture system using the polychaete *Abarenicola pusilla* (Quatrefages, 1866). *Aquaculture*, **507**, 377–384, https://doi.org/10.1016/j.aquaculture.2019.04.033

Gonen Y., Kimmel E. and Friedlander M. (1993). Effect of relative water motion on photosynthetic rate of red alga *Gracilaria conferta*. *Hydrobiologia*, **260**(261), 493–498, https://doi.org/10.1007/BF00049061

Granada L., Lopes S., Novais S. C. and Lemos M. F. (2018). Modelling integrated multitrophic aquaculture: optimizing a three trophic level system. *Aquaculture*, **495**, 90–97, https://doi.org/10.1016/j.aquaculture.2018.05.029

Granada L., Sousa N., Lopes S. and Lemos M. F. L. (2016). Is integrated multitrophic aquaculture the solution to the sectors' major challenges? – a review. *Reviews in Aquaculture*, **8**(3), 283–300, https://doi.org/10.1111/raq.12093

Guerrero S. and Cremades J. (2012). Integrated multi-trophic aquaculture (IMTA): a sustainable, pioneering alternative for marine cultures in Galicia. Xunta de Galicia.

Handå A., Min H., Wang X., Broch O. J., Reitan K. I., Reinertsen H. and Olsen Y. (2012). Incorporation of fish feed and growth of blue mussels (*Mytilus edulis*) in close proximity to salmon (*Salmo salar*) aquaculture: implications for integrated multi-trophic aquaculture in Norwegian coastal waters. *Aquaculture*, **356**, 328–341, https://doi.org/10.1016/j.aquaculture.2012.04.048

Hannah L., Pearce C. M. and Cross S. F. (2013). Growth and survival of California sea cucumbers (*Parastichopus californicus*) cultivated with sablefish (*Anoplopoma fimbria*) at an integrated multi-trophic aquaculture site. *Aquaculture*, **406**, 34–42, https://doi.org/10.1016/j.aquaculture.2013.04.022

Hargrave M. S., Nylund G. M., Enge S. and Pavia H. (2022). Co-cultivation with blue mussels increases yield and biomass quality of kelp. *Aquaculture*, **550**, 737832, https://doi.org/10.1016/j.aquaculture.2021.737832

Harrison P. J. and Hurd C. (2001). Nutrient physiology of seaweeds: application of concepts to aquaculture. *Cahiers de Biologie Marine*, **42**, 71–82.

Hoang M. N., Nguyen P. N., Le D. V. B., Nguyen D. V. and Bossier P. (2018). Effects of stocking density of gray mullet *Mugil cephalus* on water quality, growth performance, nutrient conversion rate, and microbial community structure in the white shrimp *Litopenaeus vannamei* integrated system. *Aquaculture*, **496**, 123–133, https://doi.org/10.1016/j.aquaculture.2018.07.018

Holanda M., Santana G., Furtado P., Rodrigues R. V., Cerqueira V. R., Sam Paio L. A., Wasielesky W. Jr and Poersch L. H. (2020). Evidence of total suspended solids control by *Mugil liza* reared in an integrated system with Pacific white shrimp *Litopenaeus vannamei* using biofloc technology. *Aquaculture Reports*, **18**, 100479, https://doi.org/10.1016/j.aqrep.2020.100479

Holthuijsen L. H. (2007). Waves in Oceanic and Coastal Waters. Cambridge University Press, New York.

Hossain A., Senff P. and Glaser M. (2022). Lessons for coastal applications of IMTA as a way towards sustainable development: a review. *Applied Sciences*, **12**(23), 11920, https://doi.org/10.3390/app122311920

Hughes A. D., Corner R. A., Cocchi M., Alexander K. A., Freeman S., Angel D., Chiantore M., Gunning D., Maguire J., Mendoza Beltran A., Guinée J., Ferreira J. and Rebour R. F. C. (2016). Beyond fish monoculture.

In: Developing Integrated Multi-Trophic Aquaculture in Europe, M. Cocchi, R. Corner, A. Hughes (eds.), IDREEM, ETA-Florence Renewable Energies SRL, Italy.

Huo, Y., Wu, H., Chai, Z., Xu, S., Han, F., Dong, L. and He, P. (2012). Bioremediation efficiency of Gracilaria verrucosa for an integrated multi-trophic aquaculture system with Pseudosciaena crocea in Xiangshan harbor, China. *Aquaculture*, **326**, 99–105.

Irisarri J., Cubillo A. M., Fernández-Reiriz M. J. and Labarta U. (2015). Growth variations within a farm of mussel (*Mytilus galloprovincialis*) held near fish cages: importance for the implementation of integrated aquaculture. *Aquaculture Research*, **46**(8), 1988–2002, https://doi.org/10.1111/are.12356

Irisarri, J., Fernández-Reiriz, M. J., Robinson, S. M., Cranford, P. J. and Labarta, U. (2013). Absorption efficiency of mussels Mytilus edulis and Mytilus galloprovincialis cultured under integrated multi-trophic aquaculture conditions in the Bay of Fundy (Canada) and Ría Ares-Betanzos (Spain). *Aquaculture*, **388**, 182–192.

Jiang Z., Wang G., Fang J. and Mao Y. (2013). Growth and food sources of Pacific oyster *Crassostrea gigas* integrated culture with sea bass *Lateolabrax japonicus* in Ailian Bay, China. *Aquaculture International*, **21**, 45–52, https://doi.org/10.1007/s10499-012-9531-7

Jones T. O. and Iwama G. K. (1991). Polyculture of the Pacific oyster, *Crassostrea gigas* (Thunberg) with chinook salmon, *Oncorhynchus tschawytscha*. *Aquaculture*, **92**, 313–322, https://doi.org/10.1016/0044-8486(91)90037-8

Kang Y. H., Kim S., Choi S. K., Lee H. J., Chung I. K. and Park S. R. (2021). A comparison of the bioremediation potential of five seaweed species in an integrated fish-seaweed aquaculture system: implication for a multi-species seaweed culture. *Reviews in Aquaculture*, **13**, 353–364, https://doi.org/10.1111/raq.12478

Knowler D., Chopin T., Martınez-Espineira R., Neori A., Nobre A., Noce A. and Reid G. (2020). The economics of integrated multi trophic aquaculture: where are we now and where do we need to go? *Reviews in Aquaculture*, **12**(3), 1579–1594, https://doi.org/10.1111/raq.12399

Kobayashi, M., Msangi, S., Batka, M., Vannuccini, S., Dey, M. M. and Anderson, J. L. (2015). Fish to 2030: the role and opportunity for aquaculture. *Aquaculture economics & management*, **19**(3), 282–300.

Kumar M., Clarke S. and Sierp M. (2000). Linkage between wastewater treatment and aquaculture; initiatives by the south Australian research development institute (SARDI). In: National Workshop On Wastewater Treatment and Integrated Aquaculture, 17–19 September 1999, M. S. Kumar (ed.), SARDI, West Beach, Adelaide, South Australia, pp. 153–159.

Lander T. R., Robinson S. M., MacDonald B. A. and Martin J. D. (2012). Enhanced growth rates and condition index of blue mussels (*Mytilus edulis*) held at integrated multitrophic aquaculture sites in the Bay of Fundy. *Journal of Shellfish Research*, **31**(4), 997–1007, https://doi.org/10.2983/035.031.0412

Largo D. B., Diola A. G. and Marababol M. S. (2016). Development of an integrated multitrophic aquaculture (IMTA) system for tropical marine species in Southern Cebu, Central Philippines. *Aquaculture Reports*, **3**, 67–76, https://doi.org/10.1016/j.aqrep.2015.12.006

Lawton J. R., Mata L., de Nys R. and Paul N. A. (2013). Lawton algal bioremediation of wastewaters from land-based aquaculture using *Ulva*: selecting target species and strains. *PLoS ONE*, **8**(10), e77344, https://doi.org/10.1371/journal.pone.0077344

Ledda F. D., Pronzato R. and Manconi R. (2014). Mariculture for bacterial and organic waste removal: a field study of sponge filtering activity in experimental farming. *Aquaculture Research*, **45**, 1389–1401, https://doi.org/10.1111/are.12084

Leston S., Nunes M., Viegas I., Lemos M. F. L., Freitas A., Barbosa J., Ramos F. and Pardal M. A. (2011). The effects of the nitrofuran furaltadone on *Ulva lactuca*. *Chemosphere*, **82**(7), 1010–1016, https://doi.org/10.1016/j.chemosphere.2010.10.067

Licciano M., Stabili L. and Giangrande A. (2005). Clearance rates of *Sabella spallanzanii* and *Branchiomma luctuosum* (Annelida: Polychaeta) on a pure culture of *Vibrio alginolyticus*. *Water Research*, **39**, 4375–4384, https://doi.org/10.1016/j.watres.2005.09.003

Maldonado M., Zhang X. C., Cao X. P., Xue L. Y., Cao H. and Zhang W. (2010). Selective feeding by sponges on pathogenic microbes: a reassessment of potential for abatement of microbial pollution. *Marine Ecology Progress Series*, **403**, 75–89, https://doi.org/10.3354/meps08411

Martínez-Porchas M., Martínez-Cordova L. R., Porchas-Cornejo M. and López-Elias J. A. (2010). Shrimp polyculture: a potentially, profitable, sustainable but yet uncommon aquacultural practice. *Reviews in Aquaculture*, **2**(2), 73–85, https://doi.org/10.1111/j.1753-5131.2010.01023.x

Mermillod-Blondin, F. (2011). The functional significance of bioturbation and biodeposition on biogeochemical processes at the water-sediment interface in freshwater and marine ecosystems. *Journal of the North American Benthological Society*, **30**(3), 770–778.

Milanese M., Chelossi E., Manconi R., Sara A., Sidri M. and Pronzato R. (2003). The marine sponge *Chondrilla nucula* Schmidt, 1862 as an elective candidate for bioremediation in integrated aquaculture. *Biomolecular Engineering*, **20**, 363–368, https://doi.org/10.1016/S1389-0344(03)00052-2

Milstein A. (2005). Polyculture in Aquaculture, Animal Breeding Abstracts. CABI Publishing, **73**(12), pp. 15N–41N.

Msuya F. E. and Neori A. (2008). Effect of water aeration and nutrient load level on biomass yield, N uptake and protein content of the seaweed *Ulva lactuca* cultured in seawater tanks. *Journal of Applied Phycology*, **20**, 1021–1031, https://doi.org/10.1007/s10811-007-9300-6

Nederlof M. A., Jansen H. M., Dahlgren T. G., Fang J., Meier S., Ãivind S., Sveier H., Verdegem M. C. and Smaal A. C. (2019). Application of polychaetes in (de)coupled integrated aquaculture: production of a high-quality marine resource. *Aquaculture Environment Interactions*, **11**, 221–237, https://doi.org/10.3354/aei00309

Nederlof M. A. J., Verdegem M. C. J., Smaal A. C. and Jansen H. M. (2022). Nutrient retention efficiencies in integrated multitrophic aquaculture. *Reviews in Aquaculture*, **14**(3), 1194–1212, https://doi.org/10.1111/raq.12645

Nelson E. J., MacDonald B. A. and Robinson S. M. C. (2012). The absorption efficiency of the suspension-feeding sea cucumber, *Cucumaria frondosa*, and its potential as an extractive integrated multi-trophic aquaculture (IMTA) species. *Aquaculture*, **370**, 19–25, https://doi.org/10.1016/j.aquaculture.2012.09.029

Neori A., Cohen I. and Gordin H. (1991). *Ulva lactuca* biofilters for marine fishpond effluents: II. Growth rate, yield and C:N ratio. *Botanica Marina*, **34**, 483–489, https://doi.org/10.1515/botm.1991.34.6.483

Nissar, S., Bakhtiyar, Y., Arafat, M. Y., Andrabi, S., Mir, Z. A., Khan, N. A. and Langer, S. (2023). The evolution of integrated multi-trophic aquaculture in context of its design and components paving way to valorization via optimization and diversification. *Aquaculture*, **565**, 739074.

Njiru J. M., Aura C. M. and Okechi J. K. (2019). Cage fish culture in Lake Victoria: a boon or a disaster in waiting? *Fisheries Management and Ecology*, **26**(5), 426–434, https://doi.org/10.1111/fme.12283

O'Neill E. A., Morse A. P. and Rowan N. J. (2022). Effects of climate and environmental variance on the performance of a novel peatland-based integrated multitrophic aquaculture (IMTA) system: implications and opportunities for advancing research and disruptive innovation post COVID-19 era. *Science of The Total Environment*, **819**, 153073, https://doi.org/10.1016/j.scitotenv.2022.153073

Ombwa V., Aura C., Odada E., Ogari Z., Ogik W., Ogwai K., Chelangat M., Achieng M. and Onienga P. (2018). The Socio-economic Impact of Cage Culture in Lake Victoria for informed decision making. In: Technical Report, KMF/RS/2017. Kenya Marine and Fisheries Research Institute. 57pp.

Orr L. C., Curtis D. L., Cross S. F., Gurney-Smith H., Shanks A. and Pearce C. M. (2014). Ingestion rate, absorption efficiency, oxygen consumption, and fecal production in green sea urchins (*Strongylocentrotus droebachiensis*) fed waste from sablefish (*Anoplopoma fimbria*) culture. *Aquaculture*, **422**, 184–192, https://doi.org/10.1016/j.aquaculture.2013.11.030

Osinga R., Sidri M., Cerig E., Gokalp S. Z. and Gokalp M. (2010). Sponge aquaculture trials in the East-Mediterranean Sea: new approaches to earlier ideas. *Open Marine Biology Journal*, **4**, 74–81, https://doi.org/10.2174/1874450801004010074

Page M. J., Handley S. J., Northcote P. T., Cairney D. and Willan R. C. (2011). Successes and pitfalls of the aquaculture of the sponge *Mycale hentscheli*. *Aquaculture*, **312**, 52–61, https://doi.org/10.1016/j.aquaculture.2010.12.006

Palmer P. J. (2010). Polychaete-assisted sand filters. *Aquaculture*, **306**, 369–377, https://doi.org/10.1016/j.aquaculture.2010.06.011

Palmer P. J., Wang S., Houlihan A. and Brock I. (2014). Nutritional status of a nereidid polychaete cultured in sand filters of mariculture wastewater. *Aquaculture Nutrition*, **20**, 675–691, https://doi.org/10.1111/anu.12129

Pietrak M. R., Molloy S. D., Bouchard D. A., Singer J. T. and Bricknell I. (2012). Potential role of *Mytilus edulis* in modulating the infectious pressure of *Vibrio anguillarum* 02b on an integrated multitrophic aquaculture farm. *Aquaculture*, **326–329**, 36–39, https://doi.org/10.1016/j.aquaculture.2011.11.024

Pronzato R., Cerrano C. and Cubeddu T. (1998). Sustainable development in coastal areas: Role of sponge farming in integrated aquaculture. In: Grizel H., Kesmont P. (eds), Aquaculture and Water: Fish Culture, Shellfish Culture and Water Usage, Special Publication No. 26. pp. 231–232. European Aquaculture Society, Bordeaux, France.

Purcell S. W., Williamson D. H. and Ngaluafe P. (2018). Chinese market prices of Beche-de-Mer: implications for fisheries and aquaculture. *Marine Policy*, **91**, 58–65, https://doi.org/10.1016/j.marpol.2018.02.005

Queirós A. M., Birchenough S. N., Bremner J., Godbold J. A., Parker R. E., Romero-Ramirez A., Reiss H., Solan M., Somerfield P. J., Van Colen C. and Van Hoey G. (2013). A bioturbation classification of European marine infaunal invertebrates. *Ecology and Evolution*, **3**(11), 3958–3985, https://doi.org/10.1002/ece3.769

Reid G. K., Chopin T., Robinson S. M. C., Azevedo P., Quinton M. and Belyea E. (2013a). Weight ratios of the kelps, *Alaria esculenta* and *Saccharina latissima*, required to sequester dissolved inorganic nutrients and supply oxygen for Atlantic salmon, *Salmo salar*, in integrated multi-trophic aquaculture systems. *Aquaculture*, **408**, 34–46, https://doi.org/10.1016/j.aquaculture.2013.05.004

Reid G. K., Robinson S. M., Chopin T. and MacDonald B. A. (2013b). Dietary proportion of fish culture solids required by shellfish to reduce the net organic load in open-water integrated multi-trophic aquaculture: a scoping exercise with cocultured Atlantic salmon (*Salmo salar*) and blue mussel (*Mytilus edulis*). *Journal of Shellfish Research*, **32**(2), 509–517, https://doi.org/10.2983/035.032.0230

Ridler N., Wowchuk M., Robinson B., Barrington K., Chopin T., Robinson S., Page F., Reid G., Szemerda M., Sewuster J. and Boyne-Travis S. (2007). Integrated multi-trophic aquaculture (IMTA): a potential strategic choice for farmers. *Aquaculture Economics and Management*, Vol. **11**(1), 99–110. https://doi.org/10.1080/13657300701202767

Robinson S. M. C., Martin J. D., Cooper J. A., Lander T. R., Reid G. K., Powell F. and Griffin R. (2011). The role of three dimensional habitats in the establishment of integrated multi-trophic aquaculture (IMTA) systems. *Bulletin of the Aquaculture Association of Canada*, **109**(2), 23–29.

Rosa J., Lemos M. F. L., Crespo D., Nunes M., Freitas A., Ramos F., Miguel Â. P. and Leston S. (2020). Integrated multitrophic aquaculture systems – potential risks for food safety. *Trends in Food Science & Technology*, **96**, 79–90, https://doi.org/10.1016/j.tifs.2019.12.008

Samocha T. M., Fricker J., Ali A. M., Shpigel M. and Neori A. (2015). Growth and nutrient uptake of the macroalga *Gracilaria tikvahiae* cultured with the shrimp *Litopenaeus vannamei* in an integrated multi-trophic aquaculture (IMTA) system. *Aquaculture*, **446**, 263–271, https://doi.org/10.1016/j.aquaculture.2015.05.008

Sasikumar G. and Viji C. S. (2016). Integrated multitrophic aquaculture systems (IMTA). Winter School on Technological Advances in Mariculture for Production Enhancement and Sustainability, Course Manual. ICAR, pp. 47–55, http://eprints.cmfri.org.in/10666/1/7.%20Geetha%20Sasikumar.pdf

Shi H., Zheng W., Zhang X., Zhu M. and Ding D. (2013). Ecological-economic assessment of monoculture and integrated multi-trophic aquaculture in Sanggou Bay of China. *Aquaculture*, **410–411**, 172–178, https://doi.org/10.1016/j.aquaculture.2013.06.033

Shpigel M. and Blaylock R. A. (1991). The Pacific oyster, *Crassostrea gigas*, as a biological filter for a marine fish aquaculture pond. *Aquaculture*, **92**, 187–197, https://doi.org/10.1016/0044-8486(91)90020-8

Shull D. H. (2009). Bioturbation. In: Encyclopedia of Ocean Sciences, H. S. John, K. T. Karl and A. T. Steve (eds.), Academic Press, New York, pp. 395–400.

Silva D. M., Valente L. M. P., Sousa-Pinto I., Pereira R., Pires M. A., Seixas F. and Rema P. (2015). Evaluation of IMTA-produced seaweeds (*Gracilaria*, *Porphyra*, and *Ulva*) as dietary ingredients in Nile tilapia, *Oreochromis niloticus* L., juveniles. Effects on growth performance and gut histology. *Journal of Applied Phycology*, **27**, 1671–1680, https://doi.org/10.1007/s10811-014-0453-9

Skriptsova A. V. and Miroshnikova N. V. (2011). Laboratory experiment to determine the potential of two macroalgae from the Russian Far-East as biofilters for integrated multi-trophic aquaculture (IMTA). *Bioresource Technology*, **102**, 3149–3154, https://doi.org/10.1016/j.biortech.2010.10.093

Stabili L., Licciano M., Giangrande A., Longo C., Mercurio M., Marzano C. N. and Corriero G. (2006). Filtering activity of *Spongia officinalis* var. adriatica (Schmidt) (Porifera, Demospongiae) on bacterioplankton: implications for bioremediation of polluted seawater. *Water Research*, **40**, 3083–3090, https://doi.org/10.1016/j.watres.2006.06.012

Stabili L., Schirosi R., Licciano M., Mola E. and Giangrande A. (2010). Bioremediation of bacteria in aquaculture waste using the polychaete *Sabella spallanzanii*. *New Biotechnology*, **27**(6), 774–781, https://doi.org/10.1016/j.nbt.2010.06.018

Sumbing M. V., Al-Azad S., Estim A. and Mustafa S. (2016). Growth performance of spiny lobster *Panulirus ornatus* in land-based integrated multi-trophic aquaculture (IMTA) system. *Transactions on Science and Technology*, **3**(1–2), 143–149.

Strand, Ø., Jansen, H. M., Jiang, Z. and Robinson, S. M. (2019). Perspectives on bivalves providing regulating services in integrated multi-trophic aquaculture. In: Goods and services of marine bivalves, A. C. Smaal, J. G. Ferreira, J. Grant, J. K. Petersen, Ø. Strand (eds), 209–230. Springer Cham. https://doi.org/10.1007/978-3-319-96776-9

Sukhdhane, K. S., Kripa, V., Divu, D., Vase, V. K. and Mojjada, S. K. (2018). Integrated multi-trophic aquaculture systems: A solution for sustainability. *Aquaculture Asia*, **22**(4), 26–29.

Torres M. A., Barros M. P., Campos S. C. G., Pinto E., Rajamani S., Sayre R. T. and Colepicolo P. (2008). Biochemical biomarkers in algae and marine pollution: a review. *Ecotoxicology and Environmental Safety*, **71**(1), 1–15, https://doi.org/10.1016/j.ecoenv.2008.05.009

Troell M., Joyce A., Chopin T., Neori A., Buschmann A. H. and Fang J. G. (2009). Ecological engineering in aquaculture – potential for integrated multi-trophic aquaculture (IMTA) in marine offshore systems. *Aquaculture*, **297**, 1–9, https://doi.org/10.1016/j.aquaculture.2009.09.010

Viji C. S. (2015). Studies on integrated multitrophic aquaculture in a tropical estuarine system in Kerala, India. PhD thesis, Central Institute of Fisheries Education, Mumbai, p. 128.

Viji C. S., Chadha N. K., Kripa V., Prema D., Prakash C., Sharma R., Jenni B. and Mohamed K. S. (2014). Can oysters control eutrophication in an integrated fish-oyster aquaculture system? *Journal of the Marine Biological Association of India*, **56**(2), 67–73, https://doi.org/10.6024/jmbai.2014.56.2.01811-10

Wallace J. F. (1980). Growth rates of different populations of the edible mussel, *Mytilus edulis*, in north Norway. *Aquaculture*, **19**, 303–311, https://doi.org/10.1016/0044-8486(80)90079-4

Wang H., Seekamp I., Malzahn A., Hagemann A., Carvajal A. K., Slizyte R., Standal I. B., Handå A. and Reitan K. I. (2019). Growth and nutritional composition of the polychaete *Hediste diversicolor* (OF Müller, 1776) cultivated on waste from land- based salmon smolt aquaculture. *Aquaculture*, **502**, 232–241, https://doi.org/10.1016/j.aquaculture.2018.12.047

Weisz J. B., Lindquist N. and Martens C. S. (2008). Do associated microbial abundances impact marine demosponge pumping rates and tissue densities? *Oecologia*, **155**, 367–376, https://doi.org/10.1007/s00442-007-0910-0

Wilfart A., Favalier N., Metaxa I., Platon C., Pouil S., Caruso D., Slembrouck J. and Aubin J. (2020). Integrated multi-trophic aquaculture in ponds: what environmental gain? An LCA point of view. 12th International Conference on Life Cycle Assessment of Food 2020 (LCA Food 2020), Towards Sustainable Agri-Food Systems, Berlin (virtual conference), pp. 206–208.

Winans K., Kendall A. and Deng H. (2017). The history and current applications of the circular economy concept. *Renewable and Sustainable Energy Reviews*, **68**, 825–833, https://doi.org/10.1016/j.rser.2016.09.123

Yousefi-Garakouei M., Kamali A. and Soltani M. (2019). Effects of rearing density on growth, fatty acid profile and bioremediation ability of polychaete *Nereis diversicolor* in an integrated aquaculture system with rainbow trout (*Oncorhynchus mykiss*). *Aquaculture Research*, **50**, 725–735.

Yuan X., Yang H., Meng L., Wang L. and Li Y. (2013). Impacts of temperature on the scavenging efficiency by the deposit-feeding holothurian *Apostichopus japonicus* on a simulated organic pollutant in the bivalve–macroalage polyculture from the perspective of nutrient budgets. *Aquaculture*, **406**, 97–104, https://doi.org/10.1016/j.aquaculture.2013.05.009

Zhang J. and Kitazawa D. (2016). Assessing the bio-mitigation effect of integrated multi-trophic aquaculture on marine environment through a numerical approach. *Marine Pollution Bulletin*, **110**(1), 484–492, https://doi.org/10.1016/j.marpolbul.2016.06.005

Zhang S. Y., Li G., Wu H. B., Liu X. G., Yao Y. H., Tao L. and Liu H. (2011). An integrated recirculating aquaculture system (RAS) for land-based fish farming: the effects on water quality and fish production. *Aquacultural Engineering*, **45**, 93–102, https://doi.org/10.1016/j.aquaeng.2011.08.001

Zhou Y., Yang H., Hu H., Liu Y., Mao Y., Zhou H., Xu X. and Zhong F. (2006). Bioremediation potential of the macroalga *Gracilaria lemaneiformis* (Rhodophyta) integrated into fed fish culture in coastal waters of north China. *Aquaculture*, **252**(2), 264–276, https://doi.org/10.1016/j.aquaculture.2005.06.046

doi: 10.2166/9781789064100_0097

Chapter 6

Case studies of photobioreactors employed in the treatment of wastewaters: a sustainable approach for Southeast Asian countries

*Awalina Satya[1], Tjandra Chrismadha[1], Ika A. Satya[2], Azalea D. M. Satya[3], Nina Artanti[4], Hakiki Melanie[4], Rosidah Rosidah[1], Miratul Maghfiroh[1], Eva Navisyah[1], Gunawan Gunawan[1], Nofdianto Nofdianto[1], Helen Julian[5], Ardiyan Harimawan[5] and Tjandra Setiadi[5]**

[1]Research Centre for Limnology and Water Resources-National Research and Innovation Agency-Rep. of Indonesia (BRIN), KST Soekarno, Jl. Raya Bogor-Jakarta km46, Cibinong, Bogor 16911, Indonesia

[2]Research Centre for Computation, National Research and Innovation Agency-Rep. of Indonesia (BRIN), Komplek Cisitu, Jl. Sangkuriang, Coblong, Bandung 40135, Indonesia

[3]Department of Chemical and Environmental Engineering, Faculty of Science and Engineering, University of Nottingham Malaysia, Jalan Broga, Semenyih 43500, Selangor Darul Ehsan, Malaysia

[4]Research Centre for Advance Chemistry-National Research and Innovation Agency-Rep. of Indonesia (BRIN), Gd 452, KST BJ Habibie, Setu, Tangerang Selatan 15314, Indonesia

[5]Department of Chemical Engineering, Faculty of Industrial Technology, Institut Teknologi Bandung, Jl. Ganesa No 10, Bandung 40132, Indonesia

**Corresponding author: tjandra.setiadi@itb.ac.id*

ABSTRACT

Globally, the rapidly increasing human population causes significant environmental changes such as contamination of soil, water and air. Urbanization and industrialization contribute to anthropogenic pollution, which harms ecological stability and climate change. Pollutants enter water bodies through a variety of point and non-point sources, with wastewater discharge being a major one. The existing conventional wastewater treatment methods are costly; therefore, a sustainable alternative approach to treatment costs must be obtained. Microalgae have recently been identified as a potentially cost-effective method of pollutant remediation via the mechanisms of bioaccumulation, biosorption and intracellular degradation. As a result, microalgae-based water treatment (MBWT) can be used to help with environmental resource recovery and sustainable development efforts. Photobioreactors (PBRs) play an important role in these processes because microalgae are photosynthetic organisms that are important for producing the desirable and environmentally benign products. Microalgae biomass derived from MBWT is advantageous as a potential feedstock material for a variety of industrial applications such as biofuels, bio-fertilizers, pharmaceuticals, natural colorants and so on. This review primarily focused on how microalgae are grown in open and closed PBRs. It sought to shed light on the key parameters that influence PBR performance, including irradiance, mixing, pH, temperature, nutrients, mass transfer and capital-operating costs. Along with that, a discussion on wastewater microalgae related to cultivation condition and microalgal growth, wastewater species types and properties was presented.

Keywords: anthropogenic pollution, contamination, microalgae, photobioreactor, wastewater

6.1 INTRODUCTION

The rapid growth of the global population leads to significant excessive wastewater production. Many anthropogenic activities such as domestic, agricultural, industry, mining, and so on have been major environmental issues. Those activities cause significant environmental disturbances due to pollution of soil, water and air which adversely impact ecological stability and climate change. Eliminating the pollutants (both organic and inorganic forms) from the environment is a complicated effort since those chemical structures are recalcitrant and have longer lifetimes. A global distribution of pollutants such as heavy metals, hydrocarbons, steroids, pesticides and their by-products which substantially bring about harm to the biosphere (Luu *et al.*, 2022; Rempel *et al.*, 2021).

Removal of these harmful chemicals commonly utilizes physicochemical and biological techniques. These methods are implemented in wastewater treatment systems which involve precipitation, evaporation, reverse osmosis, electrochemical treatment, ion exchange and membrane filtration. However, these conventional methods of wastewater treatment systems are not completely feasible for removing the pollutants in wastewater since they need high energy consumption and also large area during the operation which makes it expensive. On the contrary, the stakeholders (one of them is the industrial sectors), seeking more sustainable and environmentally benign eco-technologies to eliminate pollution from wastewater. Typically, the biological methods implemented in wastewater treatment aim to enhance pollutant removal performance and reduce operation costs. This method relies on bacteria, fungi and microalgae works. These microbes are capable of degrading organic and inorganic compounds through various complex metabolic pathways. Among these biotas, microalgae are preferred as it is shows dual benefits of removal of pollutants and production of biomass. The carbohydrate-rich microalga biomass can be utilized for biofuel production. Microalgae take up the pollutants as the main source of nutrition and energy from wastewater for their growth and produce biomass called phycoremediation. There are two main mechanisms of phycoremediation, which are bioaccumulation and biosorption that is then followed by biodegradation. Tolerance on organic and inorganic pollutants depends on microalgal strains. They are capable of metabolizing various organic and inorganic pollutants from wastewater to generate intermediate metabolites through enzymatic activities (Barbera *et al.*, 2018; Liang *et al.*, 2013; Siddiki *et al.*, 2022).

Microalgae are photosynthetic organisms that are largely present in our environment. They are primarily aquatic and autotrophic with pseudo leaves, stems and roots and well-defined unicellular or multicellular systems. However, both types of microalgae are cultivated using open pond systems or closed photobioreactor (PBR) systems for producing biomass for commercial purposes. The open system (low cost) is less favored than the closed system because of the lack of a controlled closed system. Therefore, for feasibility reasons, the use of PBR must be focused on energy and marketable products. Numerous advantages of using microalgae for feedstock are fast growth rate, higher biomass yield, easy handling and the ability to be cultivated in non-arable land. Broad spectrum industries are involved since microalgae biomass contains multiple high-value products (such as vitamins, enzymes, pigments, carbohydrates, lipids and proteins). It makes microalgal biomass ideal to be used as feedstock for the pharmaceutical (anti-viral, anti-microbial) industry, biodiesel, bioethanol, food supplements, biofertilizers and feed. The high-value-added products can be extracted from the biomass after downstream purification methods (Luo *et al.*, 2017; Madkour *et al.*, 2012).

A continuous increase of emerging pollutants leads to imbalance to our earth's environmental stability. Therefore, it is necessary to develop sustainable eco-technologies which can clean up the aquatic environment. The microalgae-based water treatment (MBWT) is helpful for resource recovery of the polluted sites with the production of the critical groups of molecules. Recently, MBWT has been given less attention in terms of biomass utilization for energy and fine chemicals production. Therefore, it is interesting to reveal key aspects related to the potential use of MBWT. This chapter discusses how microalgae are grown in open and closed PBRs and also sheds light on the key parameters that influence PBR performance, including irradiance, mixing, pH, temperature, mass transfer, designing models and capital and operating costs. Besides, a discussion on wastewater microalgae related to the cultivation condition and microalgal growth, wastewater species types and properties is presented.

6.2 MICROALGAE-BASED WATER TREATMENT

6.2.1 Microalgae in biological aspects

Microalgae are well known as one of the oldest life forms on earth. They are thallophytes (primitive plants) with no defined roots, stems and leaves. There is no sterile covering around the reproductive cells, and solely rely on chlorophyll-a as a primary photosynthetic pigment. Microalgae structures mainly conduct energy conversion without any development beyond their cells. This simple development allows them to adapt to any prevailing environmental conditions and survive for a long period of time. There are two types of microalgae according to its cell form that is prokaryotic cells and eukaryotic cells. Prokaryotic cells are notified as cyanobacteria which lack membrane bound organelles (plastids, mitochondria, nuclei, Golgi bodies and flagella). Cyanobacteria are more similar to bacteria than microalgae. Eukaryotic cells comprise of numerous different types of common microalgae. There have organelles that control cell function thus allowing it to survive and reproduce. Eukaryotes are classified into various classes defined by their pigmentation, life cycle and basic cellular structure. The most important classes are *Chlorophyta* (green microalgae), *Rhodophyta* (red microalgae) and *Bacillariophyta* (diatoms). Microalgae can be autotrophic or heterotrophic. Autotrophic means they only require inorganic compounds such as CO_2, salts and light energy source for growth. While heterotrophic are non-photosynthetic as they need an external supply of organic compounds as well as nutrients as energy source. However, some photosynthetic microalgae are mixotrophic (have capability to both perform photosynthesis and acquire exogenous organic nutrients). Autotrophic microalgae rely on photosynthesis as a key component for their survival. Photosynthesis allows them to convert solar radiation and absorbed CO_2 (via chloroplast) into adenosine triphosphate (ATP) and oxygen (O_2). Usable energy (forms like ATP) at the cellular level is utilized in respiration to produce energy for their growth (Chuka-Ogwude *et al.*, 2020; Udaypal Goswami *et al.*, 2024).

6.2.2 Cultivation condition and microalgal growth

Naturally, phototrophic microalgae can be grown under exposure to sunlight while assimilating carbon dioxide (CO_2) from the air and absorb nutrients from aquatic habitat. Therefore, artificial cultivation must closely attempt to replicate and enhance the optimum natural growth conditions. For commercial purposes, the use of sunlight as a free natural light resource is advantageous. But this natural light is limited by availability of sunlight in term of diurnal cycles and seasonal variations thus hinder the viability of commercial production to areas with high solar radiation (such as tropical areas). This problem was found in outdoor microalgae cultivation (Chrismadha *et al.*, 2022). To cope with this problem the artificial phototrophic cultivation was made by employing fluorescent lamps or light-emitting diodes (LED) which are utilized exclusively at a pilot scale. Artificial lighting allows continuous production, but significantly increases the energy input. It is also common that electricity for artificial lighting is generated from fossil fuels, thus negating the aim to develop a price-competitive fuel and decrease the carbon footprint of the system. The absorption spectra of major microalgae accessory pigment must be considered when determining the type of artificial light source, as it depends on the microalgae strain. These diatoms commonly have photosynthetic pigments consisting of chlorophylls a and c and fucoxanthin, while Chlorophyta contains chlorophylls a and b and zeaxanthin (Choi *et al.*, 2019; Mohapatra *et al.*, 2022).

There are three sources of carbon dioxide (CO_2) to be fixed by microalgae that is from atmosphere, industrial flue gases and soluble carbonates (Na_2CO_3 and $NaHCO_3$). Under natural growth, microalgae assimilate CO_2 from atmosphere ranging from 441–150 000 ppm. Other inorganic nutrients for microalgal growth are nitrogen (N), phosphorus (P) and silicon (Si). Some microalgae species are capable of fixing N from the air in the form of NO*x*, mostly they require it in a soluble form such as urea being the best source. During the algal growth cycle, only a small quantity of P is needed, however, this element must be supplied in excess, since PO_4^{3-} ions tend to bind with metal ions so that not all added P is bioavailable. Meanwhile, Si is required for optimum growth of certain microalgal groups such as diatoms (Choi *et al.*, 2019; Wang *et al.*, 2012).

The growth dynamics and composition of microalgae are affected by the mode growth.. Based on energy and carbon supply, there are four known major mode growth namely photoautotrophic, heterotrophic, photoheterotrophic and mixotrophic. Table 6.1 shows a summary of microalgal growth modes related to energy sources, carbon source, light requirement and metabolism variability. Mostly, microalgae are grown in a photoautotrophic mode because they are commonly photosynthetic organisms. In this mode metabolism, chemical energy is produced by assimilating CO_2 using sunlight. The overall reaction for microalgal photosynthetic growth is shown in equation (6.1).

$$nCO_2 + nH_2O \rightarrow (CH_2O)_n + nO_2 \tag{6.1}$$

Photoautotrophic cultivation plays an important role in mitigating global CO_2 emission, since it only uses CO_2 and not generates CO_2 under heterotrophic and photoheterotrophic growth conditions. Nonetheless, mixotrophic growth has possibilities for global CO_2 mitigation as it has characteristics of simultaneous utilization of carbon from both inorganic and organic sources. Heterotrophic growth of microalgae commonly uses organic carbon which exists in wastewater and independent of solar energy and facilitates scale-up. The design constraints around ensuring uniform illumination of the culture are removed which means that reactor designs do not require high surface-to-volume ratio. Compared to the photoautotrophic mode there are several advantages over heterotrophic mode, which are shown in Table 6.2. Microalgae produced via heterotrophic mode have high lipid content. The need of organic carbon substrate makes the heterotrophic microalgae to use higher energy than photoautotrophs. The input energy to ATP ratio for heterotrophs is 18% which is higher than 10% which is needed by photoautotrophs,

Table 6.1 Growth mode in microalgae cultivation.

Growth Mode	Carbon Source	Energy Source	Light Requirement	Variability in Metabolism
Photoautotrophic	Inorganic	Light	√	No alters between sources
Heterotrophic	Organic	Organic	X	Alter between sources
Photoheterotrophic	Organic	Light	√	Alter between sources
Mixotrophic	Inorganic and organic	Light and organic	X but optional	Simultaneous utilization

Modified from Perez-Garcia and Bashan (2015).

Table 6.2 Advantages and disadvantages of heterotrophic over photoautotrophic.

Growth Mode	Advantages	Disadvantages
Heterotrophic	• Higher biomass productivity • Higher growth rate • Higher lipid productivity • Low cost, simple reactor design and scalability • Ability to control certain metabolic pathways to influence biomass composition by changing culture's medium substrate • Potential being used in wastewater treatment	• Limited amount of microalgae species with high lipid contents • Utilization of glucose as substrate that can compete food and fuel • Interference of contaminant since the use of organic substrate

Modified from Perez-Garcia and Bashan (2015); Razzak *et al.* (2017); Zittelli *et al.* (2013).

thus in the economical point of view heterotrophs are more advantageous than photoautotrophs. Even so, only few microalgae are facultative heterotrophs (Chuka-Ogwude *et al.*, 2020).

There are certain microalgae that can be grown in the dark and use organic carbon from sugar, other microalgae can combine both autotrophic and heterotrophic modes (called as mixotrophic growth mode). In mixotrophic mode, microalgae utilize both organic and inorganic carbon sources for their growth. Some Chlorella species have this capability such as *Chlorella vulgaris, Chlorella protothecoides, Chlorella sorokiniana* and *Chlorella zofingiensis* (Patel *et al.*, 2022; Perez-Garcia & Bashan, 2015). However, there are some disadvantages of mixotrophic growth mode which are the requirement of O_2, light, CO_2 and organic carbon at the same time. Energy conversion efficiency is lower than heterotrophs and photoautotrophs modes. Fortunately, in mixotrophs, CO_2 is released in anaerobic respiration that could be trapped and used for photosynthesis process (Ummalyma *et al.*, 2021).

6.2.3 Waste water species types and properties

Numerous microalgal species that are reported have been used in wastewater treatment, which involved pure microalgal cultures or the natural mixed microalgae cultures–bacteria consortia. The wastewater characteristics, environmental and operating conditions determine the types of microalgal–bacterial consortia. The symbiosis favors low energy consumption and carbon footprint because the generated CO_2 during heterotrophic organic matter degradation is photosynthetically fixed, concomitantly yielded O_2 that is released (Udaypal Goswami *et al.*, 2024). The relative importance of key processes (O_2 production, CO_2 production, nutrients fixation, etc.) depends on the composition of the consortium. In this case, microorganisms such as nitrifying bacteria compete with microalgae for nutrients, affecting the N recovery of the process. Mainly, nitrifying bacteria can reduce microalgae growth by depleting the ammonium concentration in media, hence limiting N uptake, thus reduce N recovery. Moreover, nitrifying bacteria then oxidize the NH_4^+ into NO_2^- which has a negative effect on microalgae rather than being a N source for their growth. Commonly green microalgae genera (*Scenedesmus* sp., *Chlorella* sp., *Monoraphidium* sp.) has been adapted to wastewater medium (Bae & Kim, 2021; Maryjoseph & Ketheesan, 2020). Other reports also mentioned cyanobacteria instead of green microalgae since cyanobacteria are capable of producing metabolites such as phycobiliproteins and polyhydroxybutyrates. Cyanobacteria have softer cell walls which can therefore be easily digested, but nutrient uptake efficiency is lower than green microalgae (Lee *et al.*, 2022; Rehmanji *et al.*, 2022).

There are several factors that affect the MBWT system, some of them are related to environmental conditions (outdoor/indoor, light, temperature, etc.), wastewater characteristics (C:N:P ratio, nutrient levels, toxicity, etc.). It is also associated with operating conditions (batch/continuous, CO_2 addition, hydraulic and solid retention time, mixing, pH, etc.) and reactor configuration (light path, open systems, PBR, attached-growth, etc.). The main factor that affects nutrient recovery efficiency and biomass productivity is the average irradiance of the cells, which is a function of light availability (depth). Shallow depths are used to increase performance yields and greater depths to increase treatment capacity. It means a difficult trade off must be considered to optimize reactor design. Other factors that must be carefully considered are light intensity, light frequency and photoperiods (Maryjoseph & Ketheesan, 2020). Wastewater treatment in conjunction with biofuel production is considered the most plausible commercial application in the short term. This argument arises from the fact that MBWT provide a pathway for the removal of chemical and organic contaminants, heavy metals and pathogens from wastewater while producing biomass for biofuel production. The main drivers for biomass production from wastewater are to save requirements for chemicals for remediation and to minimize freshwater (Venkata Subhash *et al.*, 2022).

Pollutants removal mechanism in MBWT system substantially is biosorption, bioaccumulation, extracellular and intracellular processes. Microalgae are proven to effectively eliminate heavy metals, pesticides, antibiotics, polycyclic aromatic hydrocarbon (PAH) through their cell membrane,

Table 6.3 Microalgal heavy metal bioremediation and possible mechanism in MBWT system.

Pollutants	Microalgae Species	Initial Concentration	Possible Mechanism	References
Arsenate (V)	*Microcystis aeruginosa*	50 mg/L	Active sites binding on extracellular wall	Huang *et al.* (2014)
Arsenite (III)	*Chlamydomonas reinhardtii, Dunaliella salina*	0–20 μg/L	Polysaccharides binding, arsenic transporters	Jiang *et al.* (2022)
Chromium (VI)	*Anabaena, Oscillatoria, Phormidium, Spirogyra*	120–177 mg/L	Uronic groups binding on exopolysaccharides compound	Balaji *et al.* (2016)
Chromium (VI)	*Phaeodactylum tricornutum, Navicula pelliculosa*	1 mg/L	Extracellular polymeric substances binding	Hedayatkhah *et al.* (2018)
Mercury (Hg)	*Chlorella sorokiniana*	5 mg/L	Polypeptides binding (thiol-rich peptides)	Gómez-Jacinto *et al.* (2015)
Cadmium (II)	*Heterochlorella* sp. *MAS3 Desmodesmus* sp. *MAS1*	2 mg/L	Triglycerides binding and long-chain hydrocarbon binding in pH∼3	Abinandan *et al.* (2019)
Cadmium (II)	*Scenedesmus obliquus AS-6-1*	50 mg/L	Amide, C–N, hydroxyl and S–O binding	Zhang *et al.* (2016)
Cadmium (II)	*Aphanothece* sp. *A0, Aphanothece* sp. *A8* and *Aphanothece* sp. *A15*	1–7 mg/L	Amide, C–N, hydroxyl, carboxyl and S–S bindings	Satya *et al.* (2020)
Nickel (II)	*Chlorella sorokiniana*	20 mg/L	Hydroxyl group binding	Akhtar *et al.* (2004)

Table 6.4 Microalgal pesticides bioremediation and possible mechanism in MBWT system.

Pollutants	Microalgae Species	Initial Concentration	Possible Mechanism	References
Fluoroxypyr	*Chlamydomonas reinhardtii*	0.5 mg/L	Enhancing superoxide dismutase, catalase ascorbate peroxidase	Zhang *et al.* (2011)
Triadimefon	*Scenedesmus obliquus*	12.5 μg/L	Biodegradation favor S-enantiomer	Xu and Huang (2017)
Malathion	*Nostoc muscorum*	20 mg/L	Malathion transporter stimulation	Ibrahim *et al.* (2014)
Atrazine, Molinate, Simazine, Isoproturon, Propanil, Carbofuran, Dimethoate, Pendimethalin, Metoalcholar, Pyriprox	*Chlorella vulgaris*	2 μg/L	Biosorption	Hussein *et al.* (2017)
Chlorpyrifos	*Chlorella pyrenoidosa Merismopedia* sp.	7.63–19.64 mg/L	Enhancement of SOD and CAT activities	Chen *et al.* (2016)
Diazinon	*Chlorella vulgaris*	40 mg/L	Enhancement of SOD and CAT activities, transformation induced by phosphor hydrolytic enzymes	Kurade *et al.* (2016)

Table 6.5 Microalgal antibiotics bioremediation and possible mechanism in MBWT system.

Pollutants	Microalgae Species	Initial Concentration	Possible Mechanism	References
Carbamazepine, sulfamethazine and tramadol	*Chaetoceros muelleri Chlorella*	<40 mg/mL	Biodegradation and bioaccumulation 46	Mojiri *et al.* (2021)
Sulfamethazine	*Chlorella pyrenoidosa*	8 mg/mL	Increasing antioxidant activity ROS-scavenging enzymes SOD, CAT, and APX	Sun *et al.* (2017)
Sulfamethazine	*Spirulina platensis*	6 mg/mL	Degradation	Xu *et al.* (2018)
Levofloxacin	*Scenedesmus obliquus*	1 mg/mL	Demethylation, Dehydroxylation, side chain breakdown and ring cleavage, Decarboxylation	Xiong *et al.* (2017)
Levofloxacin	*Microcystis flos-aquae*	<10 μg/mL	Increasing antioxidant activity including superoxide dismutase and catalase and malondialdehyde activity	Wan *et al.* (2014)
Naproxen	*Cymbella* sp.	102.76 mg/L	Transformation due to hydroxylation, decarboxylation, demethylation, tyrosine conjunction and glucuronidation	Ding *et al.* (2017)
Ibuprofen	*Navicula* sp.	50 mg/L	Transformation due to hydroxylation, acylation, demethylation and glucuronidation	Ding *et al.* (2017)
Diclofenac	*Picocystis* sp. and *Graesiella* sp.	200 mg/L	Transformation due to hydroxylation, dechlorination	Ben Ouada *et al.* (2019)
Azithromycin	*Haematococcus pluvialis*	20 μg/L	Transformation by ring cleavage, demethylation and oxidation	Kiki *et al.* (2020)
Ciprofloxacin	*Chlamydomonas* sp. *Tai-03*	10 mg/L	Biodegradation and photolysis	Xie *et al.* (2020)
Enrofloxacin	*Chlorella vulgaris*	5 mg/L	Photolysis and Biodegradation by dealkylation, decarboxylation and defluorination	Chen *et al.* (2020)

extracellular and intracellular mechanisms. Though several species of the algae were discovered for their potential in wastewater treatment, *Chlamydomonas, Chlorella* and *Scenedesmus* are used extensively for wastewater treatment. The 65.05% biodegradation of ciprofloxacin was investigated with *Chlamydomonas* species Tai-03 (Xie *et al.*, 2020). Similarly, antibiotics degradation was observed with the *Haematococcus pluvialis*, *Selenastrum capricornutum*, *Chlorella vulgaris* and *Scenedesmus quadricauda* (Kiki *et al.*, 2020). The algal strains are capable of utilizing wastewater pollutants as their nutrition and producing biomass. Another way for algal biomass production is co-metabolism, in which the degradation of two compounds occurs simultaneously (Xiong *et al.*, 2021). Algae degrade toxic chemicals via their metabolic pathways. The degradation of some of the organic pollutants was investigated using microalgae. It has been proven that microalgae can degrade pesticides, dyes and antibiotics via their metabolic pathways (Tables 6.3–6.5). Biodegradation and biotransformation mechanisms mainly proceed in the removal of organic pollutants. While inorganic pollutants remediated through membrane transporter, biosorption and bioaccumulation mechanism. Figure 6.1 describes the schematic possible bioremediation route in MBWT.

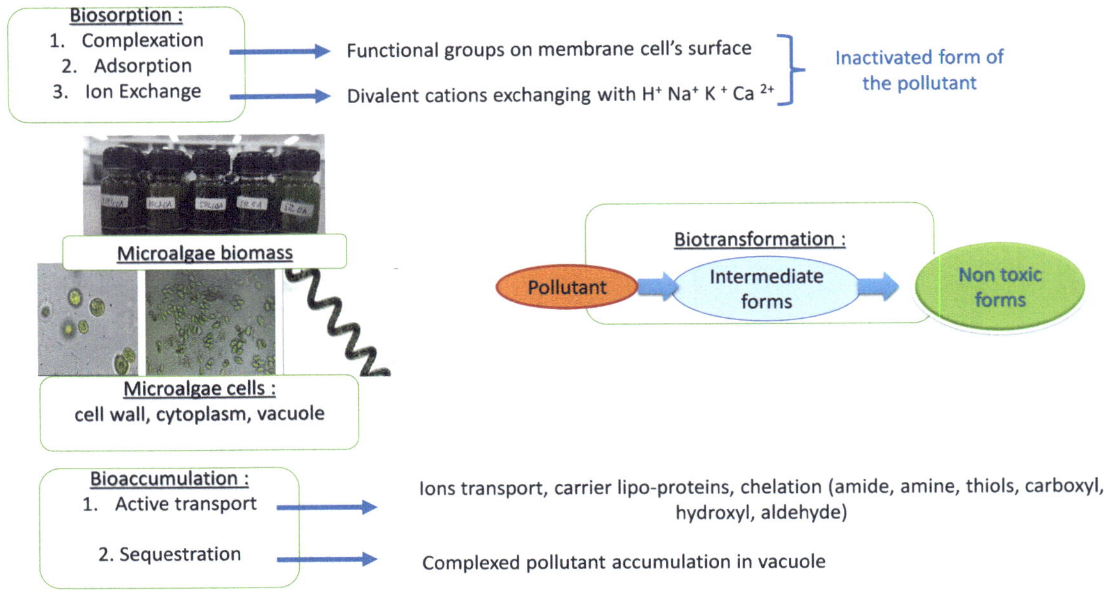

Figure 6.1 Schematic of detoxification process in MBWT system.

6.3 CULTIVATION SYSTEM

Microalgae cultivation system varies depending on the microalgae strains (type), investment cost, targeted products and nutrient source types. In general, this cultivation system belongs to two major types namely open cultivation system and the closed system.

6.3.1 Open cultivation system

Microalgae can be grown in open systems. Different open cultivation systems can be distinguished by managing growth parameters, pollution, water evaporation, productivity, downstream processing features, investment and operational costs. Open ponds have been used for large-scale outdoor microalgae culture due to their low cost and ease of construction/operation. Raceways, circular and sloped ponds are the three types of open systems, based on their size, shape, agitation type and inclination. In general, an open pond system can be divided into two types: natural pond systems (natural lakes, ponds, lagoons) and artificial ponds or containers (Chew *et al.*, 2018). Open raceway pond systems are meant to closely imitate the microalgae's native environment. Although the form of an open pond system can vary geometrically, the most common working water depths are 15–20 cm. The system is typically equipped with paddle wheels at the bottom to ensure regular distribution and recirculation of nutrients to microalgae for optimal growth, for example Figure 6.2a while Figure 6.2b is the simplest form called as an ordinary fiber pond. One type of pond system that is often used in microalgae culture is the raceway pond system, which is typically made of concrete.

Dilution of complex substrates resulted in significant microalgae output (2–7 g/(m^2.day)), with 12% nitrogen and ~100% phosphorus removal rates. In another investigation, alcohol wastewater was used for microalgae production in an outdoor system. Under optimal conditions, 405.18 ± 36.47 mg COD_{Cr}/(L.day) of nutrients were successfully removed from wastewater. Alcoholic wastewater had a higher biomass yield (35.29%), lipid content (102.68%) and productivity (227.91%) than starch

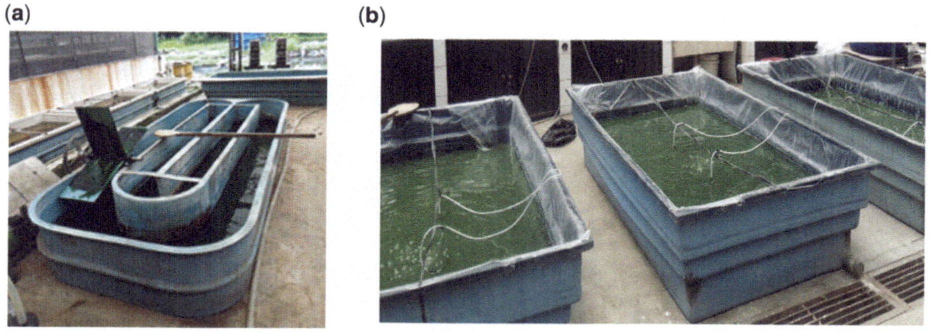

Figure 6.2 Open system for cultivating microalgae: (a) paddle wheel and fiber pond. Those are located on RC for Limnology and Water Resources-BRIN, KST Soekarno-Cibinong West of Java-Indonesia.

wastewater, while bacterial contamination increased by 4.62-fold. Open cultivation approaches are susceptible to cross-contamination by yeast and protozoa; monoculture requires a highly selective environment. Extreme habitats, such as greater salt levels (in case of *Dunaliella salina*), high nutrients (*Chlorella*) and extremely basic conditions (*Spirulina*), can be used to sustain monoculture production of desirable strains. One of the most significant issues is bacterial and other biological contamination during long-term production. Although open growing systems are cost-effective, some of its downsides include poorer yields, higher harvesting costs, high stirring energy requirements (in circular ponds), insufficient mixing, evaporation losses, uncontrolled temperature variations and CO_2 deficit. These characteristics account for the 37 tones/ha lower biomass yield of open systems compared to enclosed cultivation methods. The major advantage of an open pond system is that it is easier to build and operate than a closed system. However, it has the following limitations: (1) it requires more area, (2) climatic conditions that prevent cells from utilizing optimal illumination for their growth, (3) low biomass productivity due to poor mass transfer rate due to inefficient stirring mechanism in the system, (4) CO_2 diffusion to the atmosphere, resulting in less carbon fixation or sequestration and (5) evaporation losses (Chuka-Ogwude *et al.*, 2020; Rempel *et al.*, 2021; Zittelli *et al.*, 2013).

6.3.2 Closed PBR

The enclosed systems are appropriate for maintaining optimal growth conditions such as temperature, pH and CO_2 concentration for microalgae development. Microalgae growth has also been optimized using hybrid systems (Satya *et al.*, 2024). In these systems, microalgae are grown in enclosed PBRs to achieve the necessary cell density to restrict the growth of undesirable strains and then transferred to an open system to cultivate selected microalgae under controlled nutrition. The variation in biomass yield between microalgae species, such as *Spirulina* and *Chlorella*, could be attributed to distinct features. *Chlorella* is highly ammonia tolerant, but *Spirulina* grows rapidly. Furthermore, the pretreatment of wastewater with *Chlorella* in the first step makes the circumstances more appropriate for the growth of *Spirulina* and hence results in a higher yield. Closed culture systems have been developed, leading to the invention of flat plate, tubular, vertical column and internally illuminated PBRs. Later developments included the use of fluorescence lamps in reactors, thick transparent materials for outside panels and vertical alveolar panels for flat plate reactors. Flat plate PBRs are typically built of transparent materials that promote the maximum utilization of solar light energy. Because of its tendency to allow the maximum light route, flat plate PBRs can achieve high photosynthetic efficiency (Chew *et al.*, 2018; Maryjoseph & Ketheesan, 2020).

Most closed PBRs will contain tubes of various diameters, shapes and lengths that are made of either glass or plastic and are commonly referred to as tubular closed bioreactors. Furthermore, it may be flat, nearly horizontal, vertical, conical, or inclined. Furthermore, it gives a bigger illumination surface area and is appropriate for outdoor cultures. It also allows for plentiful biomass generation, however, dissolved oxygen accumulation is rather significant, affecting mass transfer rate. Vertical closed PBRs are inexpensive, compact and simple to operate, making them ideal for large-scale algae cultivation. Internally illuminated PBRs are a type of closed culture system that uses fluorescent lamps. In this system, CO_2 or air is fed by spargers or impellers, which are often located at the bottom of the reactor to improve algal culture agitation. It has been found that bubble column or airlift PBRs achieve higher biomass productivity than narrow tubular PBRs. This system can be customized to use both solar and artificial illumination, with the artificial lighting source turning on whenever the irradiation falls below a predetermined level. PBRs need cooling and heating systems to control the cultivation temperature. However, it demands technical effort to operate and the supply rate of illumination must be maintained uniformly by incorporating solar and artificial lighting systems (Udaypal Goswami *et al.*, 2024; Zittelli *et al.*, 2013).

Schematically, those PBR shapes are shown in Figure 6.3. PBRs appear in different configurations: vertical column reactors (bubble column (Fig. 6.3a)) or air lift (Fig. 6.3b)), flat plate reactors (Fig. 6.3c) and tubular reactors (Fig. 6.3d). The air lift reactors have great potential for industrial processes, since it has low level and homogenous distribution of hydrodynamic shear, which is a disadvantage of closed PBRs over open ponds. The medium circulates in a cyclic pattern through channels built for this purpose. The tubular design is more appropriate to the outdoor culture. It has a large illumination surface created by the disposition of the tubes. It could be configured in vertical, horizontal or inclined planes. The vertical tubular PBRs increase the contact time between gaseous and liquid phases, increases the CO_2 mass transfer. However, this disposition has a drawback on air pump costs. On the other hand, the flat-plate PBR can achieve higher cell densities than other PBRs. Further, this PBR has lower power consumption, high mass transfer capacity, no dark volumes and high photosynthetic efficiency (Mohapatra *et al.*, 2022; Razzak *et al.*, 2017). Figure 6.4 present an example of PBRs in bench-top laboratory scale (4a) and semi-pilot scale of out-door PBR (4b) for cultivating *Spirulina platensis*, respectively, in working volume of 2 and 250 L.

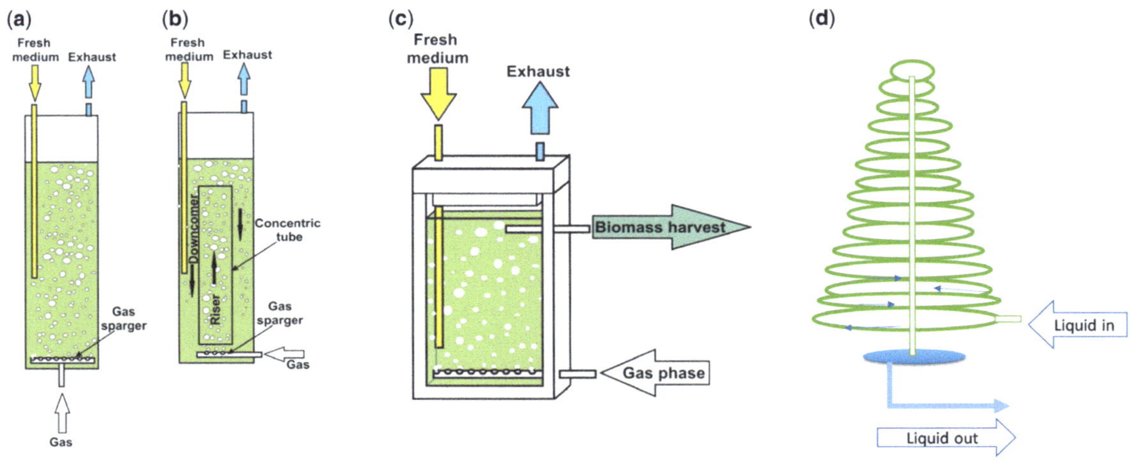

Figure 6.3 Schematic diagram of PBR: (a) bubble column, (b) internal loop airlift, (c) plate PBR, (d) coiled tubular tree.

Figure 6.4 Closed PBRs for cultivating *Spirulina platensis*: indoor-lab scale (capacity of 2 L) (a) and semi-outdoor lab scale (capacity of 250 L) (b). These PBRs located on RC for Limnology and Water Resources-BRIN, KST Soekarno-Cibinong West of Java-Indonesia.

6.4 KEY PARAMETERS OF PBR PERFORMANCE

Several environmental parameters, including irradiance, mixing, pH, temperature and mass transfer, must be modified before proceeding with microalgae cultivation. These critical characteristics determine not only the growth of microalgae, but also the biochemical content of their biomass (Mohapatra *et al.*, 2022; Razzak *et al.*, 2017). After choosing the culture mode, selecting the PBR design and preparing the cultivation media, the environment parameter must be adjusted before inoculating microalgae.

6.4.1 Irradiance

Microalgae include a variety of pigments, including chlorophyll-a (present in all microalgae), carotenoids, phycoerythrin and phycocyanin. Photosynthetic microalgae use pigments to transform light energy into chemical energy. With the exception of heterotrophic cultivation, all cultivation modes (photoautotrophic, mixotrophic and photoheterotrophic) require light supply as the predominant energy source for microalgal growth. Natural sunlight, fluorescent lamps and LEDs are commonly used as lighting systems for microalgae cultures. Ordinary fluorescent lights, which are commonly used in microalgae research, emit both growth-efficient and growth-inefficient wavelengths. LEDs have a narrower spectral range that is more compatible with the absorption bands of microalgae pigments. Regardless of whether fluorescent lights or LEDs are used to stimulate microalgae cultures, optimizing the lighting conditions is one of the most important components in achieving the highest microalgae growth rate. Light intensity, wavelength and photoperiod (lighting time) are three key light characteristics that can have a significant impact on microalgae growth in photosynthetic cultures. Light intensity refers to the amount of light received on a surface every second which is expressed in unit of $\mu mol/m^2.s$) (Perez-Garcia & Bashan, 2015; Udaypal Goswami *et al.*, 2024).

The link between light intensity and photosynthetic rate is demonstrated by photosynthetic light–response curve. This curve has three phases, namely light-limitation, light-saturation and photoinhibition phase. Light restriction and photoinhibition are the two phases, capable of lowering or even halting the microalgal growth. Light limitation occurs in cultures with insufficient light intensity or high cell density. This could emerge due to the self-shading effect in reactors with a high biomass concentration that limits the amount of light passing through the bioreactor and negatively impacts photon absorption and photosynthetic efficiency. In the light-limiting phase, increase in the light intensity stimulates microalgal growth up to the area of light saturation. Further rise in light intensity in this phase does not alter microalgal photosynthesis, but photoinhibition occurs at light levels higher than saturation point. Intensive irradiation during the photoinhibition phase damages photosystem II, drastically reducing microalgal development, or causing the culture to collapse. The estimated light intensities during light limitation, light saturation and photoinhibition phases are up to 300 μmol/(m^2. s), 300–1600 μmol/(m^2. s) and >1600 μmol/(m^2. s). The light spectrum of solar radiation contains a wide range of wavelengths of energy, the majority of which microalgae cannot utilize. Microalgae can employ visible wavelengths ranging from 400 to 700 nm for photosynthesis. This narrow spectrum is known as the photosynthetically active radiation (PAR) range and it includes 400–500 nm and 600–700 nm wavelengths (blue and red, respectively), which are the appropriate ranges for optimal microalgal photosynthesis and 500–600 nm and 700–800 nm (green-yellow and far-red, respectively) transmitted or reflected wavelengths (Mohapatra *et al.*, 2022; Patel *et al.*, 2022; Venkata Subhash *et al.*, 2022).

Each photosynthetic pigment contains main absorption bands capable of absorbing certain wavelengths of PAR. For example, chlorophyll a, chlorophyll b and carotenoids have large absorption bands at 450–475 nm (blue of blue-green), 630–675 nm (red) and 500–600 nm, respectively. Microalgae are exposed to light in accordance with the natural day–night cycle when grown outdoors. Under regulated laboratory circumstances, the length of lighting or photoperiod (light:dark cycle) which can range from 0 (for heterotrophic culture) to 24 h. Microalgal growth was assessed using various photoperiod cycles (e.g., 12:12, 14:10, 16:8 and 24:0 h). It should be mentioned that 24 h of lighting is not necessary for the continuous growth of all microalgae species and the optimal photoperiod varies on the light intensity and the microalgal strains (Ummalyma *et al.*, 2021; Zittelli *et al.*, 2013).

6.4.2 Mixing

Physical processes such as interphase, interparticle and intraparticle mass transfers occurring within a multiphase reactor depend significantly upon the mixing characteristics of the various phases involved. The mixing process is conventionally divided in macromixing (which gives information about the retention times of elementary volumes) and micromixing (describing the communication between elementary volumes). Macromixing occurs at the macroscopic scale, that is, on the scale of the vessel and corresponds to the large-scale flow processes causing distributions of the fluid elements represented by residence time distribution (RTD) or that of the local concentrations. Micromixing refers to the phenomena at the cell and molecular scales. Overall macromixing in the reactors is generally characterized by mixing time (t_m), axial dispersion coefficient (D_{az}) or the Peclet number (Pe) as given in equation (2)

$$\mathrm{Pe} = u_{\mathrm{Le}} L / D_{\mathrm{az}} \quad (6.2)$$

Mixing time is usually defined as the time required for concentration of a tracer to reach 99% of its equilibrium value from injection. The axial dispersion coefficient, D_{az}, can be calculated from empirical correlations or estimated from experimental RTD data. The RTD curves permit quantitative evaluation of the nature and degree of macromixing as well as the dynamic holdup of each fluid phase in the reactor. RTD is normally measured by the so-called 'stimulus-response' techniques that involve injecting a tracer in the inlet stream or at some point within a reactor and monitoring tracer concentration/signal in the exit stream or at some downstream point within the reactor (Maryjoseph & Ketheesan, 2020; Zittelli *et al.*, 2013).

Axial and radial mixing of the liquid phase in bubble columns is characterized by using dispersion coefficients that are analogous to the diffusion coefficient of Fick's law of diffusion. Unlike diffusion, dispersion arises from convective motion of fluid caused by the following main factors: relative movement of the gas and liquid phases; bubble coalescence and break up; the carry forward of fluid in wakes behind the rising gas bubbles and the consequent return flow generated for maintaining mass balance; and turbulency generated by any superimposed flow of liquid. In the past, radial dispersion in bubble columns has generally been lumped with the axial dispersion coefficient and the latter has been used widely and almost exclusively as an index of mixing in such reactors. When bubble columns are used as PBRs, the need arises for quantifying the radial mixing. This is because photosynthesizing microalgae and cyanobacteria suspended in column PBRs cause a radial decline in irradiance from a high value near the externally illuminated transparent wall to a low value in the centre of the PBR. Consequently, the volume of the PBR can be demarcated into a dark interior core region and a relatively better illuminated peripheral region. The frequency of radial motion of the cell-suspending fluid between the light and dark zones, or the 'flashing-light effect', influences the productivity of PBRs. Therefore, ways should be found for quantifying the extent of radial mixing and its dependence on the aeration rate, the main operational variable in a bubble column PBR. Optimization of the mixing time parameter (t_m) allows to guarantee that the photosynthetic cells will transit from a zone of maximum irradiation (close to the light source) to one of low illumination and this will generate short light/darkness cycles or micro photoperiods, which will directly affect biomass productivity and the type and proportion of metabolites. In general, the light path in a PBR is sought to maintain low values and this results in large height/light path and A/V ratios (ratio between illuminated area and working volume), with the aim of ensuring adequate light supply to microorganisms (Maryjoseph & Ketheesan, 2020; Wang *et al.*, 2012).

6.4.3 pH

The pH of cultivation medium is another factor that significantly influences microalgal metabolism and growth. The pH value controls the acid–base balance in the cultivation medium and affects the solubility and availability of different forms of inorganic carbon (CO_2, bicarbonate and carbonate) and nutrients (phosphates and ammonium/ammonia) as well as their liquid–gas transfer phenomena. For example, a high pH (>9.75) is favorable for ammonia volatilization in which ammonium (NH_4^+) is converted into ammonia (NH_3) gas. In addition, changing the pH of the medium affects the physiology and morphology of microalgae by activating the permeability of the membrane cell for certain ions and consequently affecting microalgal growth and biochemical composition. Moreover, the pH of the cultivation medium is considered as a tool for controlling biological contamination in wastewater. This is because a cultivation medium with pH higher than 9 inhibits the growth of indigenous bacteria present in wastewater. Also, during cultivation of microalgae, for example, the green alga, *Haematococcus pluvialis*, an acidic pH of 4 is recommended to avoid lethal fungal contamination of the culture. Based on microalgal growth performance, the pH of cultivation medium can be classified as fatal, tolerable, or optimal (Borowiak *et al.*, 2021; Chew *et al.*, 2018).

Extremely low (acidic) and high (basic) pH values are fatal for most microalgae species. Microalgae were reported to grow in the pH range from 5 to 8, while based on growth parameters, the optimum pH was found to be between 7.5 and 8. In another study, pH, in the range from 5 to 10, was used on the growth and lipid accumulation of microalgae, *Nannochloropsis salina*. Highest growth rates of 95.6×10^6 and 92.8×10^6 cells/mL were reported at optimum pH of 8 and 9, respectively. Therefore, most microalgal species grow well in cultivation media with pH ranging between 7 and 9, but such narrow range of pH cannot be applied for cultivation of all microalgae species under controlled conditions. The optimum, acceptable and lethal ranges of pH depend on microalgal species and cultivation conditions. It has been observed that some microalgal species can tolerate extraordinary acidic or basic pH. For instance, *Dunaliella salina* grows well in a pH close to 11.5, while the optimal pH of *Dunaliella acidophila* is between 0.0 and 3.0. Hence, the optimum pH range for cultivation of the same commercial and well-known microalgal species can thus be found from previous studies. For newly isolated microalgae strains which have less available information, the cultivation conditions need to be evaluated and optimized under controlled laboratory conditions (Patel *et al.*, 2022; Rattanapoltee *et al.*, 2021).

6.4.4 Temperature

Temperature is also an important factor for microalgal production. Temperature has a direct effect on metabolism, nutritional absorption, CO_2 bio fixation, photosynthesis and growth rate. Temperature effects not just growth but also the physiology and biochemical composition of microalgae, including the quality and amount of microalgal lipids. Temperature was discovered to have a substantial effect on the amount of carbohydrates and saturated fatty acids in microalgal biomass. For optimal microalgae growth, the temperature of the growing medium must be optimized. Microalgae's adaptation and responsiveness to varied temperatures are intimately tied to their species' origin. It should be noted that the majority of microalgal species cannot withstand excessively cold or hot temperatures. Most commercial and isolated microalgal species are cultivated at temperatures ranging from 20 to 28°C. Although microalgal strains can grow in a wide variety of temperatures, each species' greatest growth rate occurs at the optimal temperature. Temperatures above and below the optimum can have a negative impact on microalgae development and biomass output. It is also vital to understand that microalgae can handle lower temperatures better than higher temperatures. Microalgae may thrive up to 15°C below their optimal temperature, but even a few degrees more can cause cell death. Microalgae grow best at temperatures ranging from 18 to 30°C, 15 to 26°C and 20 to 30°C (Barten *et al.*, 2021; Raeesossadati *et al.*, 2014). Different optimum temperatures have been recorded for microalgae cultivation under various situations, including indoor/outdoor cultivation, open/closed systems, daytime and light intensity. Table 6.6 shows that most microalgae species may be successfully

Table 6.6 Some affecting key parameter on the microalgae cultivation performance.

Microalgae Species	Light Intensity (μmol/(m².s)	Photo-period (L:D)	Light Source	pH	Tempe-rature °C	Maximum Microalgae Yield (g/L)	References
Nannochloropsis oceanica	200–636	24:0	LED	7.8	15–30	–	Sá *et al.* (2020)
Scenedesmus obliquus	100	12:12	–	7.2	25	0.897	Qu *et al.* (2020)
Chlorella sorokiniana	1000–3000	–	Sun light	5.0–6.5	23.1–30.8	0.700	Liu *et al.* (2020)
Botryococcus braunii	0–2000	14:10	White fluorescent	5.0–10	5.0–45	3.300	Yoshimura *et al.* (2013)
Chlorella sp.	300–900	12:12	LED	6.42	25	0.532	Yan *et al.* (2020)
Chlorella sorokiniana	175	14:10	–	7.0	25	11.5	Vasconcelos Fernandes *et al.* (2015)
Chlamydomonas reinhardtii	110.3	24:0	Cool white fluorescent	7.2	23	0.19	Patel *et al.* (2015)
Tisochrysis lutea	60	24:0	LED	8.7	25	1.5	Fret *et al.* (2020)
Arthrospira (Spirulina) platensis	610.96–1205.43	12:12	Sunlight	9.5	25–45	0.4–1.44	Chrismadha *et al.* (2022)

cultivated at 25°C. This suggests that many microalgal species can thrive at normal temperature. However, for temperature-sensitive species, the culture temperature must be modified and controlled.

6.4.5 Nutrients

Culture media are solutions providing key nutrients required for microalgae to maintain a stable state, good health and growth. Nutrients are divided into macronutrients, micronutrients and trace elements based on the amount required for healthy growth. The first group includes elements such as carbon (C), hydrogen (H), oxygen (O), nitrogen (N) and phosphorous (P), which microalgae require in higher concentrations (g/L) in their culture conditions. Lower quantities (mg/L or less) of micronutrients such as cobalt (Co), zinc (Zn), manganese (Mn) and barium (Ba) in cultivation media are sufficient for microalgal growth and biomass.

Formulated media and various wastewaters (typically richer in nitrogen and phosphorus compounds) are commonly utilized as culture media to provide nutrients for microalgal development. Formulated media are synthetic broths that include the required micro- and macronutrient contents. These media have been extensively tested for the cultivation of various freshwater and marine microalgal species (e.g., the f/2 medium developed by Guillard (1975)), institutes (the Culture Collection of Algae at The University of Texas, UTEX) and commercial groups (The Culture Collection of Algae and Protozoa, CCAP). Microalgae researchers refer to formulated media by abbreviations or commercial names such as BBM (Bold's Basal Medium), f/2 (Guillard) and BG11 (Blue-Green). Each synthetic medium includes detailed instructions that describe the step-by-step solution preparation, including the names of components (macro- and micronutrients) and their concentrations (mass concentration or molarity). Formulated media may be used as non-specific media, allowing for the culture of a wide range of microalgal species, or they may be created for a specific group of microalgae. Bold's Basal Medium (BBM) and Guillard (f/2) are two popular prepared media for cultivating a wide range of freshwater and marine microalgae, respectively. Blue-Green (BG11) is an excellent medium for the culture of cyanobacteria, although it is also widely used to cultivate microalgae. Zarrouk and f/2 + Silicon (Si) (Guillard's medium for diatoms) are specialist media for growing *Spirulina* sp. (cyanobacteria) and diatoms, respectively. Another expression widely used is 'modified media,' which refers to when the composition of a medium is slightly altered by increasing or reducing the original concentrations of compound(s) in culture media. Based on the experimental design, modified media are used to promote microalgal biomass production, stimulate the creation of specific metabolites, or explore the effects of nutrient deficit or deprivation (Feng *et al.*, 2011; Michael *et al.*, 2019; Sánchez *et al.*, 2000).

Nutrient-enriched wastewaters are another low-cost and readily available medium that can deliver macronutrients and micronutrients in addition to water for microalgal development. Unlike prepared medium, the quantities of nutrients in wastewater are unknown. When determining the suitability of wastewater as a microalgal culture medium, nutrient concentrations, specifically nitrogen (N), phosphorus (P) and carbon (C), must be measured. To prevent microalgal development limitations, examine the threshold concentrations of these nutrients in wastewater. The elemental ratio of C:N:P in microalgal biomass is roughly 106:16:1, often known as the Redfield ratio. However, while the Redfield ratio is used to determine the limitations of key nutrients in microalgal culture medium, it cannot be generalized as an optimum number for all microalgal species. Depending on the species, the N:P ratio in wastewater can range from 8 to 45. Various types of home, agricultural and industrial wastewaters have been used to cultivate a variety of freshwater and marine microalgal species. The properties of the wastewater utilized for microalgal culture are typically described in research articles. The key characteristics include physical and chemical qualities (e.g., pH, total suspended solid, color and electrical conductivity), as well as nutrient concentrations (e.g., NH_4^+, NO_2^-, NO_3^- and PO_4^{3-}). The use of wastewater as a medium for microalgal growing offers various advantages, including (i) low-cost production of microalgal biomass, (ii) nutrient recovery from wastewater and (iii) wastewater treatment for safe discharge. Wastewater as a culture medium has been used primarily for wastewater treatment, bioenergy production and CO_2 sequestration. It should be noted that microalgal biomass produced in wastewater cannot be used to produce food, cosmetics, medicine and other products for human consumption due to the risk of contamination with organic and inorganic pollutants and microbes, which could be harmful to human health. Although biomass produced in formulated medium can be used for a variety of purposes, including human consumption, mass production in synthetic media is more expensive than wastewater. The various C, N and P ratios in wastewater utilized in microalgae-based wastewater treatment (MBWT) experiments have been summarized: ranging from 100/0.6/0.6 to 100/193/12 for municipal wastewater (C:N:P ratios) and from 875:1 to 0.85:1 (C:N ratio) and 20.98:1 to 0.12:1 (N:P ratio) for other wastewater. In microalgae research, both designed medium and wastewater as culture media offer advantages and disadvantages (Feng *et al.*, 2011; Raoof *et al.*, 2006).

6.4.6 Mass transfer

It is well known that the cell growth of microalgae involves three competing cellular processes: photosynthesis, photorespiration and dark respiration. The photosynthesis of microalgae makes use of light energy and CO_2 to deliver oxygen (O_2) as the byproduct. However, at high dissolved oxygen (DO) concentrations, ribulose-1,5-bisphosphate carboxylase oxygenase (RuBisCO), which is the primary carboxylating enzyme that offers CO_2 for the Calvin cycle, can also consume O_2 to produce CO_2 for photorespiration. Furthermore, much of the DO that is released as the byproduct of photosynthesis is accumulated in the culture, which can lead to a high toxic level that can menace the survival of the microalgae. Adverse conditions of high temperature and photon flux density combined with CO_2 limitation can intensify the physiological inhibitory effects of DO. It was reported that an accumulation of DO could give rise to severe inhibition of cell growth, especially in closed systems such as tubular PBRs, where the accumulated DO is removed only at the degasser through mass transfer. Since DO can not only oxidize one or more enzymes but also impact the electron transmission chain and suppress the photosynthesis process, the DO content in the culture should be limited to a certain range during the cultivation of microalgae. DO concentration should never exceed above 400% of the air saturation value in order to prevent such inhibitory effects, while a strong decrease in cell productivity for *C. sorokiniana* occurred when the DO concentration exceeded only 200% of the air saturation value (i.e., 448 μmol/L). CO_2 is usually used as the carbon source in the cultivation of microalgae in photoautotrophic cultures; it cannot be utilized directly, as the microalgae can only make use of it from the liquid phase. The CO_2 supply can be a constraining factor if the dissolved CO_2 concentration in the liquid phase is too low (e.g., when air is used as the feeding gas or when the length of a tubular PBR is too long) or if the mixing in the PBR is insufficient. It has been recommended that the partial CO_2 pressure should be higher than 0.2 kPa (i.e., 0.076 mol/m^3, which is equivalent to 3.3 mg/L). Since the partial CO_2 pressure in the atmosphere is 0.04 kPa, pure air (i.e., with a CO_2 content of 0.035% by volume) is insufficient for the CO_2 supply and a CO_2-enriched gas mixture is required (Kao *et al.*, 2014; Klinthong *et al.*, 2015).

A high concentration of dissolved carbon dioxide (DCD) can lead to a low culture pH, which would also be inhibitory to some microalgal cells. Therefore, maintaining the DCD level in a reasonable range is critically important and the optimal DCD level depends on the nature of the species. Feeding of CO_2 air volume ratios of 1% and 4% were found to be the optimal concentrations for *N. salina* and *Scenedesmus* sp., respectively (Qu *et al.*, 2020; Sá *et al.*, 2020). For the cultivation of microalgae, a dedicated space for gas exchange is usually designed in the PBR in order to maintain an optimal balance of DCD and DO. It is widely accepted that mixing is an effective method to enhance interphase mass transfer in a PBR. Mass transfer rates in a bioreactor are largely affected by the fluid properties, liquid and gas velocity and by the geometry and type of bioreactor. Mass transfer is frequently assessed by the volumetric mass transfer coefficient (k_{La}) (Maryjoseph & Ketheesan, 2020; Zittelli *et al.*, 2013). In practical terms, prediction and optimization of mass transfer by k_{La} seeks maximal mass transfer with minimal energy input.

6.5 CAPITAL AND OPERATING COST

From the economic point of view, an efficient commercial PBR should have the following features: a highly illuminated surface area-to-volume ratio to capture solar radiation, easy temperature control, good performance in mixing and mass transfer, low shear stress on the cells and low capital cost and operating cost. In practice, the requirements of a high surface area-to-volume ratio, good performance in mixing and mass transfer and low shear stress in the PBR can be easily satisfied; however, the temperature control and low capital cost and operating cost are difficult to ensure. The first consideration in designing an efficient PBR with a high surface area-to-volume ratio is the capital cost, to which the illuminated surface area contributes substantially. Currently, glass, Plexiglas, polyvinylchloride (PVC), acrylic PVC and polyethylene are the main materials applied to PBRs. These materials all meet the transparency and mechanical robustness requirements for PBR fabrication. However, each material has its own advantages and disadvantages for the construction of particular types of PBRs. For example, glass is strong, transparent and has a long-life cycle; it is the perfect material for the construction of small PBRs. However, glass requires many expensive connecting parts when used for the fabrication of large-scale PBRs and leakage usually occurs when holes for aeration are present. The transparency of materials such as Plexiglas, PVC, acrylic PVC and polyethylene decreases gradually as they oxidize in the air. It is recommended that the capital cost must fall below 43.2 USD/m^2 for the economically feasible production of bioenergy from microalgae. The second consideration is the operating cost, which includes the costs of auxiliary energy for mixing and gas transfer, temperature and pH controls, maintenance to prevent leakage and cleaning. The mixing, mass transfer and temperature-control systems that are required in commercial systems to prevent death of microalgae due to excessive solar heating are expensive elements. In addition, the mixing, mass transfer and temperature-control systems take up most of the operating cost. Cleaning is critical for a PBR to prevent contamination and cells clumping, which result in frequent reactor shut-downs for mechanical cleaning and sterilization. Hence, the cost of cleaning is also high due to the rapid fouling of microalgae cells on the illuminated surfaces of the PBR (Daneshvar *et al.*, 2021; Wang *et al.*, 2022).

Aeration rate is a significant parameter that affects the mixing, mass transfer and L/D cycles inside the PBR. Gas enriched with CO_2 is supplied to the culture to produce mixing (thereby reducing the pH and nutrients gradient, preventing cell sedimentation and preventing the occurrence of dead zones, clumping and fouling), mass transfer (thereby preventing CO_2 deficiency, DO poisoning and DCD inhibition) and an optimized L/D cycle. However, excessive aeration results in cell damage (by increasing the mechanical shear force) and a high aeration cost. The velocity of the slurry is a vital factor to prevent cell deposition and can also increase the frequency of favorable L/D cycles and the opportunity for contact with the reactant CO_2. The velocity is chosen based on the settling velocity of the cells; a velocity of 0.1–0.3 m/s was found to be effective in a PBR. A higher velocity is also preferred but would consume too much energy. Moreover, although the L/D cycles induced by aeration are beneficial for cell growth, as mentioned earlier, no significant improvement has been observed because the intermediate frequency (0.01–1 Hz) particularly prevails in PBRs. In experimental situations reported by two research groups, the maximum frequency of the L/D cycles in the PBR did not exceed 25 Hz and the frequencies, which were below 50 and 6 Hz in the plain tubes at 10 and 0.5 m/s, respectively, were also confirmed. Finally, the cell movement in a PBR is chaotic and the aeration amounts to about one-third of the cost for the largest scale systems. Therefore, a high aeration rate is not practical for mass cultivation from an economic point of view. The optimum aeration rate by the mixture of air and CO_2 for the production of microalgae in a PBR has been investigated extensively. Air with a content of 5%–10% (v/v) CO_2 at rates of 0.025–1 volume of air/(medium. time) or expressed as vvm has been verified to be cost-effective for mass cultivation. The conditions required for different styles of PBRs to reach the same mass transfer capacity have been compared: 2400–3200 W/m^3 for the tubular PBR, 40 W/m^3 in a bubble column and 53 W/m^3 for the flat-plate PBR. In a flat-panel airlift

PBR, an optimum aeration of 8 W/m^3 has been proposed as sufficient to improve the mixing and mass transfer (Chuka-Ogwude *et al.*, 2020; Udaypal Goswami *et al.*, 2024). Although many types of PBR, including plastic light-conductive structures and various types of static mixers, have been designed to increase the frequency of L/D cycles, the cleanability level is what determines the lifespan of a PBR. For example, mixers comprised of baffles or bars in the PBR have been developed to achieve rapid beneficial L/D cycles. However, it is difficult to clean and periodically repair these PBRs because the baffles or bars are fabricated directly inside the reactor wall, which is usually less than 5 cm in width, for light penetration (Daneshvar *et al.*, 2021; Ma *et al.*, 2023). Therefore, certain principles such as smoothing the internal surface, minimizing the amount and surface area of internals and bends and having a large internal space, have been put forward to increase the cleanability of such PBRs.

To summarize, rational design of a PBR is crucial and the capital cost, operating cost (including the cost of the auxiliary energy demand, cleaning and maintenance) and lifespan of the PBR should be comprehensively considered regarding commercial exploitation. The best type of PBR should have a simple structure, easy operation and temperature control, low capital cost, low operating cost (including low energy consumption and convenient cleaning and maintenance), a long lifespan and so forth. To date, many new types of PBRs have been designed aside from the types mentioned earlier. However, some of these may be less robust than they appear due to low practicability resulting from cleaning and maintenance problems (Chuka-Ogwude *et al.*, 2020). Only a few types of PBRs can be used for mass cultivation and the most promising PBRs, such as tubular PBRs, plastic bag PBRs, column airlift PBRs and flat-panel airlift PBRs, are recommended.

6.6 CONCLUSION AND PERSPECTIVES

Although microalgae-based technology provides a sustainable solution for upcycling nutrients in wastewater, there remain significant challenges to replacing conventional water treatment technologies. In the above scenario, C, N and P metabolisms in microalgae are linked and affect each other. Briefly, C metabolism provides the carbon skeleton, N is used for cellular amino acid synthesis and P is used for ATP synthesis to support the energy-required processes. It is indicated that C, N and P are the macronutrients for their growth and silicon is also included for a specific group of microalgae (diatoms) with the silicified cell walls. The nutrient requirement may be species-specific and proper nutrient ratios would promote the overall treatment efficiency. A pressing issue is that microalgae-MBWT technology does not fit all types of wastewaters and the composition of the targeted wastewater should be analyzed before being applied to estimate the potential outcomes. The inconsistent wastewater components might also largely affect system performance. Supplementation with specific nutrients (COD, N or P) might be a solution to adjust C, N and P ratios to enhance the overall performance. Other considerations for wastewater treatment using microalgae are concerns regarding the growth competition from the wastewater indigenous bacteria or the potential concomitant contamination during the treatment. Theoretically, the cooperation between microalgae and bacteria promotes the exchange of CO_2 and O_2. Besides, bacteria provide B vitamins cofactor or siderophores, which are vital for microalgal growth. However, in the pilot trials, the treatment system might fail due to bacterial overgrowth. Wastewater operation and design experience indicate that the overall performance of the microalgae treatment system can be successfully maintained through conditions such as hydraulic retention times, optimization and control of the relative ratios of microalgae to bacteria and variable CO_2 and nutrient supply. On the other hand, the interaction between microalgae with multiple bacteria could also benefit the wastewater treatment: alleviation of NH_4^+ toxicity through nitrification pretreatment improvement in nutrient removal efficiency through algae–bacteria synergistic cooperation.

The cultivation of microalgae for biofuel production requires high levels of biomass productivity per area and minimal costs. Major technical and economic challenges impede the selection of an optimal reactor type at the commercial scale. Without detailed economic considerations, closed reactors appear to perform better than open ponds because they maintain favorable growth conditions and are less vulnerable to contamination. The best reactor type, based on photosynthetic efficiency and areal productivity, appears to be column reactors, at least on the small scale used in experiments from the literature. However, technical constraints prevent the size of this reactor type from being increased to commercial scale without the use of multiple small units, which are unlikely to be economical. Combinations of open and closed reactors seem promising from a productivity perspective. However, there is not enough economic information available to assess whether the increased productivity can offset the extra capital investment required, particularly with regard to biofuel applications. Thus, at this time, no specific reactor type is optimum for the commercial cultivation of microalgae for biofuel production.

REFERENCES

Abinandan S., Subashchandrabose S. R., Panneerselvan L., Venkateswarlu K. and Megharaj M. (2019). Potential of acid-tolerant microalgae, *Desmodesmus* sp. MAS1 and Heterochlorella *sp.* MAS3, in heavy metal removal and biodiesel production at acidic pH. *Bioresource Technology*, **278**, 9–16, https://doi.org/10.1016/j.biortech.2019.01.053

Akhtar N., Iqbal J. and Iqbal M. (2004). Removal and recovery of nickel(II) from aqueous solution by loofa sponge-immobilized biomass of *Chlorella sorokiniana*: characterization studies. *Journal of Hazardous Materials*, **108**(1), 85–94, https://doi.org/10.1016/j.jhazmat.2004.01.002

Bae S. and Kim Y. M. (2021). Carbon-neutrality in wastewater treatment plants: advanced technologies for efficient operation and energy/resource recovery. *Energies*, **14**(24), 1–4, https://doi.org/10.3390/en14248514

Balaji S., Kalaivani T., Shalini M., Gopalakrishnan M., Rashith Muhammad M. A. and Rajasekaran C. (2016). Sorption sites of microalgae possess metal binding ability towards Cr(VI) from tannery effluents – a kinetic and characterization study. *Desalination and Water Treatment*, **57**(31), 14518–14529, https://doi.org/10.1080/19443994.2015.1064032

Barbera E., Sforza E., Musolino V., Kumar S. and Bertucco A. (2018). Nutrient recycling in large-scale microalgal production: mass and energy analysis of two recovery strategies by process simulation. *Chemical Engineering Research and Design*, **132**, 785–794, https://doi.org/10.1016/j.cherd.2018.02.028

Barten R., Djohan Y., Evers W., Wijffels R. and Barbosa M. (2021). Towards industrial production of microalgae without temperature control: the effect of diel temperature fluctuations on microalgal physiology. *Journal of Biotechnology*, **336**, 56–63, https://doi.org/10.1016/j.jbiotec.2021.06.017

Ben Ouada S., Ben Ali R., Cimetiere N., Leboulanger C., Ben Ouada H. and Sayadi S. (2019). Biodegradation of diclofenac by two green microalgae: *Picocystis* sp. and *Graesiella* sp. *Ecotoxicology and Environmental Safety*, **186**, 109769, https://doi.org/10.1016/j.ecoenv.2019.109769

Borowiak D., Lenartowicz P., Grzebyk M., Wiśniewski M., Lipok J. and Kafarski P. (2021). Novel, automated, semi-industrial modular photobioreactor system for cultivation of demanding microalgae that produce fine chemicals – the next story of *H. pluvialis* and astaxanthin. *Algal Research*, **53**, 1–11, https://doi.org/10.1016/j.algal.2020.102151

Chen S., Chen M., Wang Z., Qiu W., Wang J., Shen Y., Wang Y. and Ge S. (2016). Toxicological effects of chlorpyrifos on growth, enzyme activity and chlorophyll a synthesis of freshwater microalgae. *Environmental Toxicology and Pharmacology*, **45**, 179–186, https://doi.org/10.1016/j.etap.2016.05.032

Chen S., Zhang W., Li J., Yuan M., Zhang J., Xu F., Xu H., Zheng X. and Wang L. (2020). Ecotoxicological effects of sulfonamides and fluoroquinolones and their removal by a green alga (*Chlorella vulgaris*) and a cyanobacterium (*Chrysosporum ovalisporum*). *Environmental Pollution*, **263**, 114554, https://doi.org/10.1016/j.envpol.2020.114554

Chew K. W., Chia S. R., Show P. L., Yap Y. J., Ling T. C. and Chang J. S. (2018). Effects of water culture medium, cultivation systems and growth modes for microalgae cultivation: a review. *Journal of the Taiwan Institute of Chemical Engineers*, **91**, 332–344, https://doi.org/10.1016/j.jtice.2018.05.039

Choi Y. Y., Patel A. K., Hong M. E., Chang W. S. and Sim S. J. (2019). Microalgae bioenergy with carbon capture and storage (BECCS): an emerging sustainable bioprocess for reduced CO_2 emission and biofuel production. *Bioresource Technology Reports*, **7**, 100270, https://doi.org/10.1016/j.biteb.2019.100270

Chrismadha T., Satya A., Satya I. A., Rosidah R., Satya A. D., Pangestuti R., Harimawan A., Setiadi T., Chew K. W. and Show P. L. (2022). Outdoor inclined plastic column photobioreactor: growth and biochemicals response of *Arthrospira platensis* culture on daily solar irradiance in a tropical place. *Metabolites*, **12**, 1–14, https://doi.org/10.3390/metabo12121199

Chuka-Ogwude D., Ogbonna J. and Moheimani N. R. (2020). A review on microalgal culture to treat anaerobic digestate food waste effluent. *Algal Research*, **47**, 101841, https://doi.org/10.1016/j.algal.2020.101841

Daneshvar E., Sik Ok Y., Tavakoli S., Sarkar B., Shaheen S. M., Hong H., Luo Y., Rinklebe J., Song H. and Bhatnagar A. (2021). Insights into upstream processing of microalgae: a review. *Bioresource Technology*, **329**. Elsevier Ltd, pp. 1–16, https://doi.org/10.1016/j.biortech.2021.124870

Ding T., Yang M., Zhang J., Yang B., Lin K., Li J. and Gan J. (2017). Toxicity, degradation and metabolic fate of ibuprofen on freshwater diatom *Navicula* sp. *Journal of Hazardous Materials*, **330**, 127–134, https://doi.org/10.1016/j.jhazmat.2017.02.004

Feng Y., Li C. and Zhang D. (2011). Lipid production of *Chlorella vulgaris* cultured in artificial wastewater medium. *Bioresource Technology*, **102**(1), 101–105, https://doi.org/10.1016/j.biortech.2010.06.016

Fret J., Roef L., Diels L., Tavernier S., Vyverman W. and Michiels M. (2020). Combining medium recirculation with alternating the microalga production strain: a laboratory and pilot scale cultivation test. *Algal Research*, **46**, 101763, https://doi.org/10.1016/j.algal.2019.101763

Gómez-Jacinto V., García-Barrera T., Gómez-Ariza J. L., Garbayo-Nores I. and Vílchez-Lobato C. (2015). Elucidation of the defence mechanism in microalgae *Chlorella sorokiniana* under mercury exposure. Identification of Hg–phytochelatins. *Chemico-Biological Interactions*, **238**, 82–90, https://doi.org/10.1016/j.cbi.2015.06.013

Guillard R. R. L. (1975). Culture of phytoplankton for feeding marine invertebrates. In Culture Invertebrate Animals, W. L. Smith, M. H. Chanle (eds.), Plenum Press, New York, pp. 26–60.

Hedayatkhah A., Cretoiu M. S., Emtiazi G., Stal L. J. and Bolhuis H. (2018). Bioremediation of chromium contaminated water by diatoms with concomitant lipid accumulation for biofuel production. *Journal of Environmental Management*, **227**, 313–320, https://doi.org/10.1016/j.jenvman.2018.09.011

Huang W., Bi Y. and Hu Z. (2014). Effects of fertilizer-urea on growth, photosynthetic activity and microcystins production of *Microcystis aeruginosa* isolated from Dianchi Lake. *Bulletin of Environmental Contamination and Toxicology*, **92**(5), 514–519, https://doi.org/10.1007/s00128-014-1217-6

Hussein M. H., Abdullah A. M., Badr El Din N. I. and Mishaqa E. S. I. (2017). Biosorption potential of the microchlorophyte *Chlorella vulgaris* for some pesticides. *Journal of Fertilizers & Pesticides*, **08**(01), 1–5, https://doi.org/10.4172/2471-2728.1000177

Ibrahim W. M., Karam M. A., El-Shahat R. M. and Adway A. A. (2014). Biodegradation and utilization of organophosphorus pesticide malathion by Cyanobacteria. *BioMed Research International*, **2014**, 392682, https://doi.org/10.1155/2014/392682

Jiang Z., Sun Y., Guan H., Sun D., Fang S., Ma X., Wang Z., Li Z., Zhang C. and Ge Y. (2022). Contributions of polysaccharides to arsenate resistance in *Chlamydomonas reinhardtii*. *Ecotoxicology and Environmental Safety*, **229**, 113091, https://doi.org/10.1016/j.ecoenv.2021.113091

Kao C. Y., Chen T. Y., Chang Y., Bin Chiu T. W., Lin H. Y., Da Chen C., Chang J. S. and Lin C. S. (2014). Utilization of carbon dioxide in industrial flue gases for the cultivation of microalga *Chlorella* sp. *Bioresource Technology*, **166**, 485–493, https://doi.org/10.1016/j.biortech.2014.05.094

Kiki C., Rashid A., Wang Y., Li Y., Zeng Q., Yu C.-P. and Sun Q. (2020). Dissipation of antibiotics by microalgae: kinetics, identification of transformation products and pathways. *Journal of Hazardous Materials*, **387**, 121985, https://doi.org/10.1016/j.jhazmat.2019.121985

Klinthong W., Yang Y. H., Huang C. H. and Tan C. S. (2015). A review: microalgae and their applications in CO_2 capture and renewable energy. *Aerosol and Air Quality Research*, **15**(2), 712–742, https://doi.org/10.4209/aaqr.2014.11.0299

Kurade M. B., Kim J. R., Govindwar S. P. and Jeon B.-H. (2016). Insights into microalgae mediated biodegradation of diazinon by *Chlorella vulgaris*: microalgal tolerance to xenobiotic pollutants and metabolism. *Algal Research*, **20**, 126–134, https://doi.org/10.1016/j.algal.2016.10.003

Lee J. S., Sung Y. J., Kim D. H., Lee J. Y. and Sim S. J. (2022). Development of a limitless scale-up photobioreactor for highly efficient photosynthesis-based polyhydroxybutyrate (PHB)-producing cyanobacteria. *Bioresource Technology*, **364**, 128121, https://doi.org/10.1016/j.biortech.2022.128121

Liang Z., Liu Y., Ge F., Xu Y., Tao N., Peng F. and Wong M. (2013). Efficiency assessment and pH effect in removing nitrogen and phosphorus by algae-bacteria combined system of *Chlorella vulgaris* and *Bacillus licheniformis*. *Chemosphere*, **92**(10), 1383–1389, https://doi.org/10.1016/j.chemosphere.2013.05.014

Liu X., Fujiwara M., Kodera T., Watanabe K., Akizuki S., Kishi M., Koyama M., Toda T. and Ban S. (2020). Conditions for continuous cultivation of *Chlorella sorokiniana* and nutrient removal from anaerobic digestion effluent of aquatic macrophytes. *International Biodeterioration & Biodegradation*, **149**, 104923, https://doi.org/10.1016/j.ibiod.2020.104923

Luo Y., Le-Clech P. and Henderson R. K. (2017). Simultaneous microalgae cultivation and wastewater treatment in submerged membrane photobioreactors: a review. *Algal Research*, **24**, 425–437, https://doi.org/10.1016/j.algal.2016.10.026

Luu D.-N., Gachet H., Maier C.-J., Maranzana N. and Aoussat A. (2022). Eco-design and medicine: opportunities to implement eco-design in the pharmaceutical R&D process. *Journal of Cleaner Production*, **365**, 132785, https://doi.org/10.1016/j.jclepro.2022.132785

Ma M., Yu Z., Jiang L., Hou Q., Xie Z., Liu M., Yu S. and Pei H. (2023). Alga-based dairy wastewater treatment scheme: candidates screening, process advancement and economic analysis. *Journal of Cleaner Production*, **390**, 1–11, https://doi.org/10.1016/j.jclepro.2023.136105

Madkour F. F., Kamil A. E. W. and Nasr H. S. (2012). Production and nutritive value of *Spirulina platensis* in reduced cost media. *Egyptian Journal of Aquatic Research*, **38**(1), 51–57, https://doi.org/10.1016/j.ejar.2012.09.003

Maryjoseph S. and Ketheesan B. (2020). Microalgae based wastewater treatment for the removal of emerging contaminants: a review of challenges and opportunities. *Case Studies in Chemical and Environmental Engineering*, **2**, 100046, https://doi.org/10.1016/j.cscee.2020.100046

Michael A., Kyewalyanga M. S. and Lugomela C. V. (2019). Biomass and nutritive value of Spirulina (*Arthrospira fusiformis*) cultivated in cost effective medium. *Annals of Microbiology*, **69**, 1387–1395, https://doi.org/10.1007/s13213-019-01520-4

Mohapatra R. K., Padhi D., Sen R. and Nayak M. (2022). Bio-inspired CO_2 capture and utilization by microalgae for bioenergy feedstock production: a greener approach for environmental protection. *Bioresource Technology Reports*, **19**, 101116, https://doi.org/10.1016/j.biteb.2022.101116

Mojiri A., Baharlooeian M. and Zahed M. A. (2021). The potential of *Chaetoceros muelleri* in bioremediation of antibiotics: performance and optimization. *International Journal of Environmental Research and Public Health*, **18**, 1–12, https://doi.org/10.3390/ijerph18030977

Patel A. K., Huang E. L., Low-Décarie E. and Lefsrud M. G. (2015). Comparative shotgun proteomic analysis of wastewater-cultured microalgae: nitrogen sensing and carbon fixation for growth and nutrient removal in *Chlamydomonas reinhardtii*. *Journal of Proteome Research*, **14**(8), 3051–3067, https://doi.org/10.1021/pr501316h

Patel A. K., Tambat V. S., Chen C.-W., Chauhan A. S., Kumar P., Vadrale A. P., Huang C.-Y., Dong C.-D. and Singhania R. R. (2022). Recent advancements in astaxanthin production from microalgae: a review. *Bioresource Technology*, **364**, 128030, https://doi.org/10.1016/j.biortech.2022.128030

Perez-Garcia O. and Bashan Y. (2015). Microalgal heterotrophic and mixotrophic culturing for bio-refining: from metabolic routes to techno-economics. In: Algal Biorefineries, Springer International Publishing. A. Prokop, R. K. Bajpai, M. E. Zappi (eds), pp. 61–131, https://doi.org/10.1007/978-3-319-20200-6_3

Qu F., Jin W., Zhou X., Wang M., Chen C., Tu R., Han S., He Z. and Li S. (2020). Nitrogen ion beam implantation for enhanced lipid accumulation of *Scenedesmus obliquus* in municipal wastewater. *Biomass and Bioenergy*, **134**, 105483, https://doi.org/10.1016/j.biombioe.2020.105483

Raeesossadati M. J., Ahmadzadeh H., McHenry M. P. and Moheimani N. R. (2014). CO_2 bioremediation by microalgae in photobioreactors: impacts of biomass and CO_2 concentrations, light and temperature. *Algal Research*, **6**(PA), 78–85, https://doi.org/10.1016/j.algal.2014.09.007

Raoof B., Kaushik B. D. and Prasanna R. (2006). Formulation of a low-cost medium for mass production of Spirulina. *Biomass and Bioenergy*, **30**(6), 537–542, https://doi.org/10.1016/j.biombioe.2005.09.006

Rattanapoltee P., Dujjanutat P., Muanruksa P. and Kaewkannetra P. (2021). Biocircular platform for third generation biodiesel production: batch/fed batch mixotrophic cultivations of microalgae using glycerol waste as a carbon source. *Biochemical Engineering Journal*, **175**, 108128, https://doi.org/10.1016/j.bej.2021.108128

Razzak S. A., Ali S. A. M., Hossain M. M. and deLasa H. (2017). Biological CO_2 fixation with production of microalgae in wastewater – a review. *Renewable and Sustainable Energy Reviews*, **76**, 379–390, https://doi.org/10.1016/j.rser.2017.02.038

Rehmanji M., Singh R., Nesamma A. A., Khan N. J., Fatma T., Narula A. and Jutur P. P. (2022). Multifaceted applications of microalgal biomass valorization to enriched biorenewables, a review of futuristic biorefinery paradigm. *Bioresource Technology Reports*, **17**, 100972, https://doi.org/10.1016/j.biteb.2022.100972

Rempel A., Gutkoski J. P., Nazari M. T., Biolchi G. N., Cavanhi V. A. F., Treichel H. and Colla L. M. (2021). Current advances in microalgae-based bioremediation and other technologies for emerging contaminants treatment. *Science of The Total Environment*, **772**, 144918, https://doi.org/10.1016/j.scitotenv.2020.144918

Sá M., Ferrer-Ledo N., Wijffels R., Crespo J. G., Barbosa M. and Galinha C. F. (2020). Monitoring of eicosapentaenoic acid (EPA) production in the microalgae *Nannochloropsis oceanica*. *Algal Research*, **45**, 101766, https://doi.org/10.1016/j.algal.2019.101766

Sánchez S., Martínez M. E. and Espinola F. (2000). Biomass production and biochemical variability of the marine microalga *Isochrysis galbana* in relation to culture medium. *Biochemical Engineering Journal*, **6**(1), 13–18, https://doi.org/10.1016/S1369-703X(00)00071-1

Satya A., Harimawan A., Haryani G. S., Johir M. A. H., Vigneswaran S., Ngo H. H. and Setiadi T. (2020). Batch study of cadmium biosorption by carbon dioxide enriched *Aphanothece* sp. dried biomass. *Water (Switzerland)*, **12**(1), 264, https://doi.org/10.3390/w12010264

Satya I. A., Satya A., Chrismadha T., Rosadi R., Maysarah A. D., Harimawan A., Setiadi T., Tang D. Y. Y. and Show P. L. (2024). Cultivation of *Spirulina platensis* for carbon dioxide bio sequestration in hybrid photobioreactor with real-time monitoring system. *Journal of Environmental Chemical Engineering*, **12**(3), 112396, 1–11, https://doi.org/10.1016/j.jece.2024.112396

Siddiki S. Y. A., Mofijur M., Kumar P. S., Ahmed S. F., Inayat A., Kusumo F., Badruddin I. A., Khan T. M. Y., Nghiem L. D., Ong H. C. and Mahlia T. M. I. (2022). Microalgae biomass as a sustainable source for biofuel, biochemical and biobased value-added products: an integrated biorefinery concept. *Fuel*, **307**, 121782, https://doi.org/10.1016/j.fuel.2021.121782

Sun M., Lin H., Guo W., Zhao F. and Li J. (2017). Bioaccumulation and biodegradation of sulfamethazine in *Chlorella pyrenoidosa*. *Journal of Ocean University of China*, **16**(6), 1167–1174, https://doi.org/10.1007/s11802-017-3367-8

Udaypal Goswami R. K., Mehariya S. and Verma P. (2024). Advances in microalgae-based carbon sequestration: current status and future perspectives. *Environmental Research*, **249**, 118397, https://doi.org/10.1016/j.envres.2024.118397

Ummalyma S. B., Sahoo D. and Pandey A. (2021). Resource recovery through bioremediation of wastewaters and waste carbon by microalgae: a circular bioeconomy approach. *Environmental Science and Pollution Research*, **28**(42), 58837–58856, https://doi.org/10.1007/s11356-020-11645-8

Vasconcelos Fernandes T., Shrestha R., Sui Y., Papini G., Zeeman G., Vet L. E. M., Wijffels R. H. and Lamers P. (2015). Closing domestic nutrient cycles using microalgae. *Environmental Science & Technology*, **49**(20), 12450–12456, https://doi.org/10.1021/acs.est.5b02858

Venkata Subhash G., Rajvanshi M., Raja Krishna Kumar G., Shankar Sagaram U., Prasad V., Govindachary S. and Dasgupta S. (2022). Challenges in microalgal biofuel production: a perspective on techno economic feasibility under biorefinery stratagem. *Bioresource Technology*, **343**, 126155, https://doi.org/10.1016/j.biortech.2021.126155

Wan J., Guo P. and Zhang S. (2014). Response of the cyanobacterium *Microcystis flos-aquae* to levofloxacin. *Environmental Science and Pollution Research*, **21**(5), 3858–3865, https://doi.org/10.1007/s11356-013-2340-3

Wang B., Lan C. Q. and Horsman M. (2012). Closed photobioreactors for production of microalgal biomasses. *Biotechnology Advances*, **30**(4), 904–912, https://doi.org/10.1016/j.biotechadv.2012.01.019

Wang Q., Wang X., Hong Y., Liu X., Zhao G., Zhang H. and Zhai Q. (2022). Microalgae cultivation in domestic wastewater for wastewater treatment and high value-added production: species selection and comparison. *Biochemical Engineering Journal*, **185**, 108493, https://doi.org/10.1016/j.bej.2022.108493

Xie P., Chen C., Zhang C., Su G., Ren N. and Ho S.-H. (2020). Revealing the role of adsorption in ciprofloxacin and sulfadiazine elimination routes in microalgae. *Water Research*, **172**, 115475, https://doi.org/10.1016/j.watres.2020.115475

Xiong Q., Hu L.X., Liu Y.S., Zhao J.L., He L.Y., Ying G.G. (2021). Microalgae-based technology for antibiotics removal: from mechanisms to application of innovational hybrid systems. *Environment International*, **155**, 106594, pp. 1–17, https://doi.org/10.1016/j. envint.2021.106594

Xiong J.-Q., Kurade M. B. and Jeon B.-H. (2017). Ecotoxicological effects of enrofloxacin and its removal by monoculture of microalgal species and their consortium. *Environmental Pollution*, **226**, 486–493, https://doi.org/10.1016/j.envpol.2017.04.044

Xu P. and Huang L. (2017). Stereoselective bioaccumulation, transformation and toxicity of triadimefon in *Scenedesmus obliquus*. *Chirality*, **29**(2), 61–69, https://doi.org/10.1002/chir.22671

Xu X., Lu X., Ma X. and Xia H. (2018). Response of Spirulina platensis to Sulfamethazine Contamination Response of Spirulina platensis to Sulfamethazine Contamination, IOP Conference Series, Materials Science and Engineering, **301**, No.012027. https://doi.org/10.1088/1757-899X/301/1/012027

Yan C., Wang Z., Wu X., Wen S., Yu J. and Cong W. (2020). Outdoor cultivation of *Chlorella* sp. in an improved thin-film flat-plate photobioreactor in desertification areas. *Journal of Bioscience and Bioengineering*, **129**(5), 619–623, https://doi.org/10.1016/j.jbiosc.2019.12.007

Yoshimura T., Okada S. and Honda M. (2013). Culture of the hydrocarbon producing microalga *Botryococcus braunii* strain *Showa*: optimal CO_2, salinity, temperature and irradiance conditions. *Bioresource Technology*, **133**, 232–239, https://doi.org/10.1016/j.biortech.2013.01.095

Zhang S., Qiu C., Bin Zhou Y., Jin Z. P. and Yang H. (2011). Bioaccumulation and degradation of pesticide fluroxypyr are associated with toxic tolerance in green alga *Chlamydomonas reinhardtii*. *Ecotoxicology*, **20**(2), 337–347, https://doi.org/10.1007/s10646-010-0583-z

Zhang X., Zhao X., Wan C., Chen B. and Bai F. (2016). Efficient biosorption of cadmium by the self-flocculating microalga *Scenedesmus obliquus* AS-6-1. *Algal Research*, **16**, 427–433, https://doi.org/10.1016/j.algal.2016.04.002

Zittelli G. C., Biondi N., Rodolfi L. and Tredici M. R. (2013). Photobioreactors for mass production of microalgae. In: Handbook of Microalgal Culture: Applied Phycology and Biotechnology. A. Richmond, Q. Hu (eds) 2nd edn. John Wiley & Sons, Ltd, Chichester, West Sussex, UK, pp. 225–266, https://doi.org/10.1002/9781118567166.ch13

doi: 10.2166/9781789064100_0121

Chapter 7

Open and closed photobioreactors for microalgal cultivation and wastewater treatment

S. A. Laizu[1], L. H. S. Hassan[2] and I. Ahmad[3]*

[1]Institute of Forestry and Environmental Sciences, University of Chittagong, Chittagong, Bangladesh
[2]Faculty of Science, Menoufia University, Shebin El-kom 32511, Egypt
[3]ProalgaeTech, Schwabbacher Str. 15, 70437 Stuttgart, Germany
*Corresponding author Email: mustafwibinqamar@gmail.com

ABSTRACT

The utilization of microalgae in wastewater treatment (WWT) witnesses a global interest. From this sector, there are two main cultivation systems that can be used for microalgae cultivation, either open or closed systems; the choice of which of these cultivation systems to be used depends on the purpose of the produced biomass. Use of closed photobioreactors (PBRs) is better than open systems because they can provide uncontaminated monoalgal culture. Design of PBR allows controlling light intensity, light penetration ratio, with good gas exchange sectors, air/carbon dioxide supply, and mixing strategy. Different configurations of PBRs, including aerobic, horizontal tube, moving tank, and tubular aerobic PBRs, are studied in terms of their design, operation, and performance, with emphasis on their efficiency in generating microalgae biomass. Airlift pressure-feedback PBRs are energy-efficient and improve mass transfer in both internal and external loop configurations. Horizontal tube PBRs maximize light exposure but face biofouling and uneven distribution. Stirred tank PBRs have good mixing efficiency but limited light penetration and high energy requirements. PBRs are preferred for continuous cultivation systems due to their superior light penetration and high biomass productivity. The most common PBRs are flat panel reactors, tubular, and bubble column PBRs. Microalgae cultivation technology using PBRs depends on cultivation mode, factors that control the optimal growth, and harvesting techniques. PBRs are used in WWT, with suspended systems offering economical alternatives. Fixed systems, such as PBRs, improve separation but struggle with accurate biomass measurement. Hence, this chapter emphasizes the need for engineering improvements and optimization techniques to overcome operational and economic limitations of open and closed PBRs. This provides an overview of commercial WWT using microalgae, highlighting growth factors, different types and designs, and their applications.

Keywords: Microalgae, bioremediation, wastewater treatment, raceway, open ponds, photobioreactors, airlift photobioreactors, photosynthetic efficiency, tubular PBRs, bioresource recovery, biomass, biofuel

7.1 INTRODUCTION

An incredible increase in pollution in different bio-ecological systems necessitates a need for a vital biological refinery that works as a lung to reduce pollution and its harmful effects. Microalgae-based technologies have a positive environmental impact, but technical and financial challenges remain. Photosynthetic microorganisms such as microalgae and cyanobacteria are characterized by their quick growth, high biomass productivity that relies on sunlight energy, and they are promising for long-term production of highly valuable products (Touloupakis, 2023). In addition, their ability to convert organic wastes into bioenergy and their great potential for lipid accumulation in their cells (1–70% of their total mass) are also reported (Dhanasekar & Sathyanathan, 2023). Microalgae resemble an active biological tool that can be cultivated on wastewater; depending on their ability to absorb carbon dioxide (CO_2) during photosynthesis they can reduce greenhouse gas emission effects as well as global warming phenomena (Dhanasekar & Sathyanathan, 2023; Sedighi *et al.*, 2023).

More recently, microalgae wastewater treatment (WWT) technology has emerged as a supplementary approach to wastewater treatment technology and has gained increasing popularity. Microalgal biorefinery is a continuous process with the integration of simultaneous microalgae cultivation, product extraction, product processing, and purification (Chanana *et al.*, 2023). Microalgae-based WWT shows the highest biomass efficiency compared to wetland treatment, achieving 80–90% pollution removal, making it a sustainable method for obtaining renewable energy from algal biomass (Hoang *et al.*, 2022). Wastewater is a valuable resource for nutrients and energy, and has been gaining momentum for resource recovery in the context of circular economy and carbon neutrality (You *et al.*, 2023). Broadly speaking, the buildup of biomass and the absorption of nutrients, mostly influenced by photosynthetic activity, are strongly reliant on the photosynthetic effect. Nitrogen, phosphorus, and heavy metals that are present in wastewater have harmful effects on environment, but they can present a suitable culture for microalgal cultivation (Otálora *et al.*, 2023). Chemical, physical, and biological characteristics of wastewater differ according to their origin. There are different origins of wastewater, including domestic, industrial, and agricultural sources (Aditya *et al.*, 2022). Domestic wastewater has low nutrients and organic matter concentrations, which is produced from human activities, whereas industrial wastewater contains high concentrations of organic compounds, heavy metals, and dyes which are generated from various industries such as pharmaceutical, food, cloth, and mining industries. Agricultural wastewater is generated from agricultural and animal production, which contain high concentrations of organic compounds, ammonia, nitrate, and phosphorus (Sepúlveda-Muñoz *et al.*, 2023). Various microalgal and cyanobacterial genera have showed their efficiency in WWT processes, such as *Botryococcus*, *Chlorella*, *Chlamydomonas*, *Euglena*, *Limnospira*, *Phormidium*, *Scenedesmus*, *Ankistrodesmus*, and *Oscillatoria* (Hashmi *et al.*, 2023).

Contrary to traditional methods of treating wastewater, such as an activated sludge process, the use of microalgae for wastewater purification has a comparatively extended hydraulic retention time (HRT). Additionally, supplementary light and CO_2 are essential factors in this process (Whitton *et al.*, 2015). Photobioreactors (PBRs) are effective tools for outdoor and indoor microalgal cultivation, with computerized control of growth parameters and contamination inhibition (You *et al.*, 2023). PBRs promise WWT aspect, without any external containment. PBR cultivation systems are open outdoor systems such as raceway ponds and closed PBR systems such as bubble column, tubular flat plate, and spiral (Dhanasekar & Sathyanathan, 2023). Large-scale integrated processes based on industrial microalgal strains and novel PBRs require further research and innovation. Intensified microalgal cultivation processes can produce value-added biochemicals and microalgal biomass simultaneously, making them economically viable (Pavlou *et al.*, 2023).

This chapter highlights the role of microalgae in WWT in respect of their cultivation parameters in outdoor and indoor PBRs' culture, besides their various applications. In addition to focusing on the configurations and operating aspects of different types of PBRs, based on the conventional microalgae cultivation methods found in previous literature, PBRs have great potential for future application in microalgae WWT.

7.2 PBRS FOR WWT BY MICROALGAE

Algae reproduction using raceway ponds and low-cost PBRs is a promising method for producing highly valuable products such as dietary supplements, natural pigments, and aquaculture feed, which make this a viable option for commercial purposes (Abdelzaher *et al.*, 2023; Dineshbabu *et al.*, 2019; Li *et al.*, 2023). An open raceway pond, traditionally used for algal cultivation and WWT, is insufficient for CO_2 sequestration due to most CO_2 escaping into the atmosphere. A closed PBR, offering better process control and higher productivity, is more suitable for this purpose (Viswanaathan *et al.*, 2022).

7.2.1 Open ponds

Open pond systems employed for WWT can be categorized into three distinct approaches: the paddle-less open pond, the raceway pond, and the high-rate algae ponds. Microalgae are cultivated in open ponds that have sunlight exposure and access to natural water sources such as lakes and lagoons. When compared to closed ponds and bioreactors, this system is less complicated and more cost-effective, allowing wastewater to be used for biomass production (Ahmad *et al.*, 2021). The productivity of biomass can be decreased, though, by climatic changes and unmanaged growth factors. Additionally, productivity may be impacted by shading and poor light penetration. In open systems, it is impossible to maintain pure monoalgal cultures without contamination. While intensive systems, such as circular and raceway ponds, use mechanical agitators for mixing, extensive systems, such as natural lakes, use prevailing winds to homogenize the microalgal culture (Chanana *et al.*, 2023). Circular ponds, designed with a central pivot rotating agitator, form a 10,000 m^2 surface area. Similar to raceway ponds, contamination and uncontrolled growth conditions are common issues in open pond systems (Ahmad *et al.*, 2021). Raceway pond systems, ranging from 1000 to 5000 m^2, are open, shallow systems ideal for microalgae cultivation and wastewater remediation. Despite their low energy and cost requirements, their productivity can be impacted by natural contamination, temperature variation, nutrient deficiency, and illumination (Zhou *et al.*, 2020). A paddle-less open pond can also achieve a significant removal of pollutants. According to Yadavalli *et al.* (2014), open pond systems without a paddle achieved removal rates of 98.17, 96.87, and 87.93% for ammonia, total phosphorus (TP), and chemical oxygen demand (COD). This indicates that this type of basic pumped bed reactor has a significant potential for improving WWT efficiency and biomass production (Razzak *et al.*, 2024). Table 7.1 presents the characteristics of several open pond systems.

7.2.1.1 Unstirred open systems

The majority of natural water reservoirs lack a stirred component. This results in inadequate blending, but reduced expenses for large-scale commercial production. Common types of PBRs include lakes, lagoons, and ponds (Razzak *et al.*, 2024; Yu *et al.*, 2022). These culture systems offer a cost-effective, transparent, and practical method for managing and overseeing culture processes. A natural pond is typically of a depth of less than 600 mm to facilitate the penetration of light into the water and subsequent absorption by algae cells. In a prior study, it was suggested that plastic sheets can be employed to enhance temperature regulation by covering the water surface (Ahmad *et al.*, 2021).

7.2.1.2 Circular ponds

The concept of employing a circular pond with an elongated revolving arm was derived from a circular reactor designed for WWT. Consequently, a circular pond bears a striking resemblance to a WWT pond. The depth of this particular pond is consistently 20–30 cm, while its circumference ranges from 40 to 50 m. An elongated rotating arm is positioned in the middle of the pond, functioning as a clock dial and executing a paddlewheel action that is structurally similar to that of a raceway pond (Ahmad *et al.*, 2021; Razzak *et al.*, 2024).

7.2.1.3 Raceway ponds

The most prevalent species of algae that may be grown in a raceway pond are *Chlorella*, *Spirulina*, *Dunaliella*, and *Haematococcus*. These species can be cultured either as pure culture or as wastewater

Table 7.1 Comparison of different types of PBRs.

PBR Type	Wastewater Type	Scale	Design Parameters	Reference
Open pond system	Municipal and agricultural wastewater	Lab and pilot scale	Operation mode, aeration, and HRT	Yadavalli *et al.* (2014)
Membrane PBR (MPBR)	Municipal wastewater	Lab scale	Illumination, HRT, and aeration	Zou *et al.* (2022)
Pilot outdoor MPBR	Anaerobic MBR effluent	Lab scale	Reducing light path, HRT	González-Camejo *et al.* (2020)
Flat system	Sewage wastewater	Lab and pilot scale	Illumination and aeration	Pompei *et al.* (2023)
Tubular system	Pre-treated leachate landfill wastewater	Lab scale	Cultivation mode, illumination, and aeration	Porto *et al.* (2021)
Plastic bag	Malek Ashtar University of Technology wastewater	Lab scale	Aeration and size	Arabian (2024)
Cylinder system	Real domestic wastewater	Pilot scale	Illumination, HRT, and aeration	Dos Santos *et al.* (2024)
Forward osmosis (FO-PBR)	Synthetic wastewater	Lab scale	Operation mode, illumination, and bubbling air	Larronde-Larretche and Jin (2022)

culture. Additionally, they can be integrated with CO_2 capture technology in some power plants. The construction elements for a raceway pond may include either concrete or plastic. The pond may be situated either on the surface or excavated into the ground and enclosed by a wall to inhibit the infiltration of liquid from the pond into the soil. Raceway ponds exhibit a range of designs, thereby encompassing both a solitary channel and sets of channels. The typical range for the depth of the culture media in a raceway pond is between 25 and 30 cm, 10–100 m in length, and 1–10 m in width (Alazaiza *et al.*, 2023; Ali *et al.*, 2022; Tan *et al.*, 2018).

7.2.2 Closed cultivation PBRs

To overcome the challenges in open systems and enhance the production of microalgae biomass, researchers have devised closed or mostly closed vessels. These vessels prevent direct mass transfer between culture media and the atmosphere and provide a controlled environment with regulated light, CO_2, temperature, and nutrients (Ogundele *et al.*, 2023). The cost of a closed system is generally high, and the inclusion of air tightness, control procedures, and other mass transfer methods significantly contribute to the overall cost. Moreover, achieving the ideal lighting conditions requires a substantial energy consumption, which appears to be economically unviable. Hence, closed PBRs are often seen as challenging to advance to the pilot or commercial scale. Closed phase change PBRs or systems are highly valuable in the production of high value-added products. The significance of these fine chemicals (bio-pharmaceuticals, top-grade cosmetics, human health foods, and biofuels) derived from microalgae has been increasing. Consequently, the development of appropriate and sustainable closed PBRs holds a promising potential. Contemporary commonly used closed pressure bed PBRs typically consist of flat panels, vertical tubes (such as bubble columns and airlifts), horizontal tubes, stirred tanks, and their modified versions (Alazaiza *et al.*, 2023).

Open systems have problems with contamination and growth conditions; closed systems do not. Using transparent materials such as glass and polycarbonate, greenhouses are used to cover ponds and keep an eye on the ideal growing conditions (Ahmad *et al.*, 2021; Chanana *et al.*, 2023). These enclosed systems support the growth of monoalgae cultures with high concentrations of valuable and effective metabolites (Shekh *et al.*, 2022). PBRs can also prevent odour spreading of treated wastewater (Sarker, 2022). The process for secondary wastewater effluent treatment was investigated to be intensified using packed-bed biofilm PBRs and ultrafiltration membrane. The microalgal–bacterial biofilm, which evolved from a local microbial consortium, was supported by cylindrical glass carriers. Glass containers allowed the biofilm to grow properly with little suspended biomass. After a start-up period of 1000 h, stable operation was attained, and complete nitrification was observed. Supernatant biopolymer clusters were also reduced. Following that, biomass productivity attained 54 mg/L/day. Numerous strains of heterotrophic nitrification–aerobic denitrification bacteria and fungi, as well as the green microalga *Tetradesmus obliquus*, were identified. The removal rates recorded 56% for COD, 12% for nitrogen, and 20% for phosphorus. Biofilm development was the main cause of membrane fouling, and it was not effectively reduced (Díaz *et al.*, 2023). Another study by Dhanasekar and Sathyanathan (2023) tested the treatment of synthetic agrochemical wastewater by using *Chlorella vulgaris* in a bubble column PBR at 25–28°C and three light sources (blue, red, and solar) with a light intensity in the range between 1800 and 2800 lux for 21 days. The results reported that the maximum biomass obtained for blue light, which is 3.99 g/L.

7.2.2.1 Flat panel

A flat panel, shown in Figure 7.1b and Figure 7.3, is a widely used PBR model characterized by its rectangular box shape. It finds application in both pure algae cultivation and algae wastewater

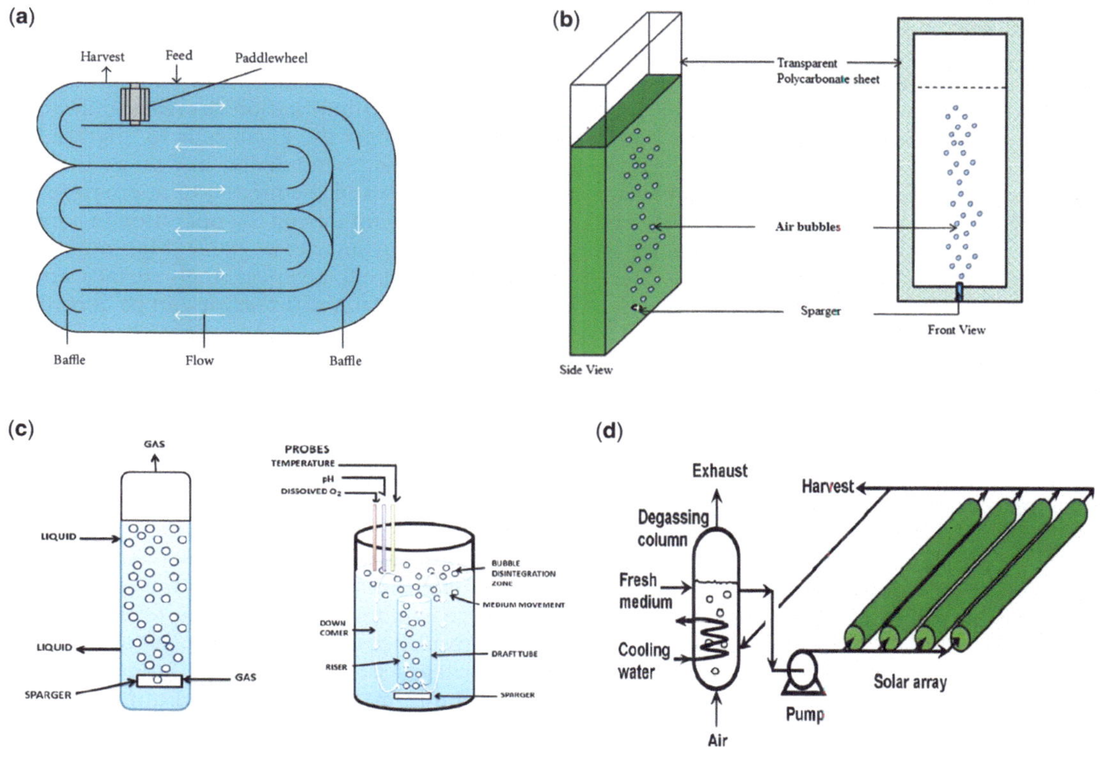

Figure 7.1 Different types of PBRs: (a) horizontal tube raceway pond, (b) flat panel PBR, (c) vertical tube PBR, and (d) horizontal tube PBR.

cultivation. It may be situated either indoors on artificially illuminated surfaces or outside on sunlight-exposed surfaces. Flat panels are constructed using transparent or semi-transparent materials such as glass, plexiglass, polycarbonate, and plastic films. Flat panels have a relatively limited propagation distance, enabling light to readily permeate the culture medium (Carone *et al.*, 2022). Air bubbles produced by an air sparger are the primary driving force for mixing. Typically, a pump is employed to augment the circulation of air bubbles along the air sparger, thereby circulating the suspension of algae cells. An exhaust gas is released at the interface between the gas and liquid phases. In order to achieve optimal intensity of incident light when the reactor is situated outside, it is slanted at a specific angle. Flat panels have several key benefits: (a) a high ratio of surface area to volume; (b) a non-significant buildup of dissolved oxygen; (c) easy cleaning; and (d) a flexible unit suitable for scaling-up. Concurrently, the primary constraints of flat panels are: (a) the cost of temperature regulation; (b) the hydrodynamic stress caused by aeration; and (c) biofouling in close proximity to the inner surface. Gabrielyan *et al.* (2022) discuss the growth of *Chlorella sorokiniana* IPPAS C-1 using a specially designed flat-panel PBR. The cultured biomass increased by 326 times in 13 days, attaining a final concentration of 2.8 g dw/L.

7.2.2.2 Vertical tube

Vertical tube PBRs can be classified into two types: bubble column and airlift. Both reactors are equipped with an affiliated air sparer located at the bottom. This spare gas is transformed into small bubbles to maintain the suspension of algae cells and improve mass transfer. The purpose of this is to control the capture of CO_2 and the emission of oxygen (Benner *et al.*, 2022). A bubble column reactor, shown in Figure 7.1c on the left, lacks an internal structure. Consequently, the fluid flow is solely propelled by bubbles, which are introduced by the air sparer located at the bottom. Typically, these cylindrical vessels have a height that is more than double the diameter and undergo external illumination. In the axial region, there is a black zone where the illumination decreases due to the presence of shielding cells. Within this arrangement, bubbles compel the algae cells to migrate from the externally lighted surface to the axially dark zone, creating a phenomenon known as the 'light flash effect'. This action enhances the efficiency of photosynthesis and the mixing of microalgae. The gas flow rate is the sole characteristic that must be taken into account during the operation of a bubble column reactor, as it considerably affects the light and dark cycle (Ahmad *et al.*, 2021; Mahmood *et al.*, 2023). A bubble column reactor can be characterized by its cost-effectiveness, high surface area to volume ratio, straightforward design, and effective mass transfer. An airlift reactor can be considered as an enhancement of the bubble column although with two interconnected zones within. The first element is referred to as the riser, whereas the second element is known as the down comer (Figure 7.1c, right). The riser zone is configured as a concentric tube, devoid of air, which propels liquid upwards and creates the dark zone. The movement of air bubbles within the concentric tube facilitates the upward movement of liquid from the riser zone to the outer zone, where algae cells can be exposed to light. This process, known as 'airlift', leads to the downward recycling of the liquid externally, ultimately returning to the riser zone. Additionally, the 'light flash effect' is observed, which serves to maximize photosynthetic efficiency. This particular type of cycle is alternatively referred to as the internal loop. Another type of airlift reactor is referred to as the external loop. In this configuration, the riser or dark zone is positioned outside the vertical tube, whereas short horizontal sections are attached at the top and bottom to divide the riser and down comer. This arrangement forces the liquid cycle and results in the 'light flash effect'. In the majority of airlift reactors, the residence duration of gas plays a crucial role in mass transfer. Several research studies on airlift reactors have asserted the following advantages: (a) excellent mass transfer efficiency; (b) effective mixing with minimal shear stress; (c) minimal energy consumption; and (d) suitable for immobilizing algae on suspended particles (Mahmood *et al.*, 2023).

7.2.2.3 Horizontal tube

Horizontal tube PBRs, sometimes known as tubular reactors, have a high surface to volume ratio, which is advantageous for maximal exposure of microalgae to light. The fundamental structure of horizontal tube PBRs consists of tubes that can be placed in several orientations, including horizontal, inclined, spiral, helicoidal, and their variations. All these classes have the same operational mechanism. The diameter of the tubes varies between 10 and 60 mm, whereas their length can extend to several hundred metres. For most tubular reactors in chemical engineering, the velocity of the flow is a crucial parameter. Another distinctive feature of tube reactors is their high efficiency in mixing in the radial direction. The diagram in Figure 7.1d illustrates a standard horizontal tube PBR. Located outside the tubes is a gas exchange system that facilitates the introduction of new culture medium, the cycling of cooling water, and the effective supplementing of CO_2 (Benner *et al.*, 2022). Furthermore, the exhaust gas is similarly segregated in this distinct system. The fundamental driving power behind the cycle between the gas exchange system and irradiation tubes is mostly attributed to a pump. The culture medium undergoes a cyclic process and circulates between these two systems, thereby inhibiting the settling of coalescent cells. Ultimately, the algal biomass can be collected after undergoing cyclic passage, therefore ensuring a continual culture process. Despite having the highest surface-to-volume ratio and employing a continuous growing technique, horizontal tube PBRs are subject to several constraints. First and foremost, there is the uneven distribution of mass in the radial directions. The presence of this constraint can easily result in uneven temperature and CO_2 distributions, thereby causing an accumulation of dissolved oxygen (Mahmood *et al.*, 2023). A further constraint of a PBR system is its difficulty to be expanded outdoors, mostly because of its substantial land use requirements and its lack of economic viability in commercial farming. Moreover, the presence of surface biological deposition induces photoinhibition, which in turn poses challenges in the cleaning process of the tubes. To prevent fouling in horizontal tube PBRs, two approaches can be employed: the use of plastic beads and the introduction of air bubbles into the tube. Furthermore, this final approach can also function as an inline degaussing. However, the expenses for upkeep will rise proportionally. Despite the aforementioned issues, horizontal tube PBRs continue to exhibit optimal performance, where microalgae growing requires an automated control system and sufficient illumination (Ahmad *et al.*, 2021; Ramasamy *et al.*, 2020).

7.2.2.4 Tubular PBRs

These structures are produced by layering plastic tubes with straight glass arranged in vertical, inclined, or helical patterns, as well as in fence-like and horizontal arrangements (Tan *et al.*, 2018). Tubular pressure-balanced reactors can be categorized into two fundamental design principles. The first category is cylindrical chambers or reactors where the microalgae are cultivated, whereas the second category consists of degausser units for gas removal and pump systems for circulation and mixing of the microalgae culture (Figure 7.2b). The key factor contributing to superior biomass productivity via photosynthesis is the optimal penetration of light. This solar array is specifically engineered with diameters of up to 0.1 m in order to enhance the penetration of light through the compact microalgal culture (Mahmood *et al.*, 2023). Continuous operation mode involves the circulation of microalgae culture from the degausser unit to the solar array and then back to the degausser unit. The efficacy of tubular PBRs has been demonstrated in several operational modes, including semi-batch, batch, turbido-static, and continuous. This makes it the most dependable form of PBR for scaling up because of its superior control and efficiency (Arun *et al.*, 2022).

7.3 BACTERIAL AND MICROALGAL CONSORTIUM

In a recent study, bacterial–microalgal consortium immobilized culture was used for the treatment of coke-oven wastewater in a closed packed-bed PBR at 20 mL/h flow rate for 24 h by using the bacterium *Bacillus* sp. NITD 19 and two microalgal strains *Chlorella* sp. and *Synechococcus* sp. The results achieved was at 10% v/v (microalgal cells to bacterial cells), with a high growth rate attained 3.08 g/L, with a removal efficiency recorded 80.32% for phenol, 47.85% for ammonia, 39.45 for nitrates, and 65.0% for fluoride (Rai *et al.*, 2023). Algal–bacterial consortium had been tested on a flat sheet of polyvinylidene fluoride membrane in a 10.35 L bioreactor, which helped in the removal of organic carbon from wastewater (Zhang *et al.*, 2020).

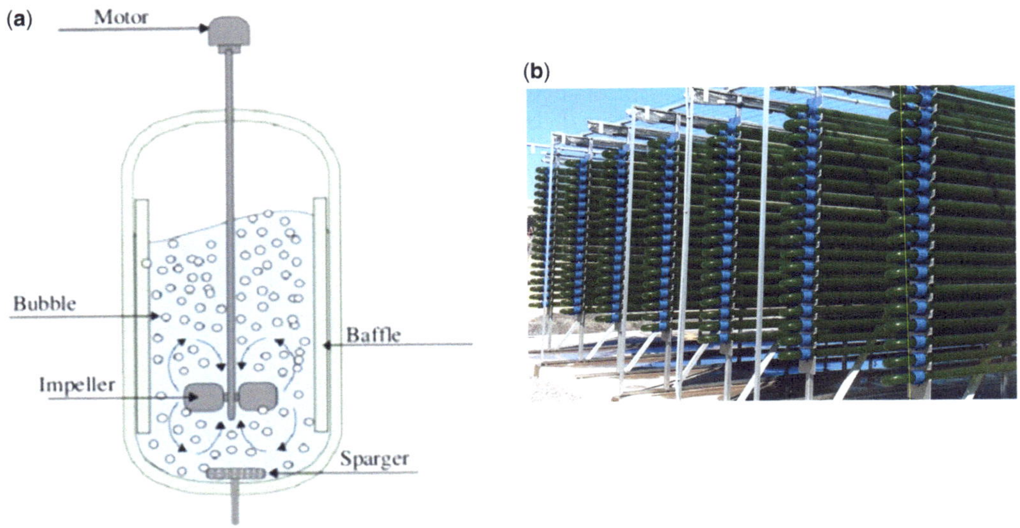

Figure 7.2 Different types of PBRs: (a) stirred tank and (b) horizontal tube PBR.

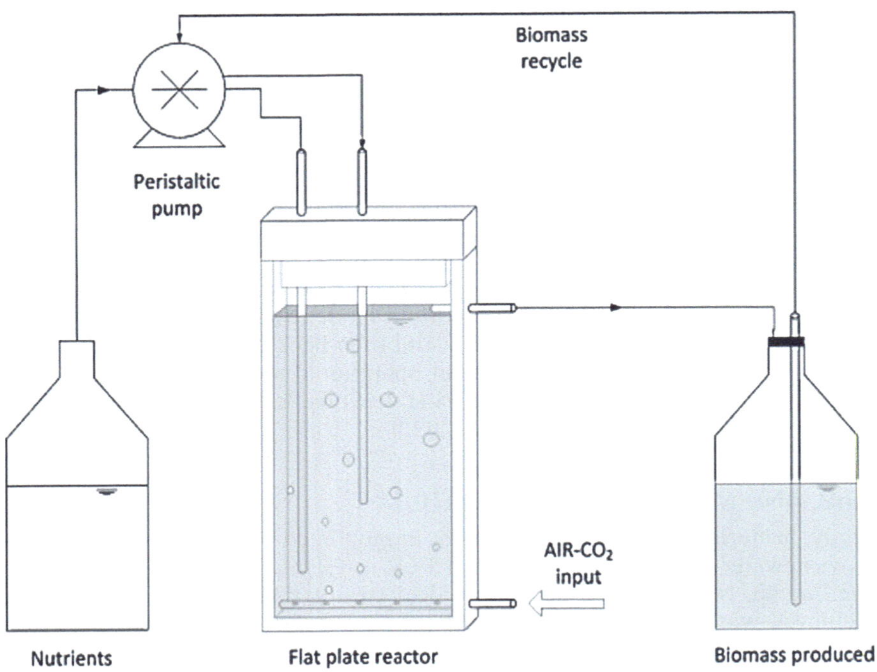

Figure 7.3 Flat panel PBR.

7.4 FACTORS THAT INFLUENCE MICROALGAL CULTIVATION IN PBRS

Microalgal cultivation in PBRs required an ideal control of chemical and physical growth factors such as temperature, pH, and dissolved oxygen (Sepúlveda-Muñoz *et al.*, 2023). The growth of algae and their ability to absorb nutrients are influenced not only by the presence of nutrients, but also by intricate interactions involving several physicochemical parameters, including pH, light intensity, temperature, and biotic factors. One of the primary biotic factors that exerts a substantial influence on the growth of algae is the initial density. It is anticipated that an increased density of algae will result in enhanced growth and improved efficiency in removing nutrients (Ogundele *et al.*, 2023). Nevertheless, the presence of a substantial concentration of algae would result in self-shading, the buildup of autoinhibitors, and a decline in the efficiency of photosynthesis (Ogundele *et al.*, 2023). The temperature will vary based on the specific species of algae. The optimal temperature range for the growing of phytoplankton is typically between 20 and 30°C. A diverse range of algal species, with a specific emphasis on green microalgae, exhibits susceptibility to mortality when exposed to temperatures over 35°C. The growth rate of algae is impeded by temperatures that fall below 16°C. The growth and development of algae are subject to the influence of light intensity, which should be maintained within an optimal range, avoiding both excessive strength and weakness. In order to flourish, algae require a mere 0.1% of direct sunlight. In the majority of water systems, the uppermost 7–10 cm of water experiences limited penetration of light. The limited penetration of light into deeper water is attributed to the abundance of bulk algal biomass. The growth of algae is also affected by the process of mixing. The process of combining algae cultures necessitates the implementation of either agitation or circulation. In the context of deep photoreactor systems, the utilization of an agitator is implemented. Paddlewheels are employed in open pond systems. The utilization of a photo-tube system capitalizes on the benefits of pump circulation. Algae, similar to all organisms, necessitate essential nutrients and an optimal pH level in order to flourish. Autotrophic development necessitates the presence of carbon, hydrogen, oxygen, nitrogen, phosphorus, sulphur, iron, as well as trace elements. Under nutrient-restricted conditions, growth is significantly impeded and there is an induction of fat accumulation. Algae exhibit a preference for a pH range that spans from neutral to alkaline (Ogundele *et al.*, 2023).

7.4.1 Temperature

In general, algae have a temperature tolerance within the range of 15–40°C. Research findings indicate that microalgae thrive most effectively within the temperature range of 25–30°C. The ideal temperature for growth differs among particular species of microalgae. Regulation of temperature in open culture systems is challenging and costly because of their increased vulnerability to overheating, medium evaporation, chilling, and lipid composition. The use of near systems, such as PBRs, is motivated by the more easier regulation of temperature in these systems (Ogundele *et al.*, 2023). Temperature adaptation of microalgae falls into three classes: psychrophiles (0–15°C), mesophiles (15–45°C), and thermophiles (>45°C) (Madigan *et al.*, 2021).

7.4.2 Light intensity

The light intensity and photosynthetic efficiency are crucial factors in determining the productivity of algae, because they impact the growth rate, biomass generation, and lipid accumulation mechanisms. The illumination intensity is influenced by the density of algal biomass, the depth of the culture medium, and the kind of PBR (Ali *et al.*, 2024). Photosynthetically active radiation is the wavelength range suitable to support the growth of photosynthetic organisms (mainly plants and microalgae), corresponding to wavelengths between 400 and 780 nm (visible spectrum), where light absorption by chlorophylls and carotenoids occurs. At high radiation intensities and low culture densities, inhibition of photosynthetic activity can occur, resulting in an increase in intracellular carotenoid concentrations or inhibiting growth (You *et al.*, 2023). Higher cell concentration and depth of the culture necessitate increased light intensity. However, it is important to note that direct sunlight and high-density artificial light can sometimes cause photoinhibition. Avoidance of system overheating is recommended. An illumination level of 1000 lux is suitable for the cultivation in Erlenmeyer flasks, but a light intensity of around 5000–10,000 lux is necessary for bigger volumes. A crucial factor is to establish the light–dark cycles in order to optimize the efficacy of photosynthesis. The photochemical phase of the light cycle is necessary for the production of adenosine triphosphate (ATP1) and nicotinamide adenine dinucleotide phosphate (NADPH2), whereas the biochemical phase requires the light cycle to synthesize the vital biomolecules for microbial development (Yousuf, 2020).

7.4.3 pH value

pH quantifies the degree of alkalinity or acidity of the aqueous medium. The optimal pH range for optimizing microalgae development and photosynthetic capacity is between 5 and 7, and 7–9 for cyanobacteria. The fluctuation in pH can modify the development of microalgae in several manners (Razzak *et al.*, 2024). The pH levels rise during the daytime as a result of the photosynthetic absorption of CO_2 by microalgae, and fall during the night-time due to the respiratory process of the respective species (Ogundele *et al.*, 2023). Elevated levels of CO_2 can result in increased biomass accumulation, but they also drive a reduction in pH, which negatively impacts the physiology of microalgae (Ali *et al.*, 2024).

7.4.4 Nutrients

Macro nutrients such as phosphorus and nitrogen, together with trace metals such as Fe, Mg, B, Mo, K, Co, Zn, and Mb, are necessary for the optimal growth of microalgae. Particularly, nitrogen and phosphorus nutrients play a crucial role in the synthesis of lipids. Research findings indicate that a nitrogen-deficient growing medium promotes increased lipid production in algae (Pérez-Mora *et al.*, 2023). In microalgae, both an abundance and a shortage of nutrients can lead to morphological and physiological alterations due to their interference with metabolic pathways (Ogundele *et al.*, 2023). Recently, researchers have adopted a two-phase approach to algal growth. In the first phase, microalgae are cultivated in a medium rich in nutrients and then moved to a nutrient-deficient media (Ali *et al.*, 2024).

7.4.5 Mixing

Effective mixing is crucial in order to avoid sedimentation and facilitate the migration of algae between the dark and light areas of cultivation systems. Alternatively, the algae located at the surface will fully absorb the light, resulting in photoinhibition, whereas the algae situated in the interior or deeper regions will be deprived of adequate light (Ekin, 2020). Moreover, it helps to preserve the uniform cell content in the medium. Mixing can be achieved by employing mechanical stirrers such as paddlewheels in ponds or by intentionally introducing bubbles in PBRs (De Souza Kirne *et al.*, 2022).

7.4.6 Predators or invasive species

They can encompass any category of living organisms that are completely undesirable and unforeseen in the microalgal production system due to their inhibitory effect on microalgae growth, imposition of nutritional deficiencies, and contamination of the media. They can be classified as fungi, insects, bacteria, or species of predatory algae (Molina-Grima *et al.*, 2022). Intensive production of extremophilic microalgae at an industrial scale is motivated by their ability to thrive in extremely harsh environments that are inaccessible to predator species. Unlike open systems, which are more susceptible to invasion by predator species, closed systems, such as PBRs, effectively prevent contamination by shielding the algae from the external environment (Ahmad *et al.*, 2021).

7.5 MECHANISMS OF NUTRIENTS UPTAKE BY ALGAE

The removal of nutrients is a crucial component of WWT, and systems based on algae provide a potential method because of their superior capacity to effectively absorb nutrients such as nitrogen and phosphorus (Mohsenpour *et al.*, 2021; Oruganti *et al.*, 2022). Specific metabolic processes are present in algae for the assimilation of nitrogen and phosphorus from their surrounding environment. Primary nitrogen intake by algae is in the form of ammonium (NH_4^+) and nitrate (NO_3^-), which are absorbed via enzymatic mechanisms. Sodium is easily absorbed by algae and used as a nitrogen source for protein production, whereas nitrate is converted into ammonium by nitrate reductase enzymes before being included into biomass (Beardall & Raven, 2020; Hussain *et al.*, 2019; Li-Beisson *et al.*, 2019). Proteins involved in phosphate transport facilitate the active movement of phosphate ions (PO_4^{3-}) across the cell membrane during phosphorus uptake in algae. Intracellularly, phosphate is employed for several metabolic activities such as nucleic acid production, energy metabolism, and membrane architecture development. Furthermore, algae may utilize phosphate storage processes, such as the buildup of polyphosphate, to retain surplus phosphorus for future utilization.

The absorption of nutrients by algae occurs via ion exchange and adsorption mechanisms, in which ions are transferred between the surface of the algal cell and the surrounding water or absorbed onto the cell surfaces. The process described enables algae to effectively sequester dissolved nutrients from the wastewater matrix, thereby promoting the removal of nutrients from the aqueous phase (Narvarte *et al.*, 2023; Sanz-Luque *et al.*, 2020; Zafar *et al.*, 2021). The nutrient concentration in the wastewater influent has a substantial effect on the rates at which algae absorb nutrients. Increased nutrient concentrations often lead to higher rates of absorption because they provide more substrate for algal biosynthesis. Nevertheless, high levels of nutrients can result in nutrient hypersaturation and impede the growth of algae, therefore restricting the effectiveness of nutrient removal (Fallahi *et al.*, 2021; Sutherland *et al.*, 2020).

The enhancement of algal biomass production and nutrient removal efficiency relies on the optimization of light conditions, including factors such as intensity, duration, and spectral quality. Optimal light availability is a critical determinant of algae growth and nutrient absorption in wastewater treatment systems. Photosynthesis in algae necessitates sufficient light energy to facilitate carbon fixation and nutrient assimilation activities. The kinetics of nutrient uptake and algal metabolism are significantly influenced by temperature and pH. The rates of algal growth and the efficiency of nitrogen uptake are typically greater at moderate temperatures and neutral pH regimes. Excessive temperatures or pH levels can impede the growth of algae and their absorption of nutrients, thereby impacting the overall effectiveness of the treatment. The stoichiometric ratio of nutrients, such as the nitrogen-to-phosphorus (N:P) ratio, can affect the preferences of algal organisms for absorbing nutrients and the composition of their biomass. Specific nutrient requirements and absorption preferences of algae are determined by their metabolic activity and growth circumstances. Deviation from balanced nutrient ratios can lead to either nutrient limitation or excess, which can affect the growth of algae and the effectiveness of nutrient removal (Fan *et al.*, 2020; Su, 2021; Zafar *et al.*, 2021).

The processes of assimilatory absorption, nitrification, and denitrification are among the key mechanisms by which algae contribute to the removal of nitrogen from wastewater (Fallahi *et al.*, 2021; Han & Zhou, 2022; Kumar & Bera, 2020). Assimilatory uptake refers to the direct integration of ammonium and nitrate into the biomass of algae, whereas nitrification is the conversion of ammonium to nitrate, which becomes available for assimilation by algae. Heterotrophic bacteria linked to algal biofilms or situated in anaerobic zones promote denitrification, resulting in the transformation of nitrate into nitrogen gas (N_2) and its subsequent emission into the atmosphere. Adsorption, precipitation, and biological assimilation processes are effective mechanisms by which algae eliminate phosphorus from wastewater. Organic phosphorus compounds are formed by the adsorption of phosphorus onto the surfaces of algal cells via electrostatic interactions or by their incorporation into algal biomass. Algae can also catalyse the formation of phosphorus as insoluble phosphate minerals, such as calcium phosphate or struvite, by releasing extracellular polymeric substances and generating alkalinity (Carrillo *et al.*, 2020; Mao *et al.*, 2021; Ummalyma *et al.*, 2023; Xing *et al.*, 2022). A comprehensive understanding of the mechanisms of nitrogen uptake by algae and the factors that affect nutrient removal efficiency is essential for optimizing the performance of

Table 7.2 Different models and studies on WWT by microalgae based on various PBRs.

Microalgae	Wastewater Origin	PBR Type	Culture Conditions	Removal Rates	Biomass Productivity	Reference
A consortium of *Chlorella* sp. and *Synechococcus* sp. with *Bacillus* sp. NITD 19	Coke-oven wastewater	Closed packed-bed PBR, 20 cm column bed height	24 h cultivation, 20 mL/h flow rate	Phenol 80.32%, NH_4-N 47.85%, NO_3-N 39.45%, fluoride 65.0%	–	Rai *et al.* (2023)
C. vulgaris	Synthetic agrochemical wastewater	Bubble column PBR	25–28°C and 3 light sources (blue, red, and solar) with light intensity in the range between 1800 and 2800 lux for 21 days	–	3.99 g/L at blue light	Dhanasekar and Sathyanathan (2023)
Mixed culture of *C. vulgaris, Scenedesmus obliquus, Botryococcus braunii, Botryococcus sudeticus,* and *Afrocarpus falcatus*		Closed system 4-L cylindrical PBR	pH = 7, 25°C	21% protein, 60% carbohydrate and 47% dissolved organic carbon (DOC)		Rafiee *et al.* (2020)
Chlorella pyrenoidosa	Food processing wastewater	Pilot-scale tubular PBR	–	15% lipid, 55% COD, 67% phosphorous, 88% nitrogen	1.8–2.1 g/L	Tan *et al.* (2021)
C. vulgaris	Piggery wastewater	MPBR	8000-lux LED lamps, CO_2 flow, 1–15 days	74.55% NH_4^+, 70.20% PO_4^{3-}, and 65.85% COD	1.1 g/L	Nguyen *et al.* (2022)
C. vulgaris (CPCC 90)	Municipal wastewater	MPBR	Illumination, HRT, and aeration, 50 days	76.7% nitrogen, and 66.2% phosphorus	3.48 g/L/day	Zou *et al.* (2022)
C. vulgaris	Urban wastewater	1.0 L PBR	5 days, dark cultivation	92.62% Phosphate	0.687	Amaro *et al.* (2023)
C. vulgaris	Anaerobic digestate of swine manure	1.5 column PBR	7 days, 25°C	95.12% NH_4^+, 76.87% phosphate	–	Shao *et al.* (2022)
Amphidinium carterae	Urban wastewater	10-L bubble column PBR	4 L/min air flow, 200 μmol photons/m²/s, 12 h/12 h light/dark (L/D) cycle, 20°C and pH 8.5	–	–	López-Rosales *et al.* (2022)
Mixed microalgae cultures' *Synedra acus, Spirulina* sp., *Euglena* sp., *Trichocerca* sp., *Paramecium* sp., and *Closteriopsis longissima*	Leachate wastewater	2.5 L cylindrical glass PBR	28 ± 0.5°C and an operating pH of 9.3 ± 0.1. 2600 lux fluorescent light source, 12/12 light/dark, 1 L/min airflow, 2% CO_2 of the airflow rate, 6 days cultivation	76.26% COD, 75.48% BOD, 74.86% nitrate, and 73.52% nitrite	117.654 mg/L	Emalya *et al.* (2023)

algae-based WWT systems. In order to improve the efficiency of nutrient removal and support the sustainable management of water resources, WWT plants can manipulate ambient conditions, nutrient concentrations, and algae species composition. Additional investigation is required to enhance our knowledge and enhance the effectiveness of algae-based nutrient removal technologies by studying the mechanisms of interactions between algae and nutrient cycling processes in wastewater treatment systems (Capodaglio & Olsson, 2019; Diaz-Elsayed *et al.*, 2019; Kurniawan *et al.*, 2021).

7.6 APPLICATION OF PBRS IN MICROALGAE WASTEWATER CULTIVATION

Many of the bioreactors used for WWT closely resemble the pure cultivation technologies. The design concepts are based on the principles of pure farming systems. In WWT, the unique properties of wastewater have led to the adoption of bioreactor designs based on WWT PBRs. Therefore, the PBRs for wastewater treatment by microalgae discussed in this chapter can be categorized into two primary types: suspended systems and fixed systems (Ali *et al.*, 2022). Within suspended systems, microalgae cells exist in a state devoid of other cells, allowing for unrestricted transport of contaminants at various dilution concentrations. This is the prevailing bioreactor employed in the treatment of microalgae wastewater. Suspended systems encompass open pond systems, tubular systems, flat bioreactor systems, and plastic bag systems. However, the isolation of the microalgae cell from wastewater can be facilitated by alternative separation techniques such as coagulation and membrane filtering, which consequently raise the overall cost of the growing system. A stationary PBR system comprises a biofilm bioreactor, an immobilized bioreactor, and a microalgae membrane bioreactor (Ali *et al.*, 2024; Patwardhan *et al.*, 2022). Although these particular PBRs have the potential to enhance separation efforts, they also have certain drawbacks in the context of WWT, notably in terms of biomass quantification (Table 7.2).

7.7 MODELLING AND OPTIMIZATION

7.7.1 Emerging PBRs for the treatment of wastewater

Several new PBRs have been developed by scientists for the purpose of treating wastewater, in addition to the conventional PBRs. An anaerobic piggy-digested effluent with a high organic load was treated by Min *et al.* (2014), who constructed a pilot scale system of PBRs that includes multi-storey trays. This particular type of PBR system can be installed in a greenhouse in order to regulate the temperature. A pH controller, a mixing tank, a sump pump, a CO_2 tank, and a stack of four trays are also included in this type of PBR system. Additionally, a supporting frame is employed. The dimensions of each PBR consisted of 4 ft in width, 32 ft in length, and 8 ft in height. Each tray had the capacity to store up to 6 in of water, which is comparable to a volume of 1800 cups. Each PBR had a total capacity of 7500 L, which included the mixing tank as well as four trays, respectively. A new supply of CO_2 was added to the mixture tank. Due to the stacked-tray bioreactor design that was utilized in the research, this particular type of PBR system possesses a number of benefits, including a high mixotrophic growth rate, low running costs, and a reduced land footprint. The adsorption rates of COD, total nitrogen (TN), and TP achieved a maximum of 19.15–23.19, 7.2, 2.65, and 3.19 g/(m^2 d), respectively, in the region of biomass productivity, COD, and TN, respectively. Using a porous sheet-shaped material, nylon filter sheets, and a reinforced glass fibre mesh as the framework, Shi *et al.* (2014) developed a prototype-scale twin-layer PBR that was employed for microalgae fixation. This PBR was produced by the researchers. In order to operate the twin-layer PBR, three modules with a single-metre-by-one-metre area were installed vertically on a metal rack. Under the influence of gravity, the wastewater proceeded to flow through the entirety of the twin-layer after being poured from the top of the twin layers. Due to this, the compounds that were present in the wastewater were uniformly distributed throughout the microalgae that were attached to the twin layers. When compared to the conventional suspended system, this particular type of PBR was able to decrease the amount of microalgae that was wasted and increase the amount of microalgae cells that were recycled. This particular type of PBR was employed for the purpose of treating municipal wastewater by *Halochlorella rubescens*. The elimination of TN and TP was achieved at a rate of 70–99%, whereas the biomass productivity was measured at 6.3 g/m^2/day.

7.7.2 Artificial neural network-genetic algorithm (ANN-GA) hybrid modelling

ANN modelling is an efficient accurate computerized tool which is used to establish the relationship between bioagents and pollutants removal. GA model helps to detect the growth optimized parameters which can maximize the removal efficiency (cultivation period, initial concentrations of pollutants, the concentration of used culture), besides reducing contamination levels (Rai *et al.*, 2023).

7.8 CONCLUSION

PBRs are promising technologies for microalgal cultivation and simultaneous wastewater treatment, with each system having unique advantages and disadvantages. Open systems are cost-efficient but face challenges such as contamination and poor light issues. Closed systems offer better control but higher costs and energy demands. Further technological advancements can improve the practicality and longevity of PBRs, contributing to improved WWT and resource recovery. However, existing limitations such as fouling, uneven resource distribution, and high operational expenses need to be addressed. Research on system architecture, scalability, and automation integration can significantly improve the practicality of PBRs for commercial applications. PBRs are essential for sustainable practices, contributing to the production of microalgal biomass, efficient WWT, and resource recovery.

REFERENCES

Abdelzaher M. A., Farahat E. M., Abdel-Ghafar H. M., Balboul B. A. A. and Awad M. M. (2023). Environmental policy to develop a conceptual design for the figure water–energy–food nexus: a case study in Wadi-Dara on the Red Sea Coast, Egypt. *Water*, **15**, 780, https://doi.org/10.3390/w15040780

Aditya L., Mahlia T. M. I., Nguyen L. N., Vu H. P. and Nghiem L. D. (2022). Microalgae–bacteria consortium for wastewater treatment and biomass production. *Science of the Total Environment*, ***838***, 155871, https://doi.org/10.1016/j.scitotenv.2022.155871

Ahmad I., Abdullah N., Koji I., Yuzir A. and Muhammad S. E. (2021). Evolution of photobioreactors: a review based on microalgal perspective. *IOP Conference Series: Materials Science and Engineering*, **1142**, 012004, https://doi.org/10.1088/1757-899X/1142/1/012004

Alazaiza M. Y. D., Albahnasawi A., Al Maskari T., Abujazar M. S. S., Bashir M. J. K., Nassani D. E. and Abu Amr S. S. (2023). Biofuel production using cultivated algae: technologies, economics, and its environmental impacts. *Energies*, **16**, 1316, https://doi.org/10.3390/en16031316

Ali S. S., Mastropetros S. G., Schagerl M., Sakarika M., Elsamahy T., El-Sheekh M., Sun J. and Kornaros M. (2022). Recent advances in wastewater microalgae-based biofuels production: a state-of-the-art review. *Energy Reports*, **8**, 13253–13280, https://doi.org/10.1016/j.egyr.2022.09.143

Ali S. S., Hassan L. H. S. and El-Sheekh M. (2024). Microalgae-mediated bioremediation: current trends and opportunities-a review. *Archives of Microbiology*, **206**, 343, https://doi.org/10.1007/s00203-024-04052-x

Amaro H. M., Sousa J. F., Salgado E. M., Pires J. C. M. and Nunes O. C. (2023). Microalgal systems, a green solution for wastewater conventional pollutants removal disinfection, and reduction of antibiotic resistance genes prevalence? *Applied Sciences*, **13**, 4266, https://doi.org/10.3390/app13074266

Arabian D. (2024). Design and fabrication of a simple and cost-effective microalgae culture photobioreactor and optimization of culture conditions. *Authorea*. Online press.

Arun S., Ramasamy S., Pakshirajan K., and Pugazhenthi G. (2022). Bioelectricity production and shortcut nitrogen removal by microalgal-bacterial consortia using membrane photosynthetic microbial fuel cell. *Journal of environmental management*, **301**, 113871. https://doi.org/10.1016/j.jenvman.2021.113871

Beardall J. and Raven J. A. (2020). Structural and biochemical features of carbon acquisition in algae. *Photosynthesis in Algae: Biochemical and Physiological Mechanisms*, **45**, 141–160, https://doi.org/10.1007/978-3-030-33397-3_7

Benner P., Meier L., Pfeffer A., Krüger K., Vargas J. A. O. and Weuster-Botz D. (2022). Lab-scale photobioreactor systems: principles, applications, and scalability. *Bioprocess and Biosystems Engineering*, **45**, 791–813, https://doi.org/10.1007/s00449-022-02711-1

Capodaglio A. G. and Olsson G. (2019). Energy issues in sustainable urban wastewater management: use, demand reduction and recovery in the urban water cycle. *Sustainability*, **12**(1), 266, https://doi.org/10.3390/su12010266

Carone M., Alpe D., Costantino V., Derossi C., Occhipinti A., Zanetti M. and Riggio V. A. (2022). Design and characterization of a new pressurized flat panel photobioreactor for microalgae cultivation and CO_2 bio-fixation. *Chemosphere*, **307**(2), 135755, https://doi.org/10.1016/j.chemosphere.2022.135755

Carrillo V., Fuentes B., Gómez G. and Vidal G. (2020). Characterization and recovery of phosphorus from wastewater by combined technologies. *Reviews in Environmental Science and Bio/Technology*, **19**, 389–418.

Chanana I., Kaur P., Kumar L., Kumar P. and Kulshreshtha S. (2023). Advancements in microalgal biorefinery technologies and their economic analysis and positioning in energy resource market. *Fermentation*, **9**, 202, https://doi.org/10.3390/fermentation9030202

De Souza Kirne P.C., Vandenberghe L., Soccol C., and De Carvalho J.C. (2022). Mixing and agitation in photobioreactors. In: Innovations in Fermentation and Phytopharmaceutical Technologies, H. Thatoi, S. Mohapatra, S.K. Das (eds.), Elsevier, Amsterdam, The Netherlands (2022), 13–35. https://doi.org/10.1016/b978-0-12-821877-8.00005-1

Dhanasekar S. and Sathyanathan R. (2023). Bioenergy potential of *Chlorella vulgaris* under the influence of different light conditions in a bubble column photobioreactor. *Global Journal of Environmental Science and Management*, **9**(4), 789–804.

Díaz O., González E., Vera L., Fernández L. J., Díaz-Marrero A. R. and Fernández J. J. (2023). Recirculating packed-bed biofilm photobioreactor combined with membrane ultrafiltration as advanced wastewater treatment. *Environmental Science and Pollution Research*, **30**, 69977–69990, https://doi.org/10.1007/s11356-023-27309-2

Diaz-Elsayed N., Rezaei N., Guo T., Mohebbi S. and Zhang Q. (2019). Wastewater-based resource recovery technologies across scale: a review. *Resources, Conservation and Recycling*, **145**, 94–112, https://doi.org/10.1016/j.resconrec.2018.12.035

Dineshbabu G., Goswami G., Kumar R., Sinha A. and Das D. (2019). Microalgae – nutritious, sustainable aqua- and animal feed source. *Journal of Functional Foods*, **62**, 103545, https://doi.org/10.1016/j.jff.2019.103545

Dos Santos T. L., Belasco G. C., Pompei C. M. E., Quartaroli L., Pessi B. A., De Silva G. R. L. and Silva G. H. R. (2024). Wastewater treatment in tubular photobioreactors: exploring the impact of support medium. In: G. Mannina, A. Cosenza and A. Mineo (eds), Resource Recovery from Wastewater Treatment. ICWRR 2024. Lecture Notes in Civil Engineering, Springer, Cham, Vol. **524**, pp. 252–257.

Ekin İ. (2020). Types of microalgae cultivation photobioreactors and production process of microalgal biodiesel as alternative fuel. *Acta Biologica Turcica*, **33**(2), 114–131.

Emalya N., Mairiza L., Bilqis P. Z., Suhendrayatna S., Munawar E. and Yunardi Y. (2023). Removal of organic and nitrogen compounds from domestic landfill leachate by microalgae. *Biointerface Research in Applied Chemistry*, **13**(2), 131.

Fallahi A., Rezvani F., Asgharnejad H., Nazloo E. K., Hajinajaf N. and Higgins B. (2021). Interactions of microalgae bacteria consortia for nutrient removal from wastewater: a review. *Chemosphere*, **272**, 129878, https://doi.org/10.1016/j.chemosphere.2021.129878

Fan X., Xu D., Wang D., Wang Y., Zhang X. and Ye N. (2020). Nutrient uptake and transporter gene expression of ammonium, nitrate, and phosphorus in *Ulva linza*: adaption to variable concentrations and temperatures. *Journal of Applied Phycology*, **32**, 1311–1322, https://doi.org/10.1007/s10811-020-02050-2

Gabrielyan D. A., Sinetova M. A., Gabel B. V., Gabrielian A. K., Markelova A. G., Rodionova M. V., Bedbenov V. S., Shcherbakova N. V. and Los D. A. (2022). Cultivation of *Chlorella sorokiniana* IPPAS C-1 in flat-panel photobioreactors: from a laboratory to a pilot scale. *Life (Basel)*, **12**(9), 1309.

González-Camejo J., Barat R., Ruano V., Seco A., and Ferrer J. (2020). Improving membrane photobioreactor performance by reducing light path: operating conditions and key performance indicators. *Water research*, **172**, 115518, https://doi.org/10.1016/j.watres.2020.115518

Han F. and Zhou W. (2022). Nitrogen recovery from wastewater by microbial assimilation – a review. *Bioresource Technology*, **363**, 127933. https://doi.org/10.1016/j.biortech.2022.127933

Hashmi Z., Bilad M. R., Fahrurrozi Z. J., Lim J. W. and Wibisono Y. (2023). Recent progress in microalgae-based technologies for industrial wastewater treatment. *Fermentation*, **9**, 311, https://doi.org/10.3390/fermentation9030311

Hoang A. T., Sirohi R., Pandey A., Nižetić S., Lam S. S., Chen W.H., Luque R., Thomas S., Arıcı M. and Pham V. V. (2022). Biofuel production from microalgae: challenges and chances. *Phytochemistry Reviews*, **22**, 1089–1126. https://doi.org/10.1007/s11101-022-09819-y

Hussain M. M., Bibi I., Shahid M., Shaheen S. M., Shakoor M. B., Bashir S., Younas F., Rinklebe J. and Niazi N. K. (2019). Biogeochemical cycling, speciation and transformation pathways of arsenic in aquatic environments with the emphasis on algae. *Comprehensive Analytical Chemistry*, **85**, 15–51.

Kumar A. and Bera S. (2020). Revisiting nitrogen utilization in algae: a review on the process of regulation and assimilation. *Bioresource Technology Reports*, **12**, 100584, https://doi.org/10.1016/j.biteb.2020.100584

Kurniawan S. B., Ahmad A., Said N. S. M., Imron M. F., Abdullah S. R. S., Othman A. R., Purwanti I. F. and Hasan H. A. (2021). Macrophytes as wastewater treatment agents: nutrient uptake and potential of produced biomass utilization toward circular economy initiatives. *Science of the Total Environment*, **790**, 148219, https://doi.org/10.1016/j.scitotenv.2021.148219

Larronde-Larretche M. and Jin X. (2022). The influence of forward osmosis module configuration on nutrients removal and microalgae harvesting in osmotic photobioreactor. *Membranes*, **12**, 892, https://doi.org/10.3390/membranes12090892

Li G., Xiao W., Yang T. and Lyu T. (2023). Optimization and process effect for microalgae carbon dioxide fixation technology applications based on carbon capture: a comprehensive review. *Journal of Carbon Research*, **9**, 35, https://doi.org/10.3390/c9010035

Li-Beisson Y., Thelen J. J., Fedosejevs E. and Harwood J. L. (2019). The lipid biochemistry of eukaryotic algae. *Progress in Lipid Research*, **74**, 31–68, https://doi.org/10.1016/j.plipres.2019.01.003

López-Rosales L., López-García P., Benyachou M. A., Molina-Miras A., Gallardo-Rodríguez J. J., Cerón-García M. C., Sánchez Mirón A. and García-Camacho F. (2022). Treatment of secondary urban wastewater with a low ammonium-tolerant marine microalga using zeolite-based adsorption. *Bioresource Technology*, **359**, 127490, https://doi.org/10.1016/j.biortech.2022.127490

Madigan M. T., Bender K. S., Buckley D. H., Sattley W. M. and Stahl D. A. (2021). Brock Biology of Microorganisms, 16th ed., Pearson Education, Harlow (UK), ISBN 978-1-292-40479-0.

Mahmood T., Hussain N., Shahbaz A., Mulla S. I., Iqbal H. M. N. and Bila M. (2023). Sustainable production of biofuels from the algae-derived biomass. *Bioprocess and Biosystems Engineering*, **46**, 1077–1097, https://doi.org/10.1007/s00449-022-02796-8

Mao Y., Xiong R., Gao X., Jiang L., Peng Y. and Xue Y. (2021). Analysis of the status and improvement of microalgal phosphorus removal from municipal wastewater. *Processes*, **9**(9), 1486, https://doi.org/10.3390/pr9091486

Min M., Hu B., Mohr M.J., Shi A., Ding J., Sun Y., Jiang Y., Fu Z., Griffith R., Hussain F., Mu D., Nie Y., Chen P., Zhou W., and Ruan R. (2014). Swine manure-based pilot-scale algal biomassproduction system for fuel production and wastewater treatment–a case study. *Applied Biochemistry Biotechnology*, **172**, 1390–1406, http://doi.org/10.1007/s12010-013-0603-6

Mohsenpour S. F., Hennige S., Willoughby N., Adeloye A. and Gutierrez T. (2021). Integrating micro-algae into wastewater treatment: a review. *The Science of the Total Environment*, **752**, 142168, https://doi.org/10.1016/j.scitotenv.2020.142168

Molina-Grima E., García-Camacho F., Acién-Fernández F. G., Sánchez-Mirón A., Plouviez M., Shene C. and Chisti Y. (2022). Pathogens and predators impacting commercial production of microalgae and cyanobacteria. *Biotechnology Advances*, **55**, 107884, https://doi.org/10.1016/j.biotechadv.2021.107884

Narvarte B. C. V., Hinaloc L. A. R., Gonzaga S. M. C., Crisostomo B. A., Genovia T. G. T. and Roleda M. Y. (2023). Nitrate and phosphate uptake of morphologically distinct calcified macroalgae. *Phycologia*, **36**(2), 1–10.

Nguyen M. T., Nguyen T. P., Pham T. H., Duong T. T., Do M. V., Trinh T. V., Nguyen Q. T. X. and Trinh VM. (2022). Removal of nutrients and COD in wastewater from Vietnamese piggery farm by the culture of *Chlorella vulgaris* in a pilot-scaled membrane photobioreactor. *Water*, **14**, 3645, https://doi.org/10.3390/w14223645

Ogundele O. D., Adewumi A. J. and Oyegoke D. A. (2023). Phycoremediation: algae as an effective agent for sustainable remediation and waste water treatment. *Environmental and Earth Sciences Research Journal*, **10**, 7–17, https://doi.org/10.18280/eesrj.100102

Oruganti R. K., Katam K., Show P. L., Gadhamshetty V., Upadhyayula V. K. K. and Bhattacharyya D. (2022). A comprehensive review on the use of algal–bacterial systems for wastewater treatment with emphasis on nutrient and micropollutant. *Bioengineered*, **13**(4), 10412–10453, https://doi.org/10.1080/21655979.2022.2056823

Otálora P., Guzmán J. L., Berenguel M. and Acién F. G. (2023). Data-driven pH model in raceway reactors for freshwater and wastewater cultures. *Mathematics*, **11**, 1614, https://doi.org/10.3390/math11071614

Pavlou A., Penloglou G. and Kiparissides C. (2023). Evaluation of tolerant to CO_2 excess microalgae for the production of multiple biochemicals in a 3G biorefinery. *Sustainability*, **15**, 3889, https://doi.org/10.3390/su15053889

Patwardhan S. B., Pandit S., Ghosh D., Dhar D. W., Banerjee S., Joshi S., Gupta P. K., Lahiri D., Nag M., Ruokolainen J., Ray R. R. and Kesari K. K. (2022). A concise review on the cultivation of microalgal biofilms for biofuel feedstock production. *Biomass Conversion and Biorefinery*, **14**(8), 7219–7236.

Pérez-Mora L. S., Doria L. M., Cezare-Gomes E. de-A., and Santo E. do-E. (2023). Phycoremediation Processes for Secondary Effluent from Sewage Treatment Plants Using Photosynthetic Microorganisms: A Review. *Applied Microbiology*, **3**(2), 400–416. http://doi.org/10.3390/applmicrobiol3020028

PompeI C. M. E., Bolzani H. R. and da Silva G. H. R. (2023). Use of native microalgae in anaerobic sewage treatment: lab and pilot-scale approaches. *Journal of Applied Phycology*, **35**, 2865–2879, https://doi.org/10.1007/s10811-023-03081-1

Porto B., Gonçalves A. L., Esteves A. F., de Souza S. M. A. G. U., de Souza A. A. U., Vilar V. J. P. and Pires J. C. M. (2021). Assessing the potential of microalgae for nutrients removal from a landfill leachate using an innovative tubular photobioreactor. *Chemical Engineering Journal*, **413**, 127546, https://doi.org/10.1016/j.cej.2020.127546

Rafiee P., Ebrahimi S., Hosseini M. and Tong Y. W. (2020). Characterization of soluble algal products (SAPs) after electrocoagulation of a mixed algal culture. *Biotechnology Reports*, **25**, e00433, https://doi.org/10.1016/j.btre.2020.e00433

Rai A., Bishayee B., Dey A., Kumar A., Lahiri S. K., Chakrabarty J. and Dutta S. (2023). Tertiary treatment of coke-oven wastewater using suspended and immobilized whole live cells of constructed bacterial–microalgal consortium: modeling and optimization using ANN-GA hybrid methodology. *Water Science and Technology: A Journal of the International Association on Water Pollution Research*, **87**(3), 509–526, https://doi.org/10.2166/wst.2023.023

Ramasamy, S., Arun, S., Pakshirajan, K. (2020). An overview of algal photobioreactors for resource recovery from waste. In L. Singh, A. Yousuf, D. M. Mahapatra (eds.), Bioreactors, Elsevier, London, UK, pp. 215–248. https://doi.org/10.1016/b978-0-12-821264-6.00013-9

Razzak S. A., Bahar K., Islam K. M. O., Haniffa A. K., Faruque M. O., Hossain S. M. Z. and Hossain M. M. (2024). Microalgae cultivation in photobioreactors: sustainable solutions for a greener future. *Green Chemical Engineering*, **5**, 418–439, https://doi.org/10.1016/j.gce.2023.10.004

Sanz-Luque E., Bhaya D. and Grossman A. R. (2020). Polyphosphate: a multifunctional metabolite in cyanobacteria and algae. *Frontiers in Plant Science*, **11**, 938, https://doi.org/10.3389/fpls.2020.00938

Sarker N. K. (2022). Exploring the potential of wastewater reclamation by means of outdoor cultivation of microalgae in photobioreactors. *Energy, Ecology and Environment*, **7**, 473–488, https://doi.org/10.1007/s40974-021-00207-4

Sedighi M., Pourmoghaddam-Qhazvini P. and Amidpour M. (2023). Algae-powered buildings: a review of an innovative, sustainable approach in the built environment. *Sustainability*, **15**, 3729, https://doi.org/10.3390/su15043729

Sepúlveda-Muñoz C. A., de Godos I. and Muñoz R. (2023). Wastewater treatment using photosynthetic microorganisms. *Symmetry*, **15**, 525, https://doi.org/10.3390/sym15020525

Shao H., Sun Y., Jiang X., Hu J., Guo C., Lu C., Guo F., Sun C., Wang Y. and Dai C. (2022). Towards biomass production and wastewater treatment by enhancing the microalgae-based nutrients recovery from liquid digestate in an innovative photobioreactor integrated with dialysis bag. *Journal of Environmental Management*, **317**, 115337, https://doi.org/10.1016/j.jenvman.2022.115337

Shekh A., Sharma A., Schenk P. M., Kumar G. and Mudliar S. (2022). Microalgae cultivation: photobioreactors, CO_2 utilization, and value-added products of industrial importance. *Journal of Chemical Technology & Biotechnology*, **97**, 1064–1085, https://doi.org/10.1002/jctb.6902

Shi J., Podola B. and Melkonian M. (2014). Application of a prototype-scale twin-layer photobioreactor for effective N and P removal from different process stages of municipal wastewater by immobilized microalgae. *Bioresource Technology*, **154**, 260–266, https://doi.org/10.1016/j.biortech.2013.11.100

Su Y. (2021). Revisiting carbon, nitrogen, and phosphorus metabolisms in microalgae for wastewater treatment. *Science of the Total Environment*, **762**, 144590, https://doi.org/10.1016/j.scitotenv.2020.144590

Sutherland D. L., Park J., Heubeck S., Ralph P. J. and Craggs R. J. (2020). Size matters – microalgae production and nutrient removal in wastewater treatment high rate algal ponds of three different sizes. *Algal Research*, **45**, 101734, https://doi.org/10.1016/j.algal.2019.101734

Tan X. B., Lam M. K., Uemura Y., Lim J. W., Wong C. Y. and Lee K. T. (2018). Cultivation of microalgae for biodiesel production: a review on upstream and downstream processing. *Chinese Journal of Chemical Engineering*, **26**, 17–30, https://doi.org/10.1016/j.cjche.2017.08.010

Tan X. B., Wan X. P., Yang L. B., Wang X., Meng J., Jiang M. J. and Pi H. J. (2021). Nutrients recycling and biomass production from *Chlorella pyrenoidosa* culture using anaerobic food processing wastewater in a pilot-scale tubular photobioreactor. *Chemosphere*, **270**, 129459, https://doi.org/10.1016/j.chemosphere.2020.129459

Touloupakis E. (2023). Photosynthetic microorganisms: cultivation and application. *Symmetry*, **15**, 742, https://doi.org/10.3390/sym15030742

Ummalyma S. B., Sirohi R., Udayan A., Yadav P., Raj A., Sim S. J. and Pandey A. (2023). Sustainable microalgal biomass production in food industry wastewater for low-cost biorefinery products: a review. *Phytochemistry Reviews*, **22**, 969–991, https://doi.org/10.1007/s11101-022-09814-3

Viswanaathan S., Perumal P. K. and Sundaram S. (2022). Integrated approach for carbon sequestration and wastewater treatment using algal–bacterial consortia: opportunities and challenges. *Sustainability*, **14**, 1075, https://doi.org/10.3390/su14031075

Whitton R., Ometto F., Pidou M., Jarvis P., Villa R. and Jefferson B. (2015). Microalgae for municipal wastewater nutrient remediation: mechanisms, reactors and outlook for tertiary treatment. *Environmental Technology Reviews*, **4**(1), 133–148, https://doi.org/10.1080/21622515.2015.1105308

Xing D., Li X., Jin C., Zhao Y., Yang X., Gao M., Ji J., Zu Zonglian S. and Guo L. (2022). Coupling of wastewater treatment and microalgae products recovery: effect of phosphorus components on the biochemical composition and phosphorus absorption of *Chlorella pyrenoidosa*. *Journal of Water Process Engineering*, **49**, 102968, https://doi.org/10.1016/j.jwpe.2022.102968

Yadavalli R., Rao C. S., Rao R. S. and Potumarthi R. (2014). Dairy effluent treatment and lipids production by *Chlorella pyrenoidosa* and *Euglena gracilis*: study on open and closed systems. *Asia-Pacific Journal of Chemical Engineering*, **9**(3), 368–373, https://doi.org/10.1002/apj.1805

You N., Deng S., Wang C., Ngo H. H., Wang X., Yu H., Tang L. and Han J. (2023). Review and opinions on the research, development and application of microalgae culture technologies for resource recovery from wastewater. *Water*, **15**, 1192, https://doi.org/10.3390/w15061192

Yousuf A. (2020). Fundamentals of microalgae cultivation. In: Microalgae Cultivation for Biofuels Production. Elsevier, New York, pp. 1–9. https://doi.org/10.1016/B978-0-12-817536-1.00001-1

Yu B. S., Lee S. Y. and Sim S. J. (2022). Effective contamination control strategies facilitating axenic cultivation of *Haematococcus pluvialis*: risks and challenges. *Bioresource Technology*, **344**, 12628.

Zafar A. M., Javed M. A., Hassan A. A., Mehmood K. and Sahle-Demessie E. (2021). Recent updates on ions and nutrients uptake by halotolerant freshwater and marine microalgae in conditions of high salinity. *Journal of Water Process Engineering*, **44**, 102382. https://doi.org/10.1016/j.jwpe.2021.102382

Zhang M., Leung K.-T., Lin H. and Liao B. (2020). The biological performance of a novel microalgal–bacterial membrane photobioreactor: effects of HRT and N/P ratio. *Chemosphere*, **261**, 128199, https://doi.org/10.1016/j.chemosphere.2020.128199

Zhou W., Lu Q., Han P. and Li J. (2020). Microalgae cultivation and photobioreactor design. In Microalgae Cultivation for Biofuels Production, A. Yousuf, (ed.), Elsevier, Amsterdam, pp. 31–50. https://doi.org/10.1016/B978-0-12-817536-1.00003-5

Zou H., Rutta N. C., Chen S., Zhang M., Lin H. and Liao B. (2022). Membrane photobioreactor applied for municipal wastewater treatment at a high solids retention time: effects of microalgae decay on treatment performance and biomass properties. *Membranes*, **12**(6), 564, https://doi.org/10.3390/membranes12060564

doi: 10.2166/9781789064100_0139

Chapter 8

The state of wastewater treatment plants and how it is affecting the effluent discharge quality within the water bodies of Gauteng

P. Jiyane[1], K. B. Hoorzook[1] and N. Janse van Rensburg[1]

1st Process Energy Environment Technology Station (PEETS), Faculty of Engineering and Built Environment, University of Johannesburg, P.O. Box 17011, Doornfontein, Johannesburg 2028, South Africa
Corresponding author: kousaro@uj.ac.za

ABSTRACT

A crucial component of both environmental sustainability and human welfare is water quality. The demand for freshwater resources has significantly increased due to growing urbanization and industrialization, which has resulted in the production of enormous amounts of wastewater. Wastewater is the water-carried waste that is removed from formal and informal residences, institutions, commercial, and industrial establishments, together with the groundwater, surface water, stormwater, and potable water as may be present. The primary objective of collection and wastewater treatment is to ensure that the effluent discharged to the environment does not pose serious threats to both human health and environment. There is an increasing concern of wastewater treatment plants (WWTPs) to perform against the required standards and expectations. The majority of WWTPs fail to meet at least three or more of the respective effluent discharge standards. Their performance infrastructure: challenges such as ageing infrastructure; historical low base budget allocation for repairs, operations, and maintenance; 'catch-up' mode to ensure optimal operation and maintenance planning with regards to the installation of new infrastructure and technologies; under investment in sewer collection and sludge handling infrastructure; low skill base to manage, operate, and maintain the specialized nature of wastewater services; meeting the demand of rapidly increasing housing developments; and inadequate design or technology decisions for the specific municipal environment. The effectiveness of WWTPs in improving water quality is evidenced by the removal of contaminants such as suspended solids, organic matter, nitrogen, phosphorus, and pathogens. Their successful operation leads to reduced eutrophication, decreased waterborne diseases, and the protection of aquatic biodiversity. Moreover, treated wastewater can be safely discharged into natural water bodies or reused for various purposes, including irrigation, industrial processes, and even potable water supply in water-stressed regions. The above paragraph serves as an overview on how WWTPs play a crucial role in safeguarding water quality and promoting sustainable water management practices. As global populations continue to grow, and environmental challenges become increasingly complex, the development and implementation of innovative technologies and sustainable practices in WWTPs are paramount. These efforts not only contribute to cleaner water but also align with broader goals of environmental conservation and public health protection.

Keywords: Effluent discharge, Gauteng, wastewater treatment.

8.1 INTRODUCTION

Water is an essential resource for various human activities, including everyday necessities, agriculture, manufacturing, and transportation. However, despite its importance, water is the most poorly managed resource in the world (Chutter, 1998). The World Health Organization (WHO) defines water pollution as any degradation of the physical, chemical, or biological parameters that has a detrimental effect on environmental organisms or renders the water resource unfit for its intended use. Every time water is used, it acquires contaminants, and its quality decreases. Nearly 80% of the used water is returned to the environment untreated, which increases freshwater scarcity worldwide.

Contaminated water may cause human diseases due to a wide range of viruses, bacteria, and protozoa which may be present in the water. In addition to biological contaminants, wastewater effluents are also contaminated by chemicals such as heavy metals (Cantinho *et al.*, 2016), nitrogen, phosphorus, and organic compounds (Barreca *et al.*, 2018, 2019) such as detergents, pesticides, hydrocarbons, and metabolites.

Most of the mentioned contaminants are present as wastewater treatment plants' (WWTP) effluents especially the heavy metals which are associated with industrial discharges (Cantinho *et al.*, 2016).

Therefore, to ensure the sustainability of the environment, it is imperative to improve the management and utilization of natural resources while reducing human pollution activities. Consequently, various legislative documents and guidelines have been developed to address these issues. These include the WHO Guidelines for the reuse of effluents, which were first developed in 1973 and revised in 1989 and 2006 (Victor *et al.*, 2008), the UN General Assembly's Millennium Development Goal for ensuring environment sustainability, which was adopted in 2000 (The Millennium Development Goals Report, 2015), the Water Framework Directive (WFD), which was adopted in 2000 (European Parliament, 2000), and its sub-directives 91/271/EEC concerning urban wastewater treatment, which establishes requirements for discharges from urban WWTPs (Directive, 1991), and 98/83/EC concerning the quality required for drinking water (Directive, 1998).

8.1.1 Contribution of this chapter

This chapter highlights the critical importance of WWTPs in Gauteng, emphasizing their role in maintaining water quality within the region's water bodies. The findings reveal that the current state of many wastewater treatment facilities is suboptimal, leading to the discharge of inadequately treated effluents into rivers and streams. This poor effluent quality is contributing to the degradation of water resources, threatening aquatic ecosystems, and posing health risks to communities reliant on these water bodies for domestic and agricultural purposes. This chapter underscores the urgent need for upgrading and better management of these facilities to protect water quality and public health.

8.2 CURRENT STATE OF WWTPS IN GAUTENG

The current state of WWTPs in Gauteng and South Africa is concerning. It has led to newly identified pollutants in our water sources which pose a threat to both environmental and human health. The pollution of these water sources is often due to the release of waste products from WWTPs.

The current WWTPs struggle to eliminate these pollutants effectively in real time because of the constant changes in process parameters, as well as the variety and levels of the pollutants (Ngila *et al.*, 2020).

The 2022 Green Drop report, the first of its kind since 2013, indicates a significant decline in the operation of wastewater treatment facilities (SA's Dismal State of Wastewater Management – Mail & Guardian, 2022). In 2022, the Department of Water and Sanitation (DWS) sent non-compliance notices to 244 wastewater treatment operations, but only half of them had responded nearly a year later (SAFACTS, 2023/2024). Most of the treatment plants evaluated from November 2022 to February 2023 were found to be unable to produce potable water that meets the SANS 241:2015 drinking water standards. Out of 824 treatment plants in South Africa, it is estimated that only 60 are discharging clean water (SA Water Crisis: Report reveals grim state of infrastructure – BizNews, 2023).

This situation signifies a notable and swift deterioration in local government performance. The management of these facilities is vital for the health and safety of the public and the environment. It is crucial that these problems are addressed quickly and efficiently (SA's Dismal State of Wastewater Management – Mail & Guardian, 2022), to be able to fulfil the fundamental needs of the populations and take care of the environment properly.

The DWS has published the complete Blue Drop Report for 2023, which evaluates the quality of drinking water; the complete No Drop Report, which concentrates on water losses and non-revenue water in all municipalities across the nation; and the Green Drop Progress Assessment Report, which offers an update on the effectiveness of wastewater management systems at the municipal level (Blue Drop Report, 2023).

The goals of the Blue, Green, and No Drop Certification programmes are to enhance the quality of municipal drinking water, manage wastewater, conserve water, and control demand. The reports provide reliable statistics and information regarding the condition of water and sanitation systems in the nation, keeping stakeholders and the public informed. Institutions that provide water services and attain excellence in doing so are acknowledged. Infrastructure owners receive direction and advice on how to enhance their water and sanitation services. The presentation that goes with the reports describes the actions being taken to tighten regulations and assist towns that are having difficulty meeting national standards and norms for services (Blue Drop Report, 2023).

These incentive-based regulatory programmes were implemented by DWS in 2008, as the water sector regulator under the National Water Act and the Water Services Act. DWS discontinued the programmes in 2014. In 2021, the programmes were resumed by Senzo Mchunu, Minister of Water and Sanitation. DWS published progress reports for Blue Drop and No Drop as well as the complete Green Drop Report in 2022. Given that wastewater treatment works (WWTWs) are the focus of the Green Drop Progress Assessment report. It was discovered that 64% of WWTWs are very or critically vulnerable to leakage of partially treated or untreated water into the environment, including rivers. Since 2013, there have been more WWTWs in the high- and critical-risk categories (Blue Drop Report, 2023).

Wastewater treatment is a crucial process that helps to reduce the number of pollutants in wastewater before it is discharged into water bodies. The discharge of raw and improperly treated wastewater into water bodies can have adverse effects on the environment and human health (Edokpayi *et al.*, 2017). Wastewater effluents are major contributors to a variety of water pollution problems, and the degradation of surface water bodies (Edokpayi *et al.*, 2017). Most cities generate on average 30–70 mm^3 of wastewater per person per year and owing to the lack of improper wastewater treatment facilities, wastewater and its effluents are often discharged into surface water sources, which are receptacles for domestic and industrial wastes, resulting in pollution (Edokpayi *et al.*, 2017). The release of raw and improperly treated wastewater onto water courses has both short- and long-term effects on the environment and human health. Hence, there should be proper enforcement of water and environmental laws to protect the health of inhabitants of both rural and urban communities (Edokpayi *et al.*, 2017).

Taking for example the Nancefield region receives sewage discharge daily from Olifantsvlei (182 ml/day), Bushkoppies (186 ml/day), and Goudkoppies (126 ml/day) WWTWs (Vermaak, 2010). The raw sludge thickened by gravity is brought to the Goudkoppies WWTW, where it is processed, dewatered, dried in the sun, and disposed of. The Regional Sludge Handling Facility at Olifantsvlei receives gravity-thickened waste-activated sludge for dewatering and composting. The Goudkoppies WWTW collects and treats sewage from Johannesburg's southeast and City centres (Marara & Palamuleni, 2019).

Beginning in 1978, construction of the Goudkoppies WWTW included installing a new head of works that included the following processes: screening, degritting, primary sedimentation, activated sludge incorporating the five-stage Phoredox process, activation, chlorination, final clarification, digestion, dewatering, and solar sludge drying (Marara & Palamuleni, 2019). The treatment process at Bushkoppies Works involves mixing raw thickened sludge with waste-activated sludge, which is then anaerobically broken down (Marara & Palamuleni, 2019). Following this, solar drying and mechanical dewatering through linear screen or belt presses are employed. The final dried sludge is disposed of on private farmland. In the context of sewage treatment, phosphates (P) and nitrogen (N) are abundant. Nitrogen, abundant in the atmosphere as N_2, is essential for life but can lead to eutrophication when present in excess in water bodies, primarily due to nitrate (NO_3) forms (Adelyn *et al.*, 2022).

Human activities, notably fertilizer use and sewage discharge, significantly contribute to elevated nitrate levels in water, promoting excessive algae and plant growth (Marara & Palamuleni, 2019). Phosphorus (P) is a necessary component for all life; 5–8 parts per million of phosphate are present in wastewater undergoing primary and secondary treatment; most of this phosphate is attributed to household soaps and detergents (Correll, 1999). A high phosphorus concentration in a water body is detrimental due to eutrophication, a process where excessive nutrients, particularly phosphorus, promote the rapid growth of algae. This algal bloom can deplete oxygen levels in the water as the algae die and decompose, creating hypoxic or anoxic conditions that harm aquatic life. The resulting oxygen depletion can lead to dead zones where aquatic organisms cannot survive, disrupting ecosystems and harming fish populations. Additionally, some algal blooms produce toxins that are dangerous to both aquatic organisms and humans, affecting water quality and posing health risks (Correll, 1999). Due to poor infrastructure maintenance and overcapacity, raw sewage is pumped from the Olifantsvlei and Goudkoppies WWTPs directly into the Klip River and its headwaters (Figure 8.1).

8.2.1 Gauteng: municipal wastewater corrective action plan status

8.2.1.1 Overview of critical wastewater systems

The 2022 Green Drop report and the Gauteng provincial chapter identified nine systems to be in a critical state, compared to zero (none) in 2013 (Green Drop 2022 Report Release, 2022). Most of these systems are managed by Merafong and Rand West, with one system each in the City of Tshwane and Lesedi (Green Drop 2022 Report Release, 2022). The provincial risk ratio for treatment plants remained constant in medium risk space, at 58.8% in 2021, compared to 58.5% in 2013, which suggests limited risk movement since 2013 (Green Drop 2022 Report Release, 2022). The most prominent risks were associated with WWTWs that exceeded their design capacity, dysfunctional processes,

Figure 8.1 Raw sewage outflow into the Olifantsvlei Nature Reserve at the Olifantsvlei Wastewater Treatment Works. Taken by Kyle van Heyde, May 2024.

Table 8.1 Gauteng 2021 Green Drop Summary.

WSA Name	2013 GD Score (%)	2021 GD Score (%)	2021 Critical State (<31%)
Lesedi LM	78	79 (up) 60 (down)	Devon
City of Tshwane	82	24 (down)	Klipgat
Rand West LM		8.27	
Randfontein LM	67	20	Randfontein, Hannes van Niekerk
Westonaria LM	40		
Merafong LM	54		Khutsong, Kokosi-Fochville, Murray & Roberts, Wedela, Welverdiend
Total	–	–	9

Source: Green Drop 2022 Report Release (2022).

and equipment (especially disinfection), and effluent and sludge non-compliance. Opportunities are presented in terms of reducing cost through process optimization, improved energy efficiency, and beneficial use of sludge, nutrients, biogas, and other energy resources (Green Drop 2022 Report Release, 2022) (Table 8.1).

8.2.2 Provincial Green Drop analysis

The 100% response from the nine municipalities audited during the 2021 Green Drop process demonstrated a firm commitment to wastewater services in the province. Local Government reforms resulted in the merging of Randfontein LM and Westonaria LM into Rand West LM. Therefore, nine Water Services Authority/Water Services Act (WSAs) were audited in 2021 compared to the 10 WSAs in 2013. The analysis for the periods 2009, 2011, 2013, and 2021 indicated that most of the system scores were in the 50–80% (Average Performance) category, with the 80–90% (Good Performance) being the next largest category. The most concerning finding was that nine systems resided in critical state (<31%) compared to zero systems in this category in 2013 (Green Drop 2022 Report Release, 2022).

8.2.3 Regulatory enforcement

WWTPs which failed to achieve the minimum Green Drop target of 31% were placed under Regulatory focus. The Regulator required these municipalities to submit a detailed corrective action plan within 60 days from publishing of the Green Drop 2022 report. Four municipalities and nine wastewater systems that received Green Drop scores below 31% were placed under regulatory surveillance, in accordance with the Water Services Act (108 of 1997). In addition, these municipalities were advised to ringfence water services grant allocation to rectify/restore wastewater collection and treatment shortcomings identified (Green Drop 2022 Report Release, 2022) (Table 8.2).

The following municipalities and their associated WWTPs were found to be in high Cumulative Risk Rating (CRR) risk positions, which means that some or all the risk indicators were in a precarious state, that is, operational flow, technical capacity, and effluent quality. WWTWs in high risk and critical risk positions pose a serious risk to public health and the environment. The following municipalities were required to assess their risk contributors and develop corrective measures to mitigate these risks (Green Drop 2022 Report Release, 2022).

The Green Drop Watch Report monitors the actions and progress taken by 334 critical-state wastewater systems since April 2022, as well as the enforcement measures initiated by the Department's regional offices to address defective services. The Watch report reveals vulnerabilities and deficiencies by the responding municipalities, with some being more despondent than others. Despite the implementation of support and enforcement measures by DWS (and relevant sector

Table 8.2 WWTWs with (<31%) Green Drop scores.

WSA Name	2013 GD Score (%)	WWTWs with <31% score
Lesedi LM	79	Devon
City of Tshwane	60	Klipgat
Rand West LM	24	Randfontein, Hannes van Niekerk
Merafong LM	21	Khutsong, Kokosi-Fochville, Murray & Roberts, Wedela, Welverdiend

Source: Green Drop 2022; Report Release (2022).

partners), the interventions appear to lack the desired impact to halt declining services or ensure timely and successful turnaround. This begs a rethink of the type and nature of interventions required to deepen and fast-track intercessions to match the magnitude and pace of decline and lack of responsiveness by municipalities who persist in critical state.

8.2.4 Highlight the issues related to poor operation and maintenance of wastewater and sewage treatment infrastructure

The lack of improper wastewater treatment facilities, wastewater and its effluents are often discharged into surface water resources, which are receptables for domestic and industrial wastes, resulting in pollution (Mema, 2010). The poor quality of wastewater effluents is responsible for the degradation of the receiving surface water body. Wastewater effluents should be treated efficiently to avert adverse health risks of the user of surface water resources and the aquatic ecosystem. The release of raw and improperly treated wastewater onto water courses has both short- and long-term effects on the environment and human health. Hence there should be proper enforcement of water and environmental laws to protect the health of inhabitants of both rural and urban communities (Mema, 2010).

Wastewater comprises of all used water in homes and industries including storm water and runoffs from lands, which must be treated before it is released into the environment to prevent any risk it may have on the environment and human health. Wastewater is treated in wastewater treatment facilities where the conventional pollution parameters such as biochemical oxygen demand, chemical oxygen demand, total suspended solids, and nutrients are removed from the wastewater stream so that the effluent do not constituent new sources of pollution (Ratola *et al.*, 2012). However, it has been discovered that the wastewater organic load contains high levels of a variety of hazardous organic pollutants, and thus, additional treatment steps and control measures become very necessary (Ratola *et al.*, 2012; Weber *et al.*, 2006).

The quality of wastewater varies according to the types of effluents WWTPs receive such as domestic wastewater, dry and wet atmospheric deposition, urban runoff containing traffic-related pollution, or agricultural runoff (Ratola *et al.*, 2012). The range of contaminants becomes broader when industrial wastewater is included into the raw water stream that enters a WWTP (Katsoyiannis & Samara, 2004, 2005; Ratola *et al.*, 2012). It has been shown that wastewater effluents contain emerging organic contaminants such as persistent organic pollutants, brominated flame retardants, perfluorinated compounds, and pharmaceuticals, which may not be removed during the treatment process (Müller *et al.*, 2007; Ratola *et al.*, 2012). This has fostered the technology used for wastewater treatment to evolve so quickly to meet the demands of the modern world.

In many countries, urbanization is growing at an unprecedented rate, and such development is often unbalanced with much of the disposable municipal expenditure devoted to high-profile infrastructure with waste disposal and management coming well down the list of priorities in terms of allocation of funds (Klinch & Stuart, 1999).

8.3 IMPACT OF WASTEWATER DISCHARGE ON WATER BODIES

Research conducted by various bodies such as Saving Water South Africa (Saving Water SA, 2011) has indicated that less than half of the South African WWTPs treat wastewater they receive to a safe and acceptable level (Edokpayi *et al.*, 2015). The health risk from wastewater usually comes from microbial pathogens, nutrient loads, heavy metals, and some organic chemicals (Edokpayi *et al.*, 2015; Toze, 1997). Bacteria are the most common pathogens usually found in treated wastewater and cause several infections and diseases particularly to young, pregnant, immune-compromised, and aged people (Ashbolt *et al.*, 1997; Edokpayi *et al.*, 2015).

Most wastewater treatment facilities in South Africa dispose their effluents directly to nearby rivers or streams, which are used by the surrounding villages for their various water needs. A case study of the Olifantsvlei WWTP in the Gauteng province, South Africa found that effluents from the plant was not meeting the required standards for discharge into the environment, which could lead to negative impacts on the receiving environment (Manus and van der Merwe-Botha, 2009). The study also found that the plant was not operating at full capacity, which could be due to a lack of maintenance and management (Manus and van der Merwe-Botha, 2009). This was further confirmed in another study where it was illustrated that even though the treatment plant can remove some pathogens from the influent, effluent discharges were only occasionally devoid of the organisms, a condition that posed a health threat from infectious diseases (Manus and van der Merwe-Botha, 2009). These findings suggest that inadequate investment in wastewater treatment infrastructure, shortage of skilled manpower, poor planning, or corruption (Edokpayi *et al.*, 2015) could have resulted in the poor performance of the WWTP.

Wastewater needs to be adequately treated prior to its disposal or reuse to protect receiving water bodies from contamination (Edokpayi *et al.*, 2015). The discharge of poorly treated wastewater usually affects water users downstream and contaminates groundwater (Edokpayi *et al.*, 2015). Waste stabilization ponds (WSPs) are usually used to provide an effective and low-cost means of handling domestic wastewater for smaller towns and communities (Ogola *et al.*, 2009). The use of WSPs is advantageous over the conventional WWTPs because they are very simple to design, operate, and most wastewater, dissolved oxygen (DO), is consumed by bacteria. When this happens the DO level drops a threshold point, with a negative impact on life as they are unable to continue their normal life sustaining processes such as growth and reproduction. Such decrease affects fish and other aquatic life.

The health impacts of the contamination of the surface water with pathogenic organisms in wastewater could result in the transmission of waterborne diseases for people who use the water for domestic and other purposes downstream (Chigor *et al.*, 2013; World Health Organization, 2006). Approximately 25% of all the deaths worldwide are the result of infectious diseases caused by pathogenic microorganisms (UNEP (United Nations Environment Programme), 2006). Scientists have identified about 1400 species of microorganisms that can cause ill health, including bacteria, protozoa, protozoan parasites, parasitic worms, fungi, and viruses (Blue Drop Report, 2023). The major concern of wastewater discharge of various classes of pathogens is that they can cause diseases of various magnitudes in human beings. Unlike some of the environmental impacts that can take a long time before they manifest, pathogens cause immediate negative health impacts on people who use contaminated surface water resources for domestic, agricultural, and recreational purposes. Some common pathogens found in untreated and ill-treated wastewater are listed in Table 8.3.

According to the WHO, some of the waterborne diseases that pose a high risk to South Africans include gastroenteritis, cholera, viral hepatitis, typhoid fever, bilharziasis, and dysentery (Kotzé, 2023).

To further validate the pathogens found in untreated wastewater listed in Table 8.3, a study conducted in Gauteng province, South Africa, assessed the effectiveness of four WWTPs (Baviaanspoort, Zeekoegat, Rayton, and Refilwe) in the removal of *Cryptosporidium* and *Giardia* (Dungeni & Momba, 2010). The study discovered that although the treatment plants were able to remove some pathogens from the influent, effluent discharges were only occasionally devoid of the organisms, a condition that posed a health threat from infectious diseases (Mema, 2010). Another study found that the

Table 8.3 Pathogens found in untreated wastewater.

Agent	Species	Disease
Bacteria	*Campylobacter jejune*	Gastroenteritis (possibly long-term sequelae, e.g. arthritis)
	E. coli	Gastroenteritis
	E. coli O157:H7	Bloody diarrhoea, haemolytic uremic syndrome
	Helicobacter pylori	Abdominal pain, peptic ulcers, and gastric cancer
	Salmonella spp.	Salmonellosis, gastroenteritis, diarrhoea (possibly long-term sequelae, e.g. arthritis)
	Salmonella typhi	Typhoid fever
	Shigella spp.	Dysentery (possibly long-term sequelae, e.g. arthritis)
	V. cholerae	Cholera
Helminths	*Ascaris lumbricoides* (roundworm)	Ascariasis
	Ancylostoma duodenal and *Necator americanus* (hookworm)	
	Clonorchis sinensis (liver fluke)	Clonorchiasis
	Fasciola (liver fluke)	Fascioliasis
	Fasciolopsis buski (intestinal fluke)	Fascioloidiasis
	Opisthorchis viverrine	Opisthorchiasis
	Schistosoma (blood fluke)	Schistosomiasis (bilharzia)
	Trichuris (whipworm)	Trichuriasis
	Taenia (tapeworm)	Taeniasis
Protozoa	*Balantidium coli*	Balantidiasis (dysentery)
	Cryptosporidium parvum	Cryptosporidiosis
	Cyclospora cayetanensis	Persistent diarrhoea
	Entamoeba histolytica	Amoebiasis (amoebic dysentery)
	Giardia lamblia	Giardiasis
Viruses	*Adenovirus*	Respiratory disease and eye infections
	Astrovirus	Gastroenteritis
	Calivivirus	Gastroenteritis
	Enteroviruses	Gastroenteritis
	Echovirus	Fever, rash, respiratory, heart disease, and aseptic meningitis
	Poliovirus	Gastroenteritis
	Hepatitis A and E	Infectious hepatitis
	Parvovirus	Gastroenteritis
	Norovirus	Gastroenteritis
	Rotavirus	Gastroenteritis
	Coxsackie viruses	Herpangina, aseptic meningitis, respiratory illness, fever, paralysis, heart, and kidney disease

Source: World Health Organization (2006).

abundance of *Cryptosporidium* and *Giardia* spp. in treated effluents produced by the four WWTPs in the Gauteng province, South Africa was high, and the effluent from the plants was not meeting the required standards for discharge into the environment, which could lead to negative impacts on the receiving environment (Dungeni & Momba, 2010).

In another study presented by a WRC Research Manager, Dr Eunice Ubomba-Jaswa, water samples were collected and analysed from various points in natural bodies within the Hammanskraal area in the Tshwane Metro Council, which revealed a very high degree of faecal contamination. This was confirmed by the presence of excessive numbers of *Escherichia coli* and the presence of other known enteric pathogens such as *Salmonella* (a common bacterial disease that affects the intestinal tract) and *Shigella* (a bacterium that causes a diarrhoeal illness). The results indicated that the quality of the river water was not acceptable for recreational use including religious ritual baptisms, swimming, and bathing. Samples collected along the Apies River downstream from the Temba Wastewater Treatment Works tested positive for the presence of the toxigenic *Vibrio cholerae* strain. In the study, Ubomba-Jaswa also added that a cholera strain was detected in water samples collected from the Kaalplaasspruit, which flows into the Apies River further downstream in the Hammanskraal area. Even though the study was not able to pinpoint the source of the Hammanskraal cholera outbreak it highlighted some important public health risks which may well contribute towards causing future cholera and/or other waterborne disease outbreaks (Anon, 2023).

The test results of the Rooiwal Wastewater Treatment Works and Temba (drinking) Water Treatment Works indicated that both plants as dysfunctional with the final treated wastewater and final treated drinking water were found not to be compliant with the regulated water quality standards. Additionally, excessive pollution levels were observed in both the Pienaars and the Apies Rivers systems, manifested by the dominant visual presence of blue-green algae and water hyacinth (Anon, 2023).

8.4 CASE STUDIES

Consider a case study titled the evaluation of the Impacts of Municipal Wastewater Treatment on the Receiving Environment on the Olifantsvlei Wastewater Treatment Plant in the Gauteng province, South Africa (Mothetha, 2016). The study highlighted the challenges and successes of the WWTP. The identified challenges included the effluents from the plant not meeting the required standards for discharge into the environment, which has a potential negative impact on the receiving environment (Mothetha, 2016). In addition to the effluents, it was identified that the plant was not operating at full capacity which could be due to a lack of maintenance and management (Mothetha, 2016). Besides the challenges identified from the Olifantsvlei WWTP, the plant was able to remove a significant number of pollutants from the wastewater, which would have otherwise been discharged into the environment (Mothetha, 2016). The study found that the plant was able to generate biogas from the wastewater treatment process, which was used to generate electricity for the plant (Mothetha, 2016).

8.5 STRATEGIES FOR IMPROVEMENT

Several studies conducted on most of the non-compliant plants have indicated that their root cause for the poor performance does not seem to be the need for additional or upgraded plant infrastructure or the need for additional funding, but the challenge is that the available plant infrastructure and equipment are not well operated and/or sufficiently maintained. The most pressing deficiency is the critical shortage of trained, skilled, and experienced process controllers and mechanical/electrical maintenance staff. It has been found that most plants operate with insufficient information to guide optimized operations.

The most problematic unit process is disinfection. More than two-thirds of the plants equipped with disinfection or attempting to disinfect the treated wastewater are experiencing problems. Inadequate disposal and use of sludge was found at 81% of the plants. None of the plants visited would comply with the new South African Sludge Guidelines. In terms of electrical and mechanical maintenance of the plant infrastructure, immediate intervention is required at approximately 30% of the plants while some kind of intervention would be required at more than 60% of the plants. Only 10% of plants are maintained to ensure acceptable functionality. The underlying cause and reason for the overall poor track record related to plant and equipment is the lack of capable and trained maintenance staff. This was also reflected in the resources available for wastewater plant operations. The compounding effect of all the evaluated aspects including the lack of maintenance becomes clear when the compliance is assessed. The Department of Water Affairs and Forestry requires a 95-percentile compliance with the conditions stipulated in the authorization. If these criteria were strictly enforced, only 4% of the surveyed plants are adequately operated and maintained. Immediate intervention is required at approximately 30% of the plants to avoid crisis situations such as an outbreak of waterborne diseases. In the short-to-medium term, intervention would be required at more than 66% of the plants. The overall conclusion of the national survey is that the majority of micro-, small-, and medium-sized WWTPs in South Africa are in trouble and do not comply with the regulatory standards. The key aspects to resolve this situation relate to the availability of trained and competent process controllers and skilled mechanical/electrical maintenance crews (Snyman *et al.*, 2006).

Even though a one-size-fits-all approach does not work for municipalities but rather a differentiated approach, based on the specific circumstance and base information for a particular municipality, may be more effective in overall turnaround of wastewater services in the municipal sector. Prioritizing the specific WWTWs that are regarded to have the highest impact on the receiving environment and those that do not have the minimum of critical data readily available, by applying the following rationale:

- Select the WWTWs with Cumulative Risk Rating (CRR) with >18 within a specific province. Develop and implement risk-reduction action plans for those plants by targeting a specific (reduced) CRR target to be achieved within a specific period.
- After achieving target and deducing the CRR >18 category risks to their target risk, take the same approach by targeting CRR 10–18, and so forth (Manus & van der Merwe-Botha, 2009).

The other response to these challenges has been an effort to develop more robust but low-cost treatment methods. To achieve this balance between cost and robustness, a lot of scientific and engineering research needs to be carried out to exploit the competencies at the interface of the relevant disciplines.

There are two primary approaches that have been used recently in which the first involves combining cutting-edge techniques (enhanced oxidation processes) with traditional ones (biodegradation). The second one is the process intensification. Earlier research has shown that integrating biological processes with heterogeneous photocatalysis improves the performance of anaerobic digesters. Furthermore, the performance of the integrated systems can be improved through reactor choice and process intensification, in which individual reactors are optimized. Fluidized-bed reactors are of special interest because they have high performance for multiphase operations such as heterogeneous photocatalytic degradation of bio-recalcitrant wastewater.

Photocatalysis has been considered as one of the quick methods for breaking down resistant organic pollutants in wastewater with titanium dioxide (TiO_2) being widely applied as an appropriate photocatalyst due to its high activity and in slurry or suspension during photocatalysis. However, the problem of applying TiO_2 in slurry form has been the difficulty in separating it from the treated effluent because it does not settle but remains in suspension.

Another aspect that is essential in improving the state of WWTPs and the quality of effluent discharge is addressing the skill set needed to implement cutting-edge technologies to address emerging contaminants such as acid mine drainage (AMD), pharmaceutical wastes, and other bio-recalcitrant wastes, which pose problems to both water and wastewater treatment systems. The skill requirements for the processes that manage such wastes include the use of membranes, photocatalysis, and adsorption techniques. Some South African universities including the University of Stellenbosch, University of the Witwatersrand, and the Vaal University of Technology have developed membrane techniques for AMD remediation and advanced oxidation processes. However, these skills have not been transferred to the people who need them most, local government, and industries (Ochieng Aoyi *et al.*, 2015).

When it comes to progress integration of addressing the problem of emerging contaminants, there are two approaches that define the trajectory of the research path that adds value to life. The first one is the use of waste or low-cost materials as feedstock for energy production. This approach is often described as waste beneficiation. In recent years, the efficiency of many processes has been enhanced using nano-modified materials. The second approach involves the integration of conventional energy production processes with advanced ones to improve production. The application of process integration in wastewater treatment is an area that has not been developed in many engineering curricula in South Africa.

When it comes to water management, sustainable process engineering (SPE) is one of the emerging technologies, which can optimize the use of water thus improving the state of WWTPs and the quality of effluent discharge; however, this technology still needs a lot of enhancement. SPE technology aims at reducing water consumption and hence reduces water demand thereby lifting off the burden of waste treatment plants (Ochieng Aoyi *et al.*, 2015).

8.6 FUTURE PROSPECTS

Water is an essential resource for various human activities, including everyday necessities, agriculture, manufacturing, and transportation. When it comes to the future aspects of addressing the state of WWTPs and how it is affecting the effluent discharge quality within the water bodies of Gauteng, a more structured programme will be put forward on one of the programmatic areas of interventions. Effective coordination will be a critical success factor. It will also be necessary to prioritize actions, to make the best possible use of limited institutional, technical, and financial resources.

The Department of Water Affairs on the contrary has embarked on the following targeted, risk-based regulatory action:

Risk-based, plant-size specific: Prioritizing the specific WWTWs that are regarded to have the highest impact on the receiving environment and those that do not have the minimum of critical data readily available, by applying the following rationale:

- Seven or higher (micro- and small-sized WWTWs) to be the indicative point where attention needs to be focused. In this event it would imply that 197 of the WWTWs are experiencing serious to critical problems and need urgent to critical attention to reduce their risk profile.
- Ten or higher (medium-sized WWTWs) to be the indicative point where attention needs to be focused. In this event it would imply that 147 of the WWTWs are experiencing serious to critical problems and need urgent to critical attention to reduce their risk profile.
- Fourteen or higher (large-sized WWTWs) to be the indicative point where attention needs to be focused. In this event it would imply that 48 of the WWTWs are experiencing serious problems and need urgent attention to reduce their risk profile.
- Twenty-eight or higher (macro-sized WWTWs) to be the indicative point where attention needs to be focused. In this event it would imply that that two of the WWTWs are experiencing serious problems and need urgent attention to reduce their risk profile.

Risk-based, plant-size non-specific: Prioritizing the specific WWTWs that are regarded to have the highest impact on the receiving environment and those that do not have the minimum of critical data readily available, by applying the following rationale:

- Select the WWTWs with CRR with >18 within a specific province. Develop and implement risk-reduction action plans for those plants by targeting a specific (reduced) CRR to be achieved within a specific time period.

- After achieving the aforementioned target and deducing the CRR >18 category risks to their target risk, take the same approach by targeting CRR 10–18, and so forth.

Risk-based, geo-graphical spread: Prioritizing a specific water management area or municipal area that represent WWTWs with high impact on the receiving environment and with insufficient operational/compliance data and apply the same rationale as with risk-based plant-size non-specific approach (Manus & van der Merwe-Botha, 2009).

8.7 CONCLUSION

The general finding of this chapter indicated that most of Gauteng's micro-, small-, and medium-sized wastewater treatment facilities are in crises and do not adhere to the regulatory standards. The key aspects to resolve this situation relate to the availability of trained and competent process controllers and skilled mechanical/electrical maintenance crews. There is also a need to prioritize the specific WWTWs that are regarded to have the highest impact on the receiving environment and those that do not have the minimum of critical data readily available. This will enable an effort to develop more robust but low-cost treatment methods. This balance between cost and robustness can be achieved through a lot of scientific and engineering research to exploit the competencies at the interface of the relevant disciplines.

The other factor for the failing state of wastewater treatment facilities in Gauteng as well as other provinces includes inadequate coverage of wastewater treatment facilities in both urban and rural areas, poor operational state of wastewater infrastructure, design weaknesses, expertise, corruption, insufficient funds allocated for wastewater treatment, overloaded capacities of existing facilities, and inefficient monitoring for compliance with recommended guidelines.

According to reports, 63% of South Africans live in cities, with 40% of them residing in metropolitan towns. According to demographic predictions, 7.8 million more South Africans will live in cities after 15 years, and an additional 6 million will do so after 20 more years, in 2050 (Department of Cooperative Governance and Traditional Affairs (DCOG), 2014; National Planning Commission, 2012; South African Cities Network (SACN), 2011).

Given these figures, some of the recommendations are that there is a need to:

(a) coordinate advisory functions and apply differentiated regulations;
(b) inculcate in the public a sense of deep reverence for water as is the case in the Middle East;
(c) generate ideas that can be converted into content, which result in the development of relevant curriculum to equip people with skills relevant to our situation;
(d) develop more skills in science and engineering;
(e) collect and exchange data; identify priorities, pollution sources, and hotspots;
(f) generate region-specific data and extrapolate to other regions, only where applicable;
(g) implement low-cost but efficient technologies and management strategies such as intelligent pressure regulation.

ACKNOWLEDGEMENT

The authors wish to acknowledge Dr Marlene van der Merwe-Botha, Director of WaterGroup Holdings (PTY) Ltd, and Kyle van Heyde, UJ PEETS, for the drone picture, figure 8.1.

REFERENCES

Adelyn A. O., Ndlovu S. S., Ngwagwe L. M., Mudau M., Alabi M. A. and Edokpayi J. N. (2022). Wetland resources in South Africa: threats and metadata study. *Resources*, **11**(6), 54, https://doi.org/10.3390/resources11060054

Anon. (2023). sanews. https://www.sanews.gov.za/south-africa/study-unable-pinpoint-source-hammanskraal-cholera-outbreak (accessed 03/01/2024).

Ashbolt N. J., Dorsch M. R., Cox P. T. and Banens P. (1997). Blooming *E. coli*, what do they mean? *Special Publications of the Royal Society of Chemistry*, **191**, 78–88.

Barreca S., Busetto M., Vitelli M., Colzani L., Clerici L. and Dellavedova P. (2018). Online solid-phase extraction LC-MS/MS: a rapid and valid method for the determination of perfluorinated compounds at sub ng· L^{-1} level in natural water. *Journal of Chemistry*, **2018**, 1–9, https://doi.org/10.1155/2018/3780825

Blue Drop Report (2023). Department of Water and Sanitation. https://ws.dws.gov.za/iris/releases/BDN_2023_Report.pdf

Barreca S., Busetto M., Colzani L., Clerici L., Daverio D., Dellavedova P., Balzamo S., Calabretta E. and Ubaldi V. (2019). Determination of estrogenic endocrine disruptors in water at sub-ng L^{-1} levels in compliance with Decision 2015/495/EU using offline–online solid phase extraction concentration coupled with high performance liquid chromatography-tandem mass spectrometry. *Microchemical Journal*, **147**, 1186–1191, https://doi.org/10.1016/j.microc.2019.04.030

Cantinho P., Matos M., Trancoso M. A. and dos Santos M. C. (2016). Behaviour and fate of metals in urban wastewater treatment plants: a review. *International Journal of Environmental Science and Technology*, **13**, 359–386, https://doi.org/10.1007/s13762-015-0887-x

Chigor V. N., Sibanda T. and Okoh A. I. (2013). Studies on the bacteriological qualities of the Buffalo River and three source water dams along its course in the Eastern Cape Province of South Africa. *Environmental Science and Pollution Research*, **20**, 4125–4136, https://doi.org/10.1007/s11356-012-1348-4

Chutter F. M. (1998). Research on the rapid biological assessment of water quality impacts in streams and rivers: Final report to the Water Research Commission. WRC.

Correll D. L. (1999). Phosphorus: a rate-limiting nutrient in surface waters. *Poultry Science*, **78**(5), 674–682, https://doi.org/10.1093/ps/78.5.674

Department of Cooperative Governance and Traditional Affairs (DCOG) (2014). Integrated urban development framework: Draft for discussion. Pretoria.

Directive E. U. W. (1991). Council Directive of 21. May 1991 concerning urban wastewater treatment (91/271/EEC). *Journal of European Communications*, **34**, 40.

Directive C. (1998). On the quality of water intended for human consumption. *Official Journal of the European Communities*, **330**, 32–54.

Dungeni M. O. M. B. A. and Momba M. N. B. (2010). The abundance of *Cryptosporidium* and *Giardia* spp. in treated effluents produced by four wastewater treatment plants in the Gauteng Province of South Africa. *Water SA*, **36**(4).

Edokpayi J. N., Odiyo J. O., Msagati T. A. and Popoola E. O. (2015). A novel approach for the removal of lead(II) ion from wastewater using mucilaginous leaves of *Diceriocaryum eriocarpum* plant. *Sustainability*, **7**(10), 14026–14041, https://doi.org/10.3390/su71014026

Edokpayi J. N., Odiyo J. O. and Durowoju O. S. (2017). Impact of wastewater on surface water quality in developing countries: a case study of South Africa. Water Quality, **10**(66561), pp. 10–5772.

European Parliament, C. (2000). Directive 2000/60/EC of the European Parliament and of the Council of 23 October 2000 establishing a framework for Community action in the field of water policy. *Official Journal of the European Communities*, **327**, 1–72.

Green Drop 2022 Report (2022). Water Institute of Southern Africa. https://wisa.org.za/2022/04/01/green-drop-2022-report-release/

Katsoyiannis A. and Samara C. (2004). Persistent organic pollutants (POPs) in the sewage treatment plant of Thessaloniki, northern Greece: occurrence and removal. *Water Research*, **38**(11), 2685–2698, https://doi.org/10.1016/j.watres.2004.03.027

Katsoyiannis A. and Samara C. (2005). Persistent organic pollutants (POPs) in the conventional activated sludge treatment process: fate and mass balance. *Environmental Research*, **97**(3), 245–257, https://doi.org/10.1016/j.envres.2004.09.001

Klinch B. A. and Stuart M. E. (1999). Human health risk in relation to landfill leachate quality; British Geological Survey. Technical Report, pp. 1–55.

Kotzé P. (2023). New technology to transform water quality testing in South Africa. *Water Wheel*, **22**(3), 16–19.

Manus L. and van der Merwe-Botha M. (2009). Raising wastewater treatment performance through incentive- and risk-based targeted regulation. Proceedings, Water Institute of South Africa (WISA) 2010.

Marara T. and Palamuleni L. G. (2019). An environmental risk assessment of the Klip River using water quality indices. *Physics and Chemistry of the Earth, Parts A/B/C*, **114**, 102799, https://doi.org/10.1016/j.pce.2019.09.001

Mema V. (2010). Impact of poorly maintained wastewater sewage treatment plants – lessons from South Africa: wastewater management. *ReSource*, **12**(3), 60–65.
Mothetha M. L. (2016). Evaluation of the impacts of municipal wastewater treatment on the receiving environment: A case study of the Olifantsvlei wastewater treatment plant in the Gauteng Province, South Africa, University of South Africa, Pretoria, https://uir.unisa.ac.za/handle/10500/22059
Müller K., Magesan G. N. and Bolan N. S. (2007). A critical review of the influence of effluent irrigation on the fate of pesticides in soil. *Agriculture, Ecosystems & Environment*, **120**(2–4), 93–116, https://doi.org/10.1016/j.agee.2006.08.016
National Planning Commission (2012). National Development Plan 2030. Our future – make it work. http://www.npconline.co.za/MediaLib/Downloads/Downloads/NDP%202030%20-%20Our%20future%20-%20make%20it%20work.pdf (accessed 19 November 2014).
Ngila J. C., Matheri A. N., Muckoya V., Ngigi E. and Ntuli F. (2020). Mathematical modelling for biological wastewater treatment plants, Gauteng, South Africa.
Ochieng Aoyi M. O., Majozi T., Seid E., Leswifi T., Rwanga S. and Kesi J. (2015). Water and wastewater management in local government: Skills needs and development.
Ogola J. S., Chimuka L. and Maina D. (2009, July). Occurrence and fate of trace metals in and around treatment and disposal facilities in Limpopo Province, South Africa: A case of two areas. Proceedings of the IASTED International Conference on Environmental Management and Engineering.
Ratola N., Cincinelli A., Alves A. and Katsoyiannis A. (2012). Occurrence of organic microcontaminants in the wastewater treatment process. A mini review. *Journal of Hazardous Materials*, **239**, 1–18, https://doi.org/10.1016/j.jhazmat.2012.05.040
SAFACTS (2023/2024). List of wastewater treatment plant in South Africa. https://safacts.co.za/list-of-wastewater-treatment-plant-in-south-africa/.
SA's Dismal State of Wastewater Management – Mail & Guardian (2022). https://mg.co.za/the-green-guardian/2022-04-06-only-3-of-sas-wastewater-treatment-plants-function-properly/.
Saving Water, S.A. (2011). Over half of wastewater treatment plants are well below standard. [Internet].
SA Water Crisis: Report reveals grim state of infrastructure – BizNews (2023). https://www.biznews.com/news/2023/06/12/water-report-infrastructure.
Snyman H. G., Van Niekerk A. M. and Rajasakran N. (2006, May). Sustainable wastewater treatment – what has gone wrong and how do we get back on track. Water Institute of Southern Africa Conference, Durban, May.
South African Cities Network (SACN). (2011). 2011 State of SA Cities Report. Johannesburg: SACN.
The Millennium Development Goals Report (2015). The United Nation. New York, NY, USA.
Toze R. S. G. (1997). Microbial pathogens in wastewater: literature review for urban water systems multi-divisional research program.
UNEP (United Nations Environment Programme) (2006). Water quality for ecosystem and human health. United Nations Environmental Monitoring System (GEMS)/Water Programme.
Vermaak V. (2010). A geomorphological investigation of the Klip River wetland, south of Johannesburg. Doctoral dissertation, University of the Witwatersrand.
Victor R., Kotter R., O'Brien G., Mitropoulos M. and Panayi G. (2008). WHO guidelines for the safe use of wastewater, excreta and greywater, volumes 1–4.
Weber S., Khan S. and Hollender J. (2006). Human risk assessment of organic contaminants in reclaimed wastewater used for irrigation. *Desalination*, **187**(1–3), 53–64, https://doi.org/10.1016/j.desal.2005.04.067
World Health Organization. (2006). France. WHO guidelines for the safe use of wastewater excreta and greywater, (Vol. **4**). World Health Organization.

doi: 10.2166/9781789064100_0153

Chapter 9

Microalgal wastewater treatment: a circular economy approach

*Nilay Kumar Sarker and Prasad Kaparaju**

School of Engineering and Built Environment, Griffith University (Nathan Campus), Queensland, Australia
**Corresponding author: Prasad Kaparaju, p.kaparaju@griffith.edu.au*

ABSTRACT

In this chapter, we explore the application of microalgae in wastewater treatment and the advantages of this integrated approach in fulfilling various sustainable development goals via circularity. We highlight how this integration contributes to environmental protection, reduction in chemical usage, and decrease in carbon emissions while fostering a green system that supports the energy, aquaculture, and agriculture sectors. This is achieved by converting the algal biomass generated from wastewater treatment into algae-based biofuel, fish feed, and fertilizer. Additionally, we examine the obstacles and necessary policy support needed to commercialize and expand this field. The role of microalgae in nutrient assimilation, pollutant removal, and biomass production is discussed alongside the technological advancements in cultivation methods, including photobioreactors and hybrid treatment systems. The chapter also delves into the economic and environmental impacts of adopting a circular economy model, highlighting the potential for resource recovery and sustainable development.

Keywords: Bioremediation, circular economy, microalgae, wastewater treatment

9.1 INTRODUCTION

Population growth and societal development increase the demand for natural resources, such as clean water, beyond natural replenishment capacities. This growth also escalates municipal wastewater production, potentially causing environmental issues if not properly treated. Eutrophication, driven by excess nitrogen and phosphorus, is a significant environmental concern, leading to anoxic conditions and ecosystem degradation. Proper municipal wastewater treatment is essential to mitigate environmental impacts and public health risks from untreated components (Nobre *et al.*, 2024).

The future will see a shift towards circular economy principles, emphasizing reuse and recycling. Wastewater will be seen as a resource, transforming treatment plants into resource recovery facilities. This will address water stress and nutrient scarcities, such as phosphate and zinc, which are crucial for agriculture and other applications (Fernandes *et al.*, 2022).

Microalgae cultivation offers a promising alternative for traditional effluent treatment, supporting resource and energy recovery. First proposed decades ago, microalgae-based treatment has been extensively studied. Microalgae remove nitrogen and phosphorus through growth, complying with legislative limits, and can also uptake carbon dioxide (CO_2) (Nobre *et al.*, 2024). The resulting biomass can be valorized for biofuels and high-value products. This chapter will explore the circular approach of microalgal wastewater treatment. Our discussion will focus on the applications of microalgal biomass in aquaculture, agriculture, and biofuel production. This is due to the technical challenges and public perception issues associated with using wastewater-derived algal biomass in food, feed, pharmaceuticals, and cosmetics.

9.2 OVERVIEW OF CURRENT WASTEWATER GENERATION AND TREATMENT

9.2.1 Wastewater generation

The generation of wastewater is a growing global concern due to rapid industrialization and urbanization. In many regions, the volume of wastewater produced is increasing significantly, posing severe challenges for treatment and management. According to a UN-Water Analytical Brief, approximately 80% of the world's wastewater is discharged untreated into the environment, resulting in widespread pollution and health risks (Reno *et al.*, 2020).

Industrial sectors such as the agro-industrial and pharmaceutical industries are major contributors to wastewater generation. For instance, the palm oil industry, particularly in countries such as Malaysia and Indonesia, generates large volumes of palm oil mill effluent (POME). POME is characterized by its high strength and significant organic content, which includes high levels of chemical oxygen demand (COD) and biochemical oxygen demand (BOD) (Maaroff *et al.*, 2019). Similarly, the ayurvedic pharmaceutical industry produces wastewater with high concentrations of COD, BOD, and total solids, necessitating effective treatment methods (Rao *et al.*, 2011).

In the context of agricultural wastewater, dairy farms produce effluents rich in nutrients and organic matter. The management and treatment of these effluents are critical to prevent environmental degradation and comply with regulatory standards (Othman *et al.*, 2013). Moreover, wastewater from soy sauce production and other food processing industries also contributes to the overall wastewater burden, highlighting the need for innovative treatment solutions (Harun *et al.*, 2020).

9.2.2 Wastewater treatment

Effective wastewater treatment is essential to mitigate the environmental and health impacts of wastewater discharge. Various technologies and processes have been developed to treat wastewater from different sources. Conventional methods include physical, chemical, and biological treatments, each with specific advantages and limitations.

9.2.2.1 Physical and chemical treatments

Electrocoagulation is an example of an effective electrochemical technique used to treat industrial wastewater. It is praised for its simplicity, low operating costs, and high efficiency in removing pollutants such as COD and suspended solids. However, it is often used as a pretreatment step due to its limitations in achieving complete pollutant removal (Gerardi, 2006).

9.2.2.2 Biological treatments

Biological treatment methods, particularly those involving microalgae, are gaining attention for their dual benefits of wastewater treatment and biomass production. Microalgae can uptake nutrients and heavy metals from wastewater, converting them into valuable biomass that can be used for biofuel production and other applications. This approach not only treats wastewater effectively but also contributes to the circular economy by generating renewable resources (Milhazes-Cunha & Otero, 2017).

For instance, the integration of microalgal cultivation with wastewater treatment has shown promising results. Species such as *Chlorella vulgaris* have demonstrated high efficiency in removing nutrients from wastewater while producing significant amounts of biomass (Cheah *et al.*, 2018).

Additionally, the use of microalgae in conjunction with other treatment methods, such as anaerobic digestion (AD), can enhance the overall efficiency of wastewater treatment processes by providing a secondary treatment step and generating bioenergy.

9.2.2.3 *Innovative approaches*

Recent advancements in wastewater treatment include the use of aerobic granular sludge and bioelectrochemical systems (BESs). Aerobic granular sludge systems are effective in treating high-strength industrial effluents due to their compact structure and high biomass retention, which enhance the degradation of organic matter and nutrient removal (Othman *et al.*, 2013). BESs, including microbial fuel cells and microbial electrolysis cells, offer the potential for simultaneous wastewater treatment and energy generation, making them attractive for sustainable wastewater management (Milhazes-Cunha & Otero, 2017).

9.2.3 Limitations of conventional wastewater treatment processes

Conventional wastewater treatment methods are aimed at the contamination removal process rather than by-product recovery. As these techniques aim to remove such contamination, the traditional division of these methods applies to physical, chemical, and biological approaches. The latter methods are arranged based on their mechanisms and efficiencies, and some of the most remarkable examples include constructed wetlands, activated sludge processes, bioreactor-based bioprocesses, and BESs (Kumar *et al.*, 2022). However, traditional approaches face considerable issues in terms of products and energy requirements. Specifically, the challenges of upscaling traditional methods to promote efficacious technologies for carbon sequestration and wastewater treatment are primarily tackled in terms of heavy energy penalties and significant operational costs. On the contrary, emerging techniques such as microalgae cultivation present viable alternatives. Indeed, this biological method is distinguished by optimal biological features. These aspects facilitate efficient CO_2 and nutrient utilization from wastewater, drive rapid biomass generation, and concurrently precipitate valuable biomass components suitable for biorefinery applications. However, microalgae-enabling carbon capture and utilization are negatively impacted by high reliance on nutrients and water for decent microalgal growth, as well as energy-consumption issues during downstream processes such as centrifugation, flocculation, extraction, and dehydration (Razzak *et al.*, 2017).

In recent years, the massively changing socioeconomic and environmental context resulted in a shift towards the so-called circular economy, which stands for a viable solution against the intensively utilizing consumption of finite resources and hazardous environmental pollution emission. The priorities of the circular economy are directed at reducing the materials' use and waste generation, and these rules are characterized by the notions of reducing, redesigning, reusing, recycling, and recovering (Sheldon & Norton, 2020). Microalgae cultivation is also deeply associated with the generation of waste streams usable in other types of cultivation processes. For instance, CO_2 and heavy levels of nutrients in wastewaters precipitate as ideal materials for promoting the growth of microalgal cells. This approach enables the initiation of microalgae cultivation processes in a loop, leading to good bioproducts and fostering low-carbon and circular practices in the bioeconomy (Lim *et al.*, 2021).

9.3 ROLE OF MICROALGAE IN WASTEWATER TREATMENT

Microalgae thrive in wastewater despite its differing chemical and physical properties compared to freshwater, marine water, and culture media. These organisms efficiently utilize the organic and inorganic compounds present in wastewater – ranging from proteins and carbohydrates to potassium and phosphate – coupled with natural solar energy and carbon dioxide. This capability not only supports their growth but also facilitates wastewater treatment by recovering nutrients and aligning with circular bioeconomy principles (Díaz *et al.*, 2022).

The treatment process capitalizes on the natural ability of microalgae to utilize inorganic nutrients and produce oxygen, which increases the pH and thereby creates a disinfectant effect. This also hinders the growth of pathogenic microorganisms such as coliforms. However, the cultivation process must be carefully monitored due to the presence of harmful pathogens that could impact the microalgae. Despite these challenges, the robust adaptability of microalgae to various environmental conditions such as pH and temperature makes them an effective solution for wastewater treatment (Díaz *et al.*, 2022).

Use of microalgae not only mitigates the high costs associated with tertiary biological treatments by removing harmful compounds but also reduces greenhouse gas emissions by absorbing carbon during treatment processes. Furthermore, the biomass produced can be converted into value-added products such as biofuels and biogas, enhancing the economic viability of this approach. The cultivation in raceways and photobioreactors (PBRs), preferred for their low operational costs and effectiveness in processing large volumes, supports the scale-up of this technology. Overall, microalgae offer a promising, cost-effective strategy for wastewater treatment, nutrient removal, and the production of useful bioproducts, leveraging their unique biochemical properties and growth dynamics (Díaz *et al.*, 2022).

9.4 ALGAL CULTIVATION METHODS FOR WASTEWATER TREATMENT

The cultivation of algae and the systems used for wastewater treatment are crucial for optimizing the efficiency of microalgae in removing contaminants and producing valuable biomass. The choice of cultivation system significantly impacts the growth rates of microalgae, their ability to remove pollutants, and the overall sustainability of the treatment process.

9.4.1 Traditional microalgae wastewater treatment systems

Microalgae can be cultivated in various systems, each offering different advantages and challenges. The most common systems are open ponds and closed PBRs.

9.4.1.1 Open ponds

Open ponds are the simplest and most cost-effective method for algal cultivation, particularly suitable for large-scale operations. They are generally shallow and allow natural sunlight to drive the photosynthetic process. Raceway ponds, a type of open pond system, are widely used due to their ease of construction and operation. These systems, however, are highly dependent on environmental conditions such as light, temperature, and contamination from external sources (Razzak *et al.*, 2013).

9.4.1.2 Closed PBRs

PBRs offer controlled environments for algae cultivation, minimizing the risks of contamination and allowing precise control over growth conditions. Various configurations of PBRs, including tubular, flat-panel, and airlift reactors, are used depending on the specific needs of the algal species and the pollutants being treated. PBRs typically achieve higher biomass densities and can operate effectively in areas with limited space, but they require higher initial and operational costs (Razzak *et al.*, 2013).

9.4.2 Hybrid treatment systems

To enhance the efficiency of wastewater treatment, hybrid systems that combine traditional algal cultivation with other treatment technologies have been developed.

9.4.2.1 Activated sludge and microalgae

Integrating microalgae with activated sludge processes enhances nutrient removal from wastewater. This hybrid system utilizes the symbiotic relationship between microalgae and bacteria, where microalgae supply oxygen for bacterial decomposition processes while consuming the carbon dioxide produced by bacteria (Zhang *et al.*, 2020b).

9.4.2.2 Advanced oxidation processes and microalgae

Combining advanced oxidation processes with microalgae cultivation allows for the effective breakdown of persistent organic pollutants (POPs) before they are biologically treated by algae. This system can significantly reduce the presence of complex chemicals in wastewater, making the treatment process more efficient (Urbina-Suarez *et al.*, 2021).

9.4.2.3 Constructed wetlands and microalgae

The use of constructed wetlands in conjunction with algal cultivation is another promising hybrid approach. In this system, microalgae enhance the oxygenation of wetlands, improving the removal efficiencies of nitrogen and phosphorus, and thereby supporting healthier aquatic ecosystems (Pahunang *et al.*, 2021).

9.5 RECENT ADVANCES IN ALGAL WASTEWATER TREATMENT SYSTEMS

Recent studies have shown the effectiveness of various microalgae cultivation systems in treating wastewater, with a focus on nutrient recovery and biomass production. This section provides in-depth insights into the different culturing systems, their operational conditions, and the outcomes in terms of pollutant removal and biomass yields.

9.5.1 Photobioreactors

PBRs offer a controlled environment for microalgae cultivation, enhancing nutrient removal efficiency. Foladori *et al.* (2018) demonstrated high removal rates of COD and nitrogen (>80%) using a photo-sequencing batch reactor with a microalgae–bacterial consortium treating primary effluent from a wastewater treatment plant. However, phosphorus removal was relatively low in this system. Purba *et al.* (2021) also employed a photo-sequencing batch reactor to treat domestic wastewater, developing microalgae–bacteria granular sludge from the initial reactor inoculum. Although COD removal was efficient, nitrogen and phosphorus removal yields were modest, with ammonia nitrogen removal being the exception. Wiley (2013) operated an offshore membrane tubular PBR to treat secondary effluent using a mixed microalgae culture, achieving efficient nitrogen removal and biomass production. Similarly, Novoveská *et al.* (2016) conducted a yearlong operation of a larger-scale offshore system composed of 12–48 PBRs, which consistently removed nutrients efficiently. Ruiz *et al.* (2013) used a flat-panel PBR to cultivate *Scenedesmus obliquus* in secondary effluents with different hydraulic retention times (HRTs). They found that nitrogen and phosphorus removal was above 87% for HRT values over 1.1 days, but an HRT of 0.5 days led to the washout of microalgae and no nutrient removal. Ling *et al.* (2019) also worked with *Scenedesmus obliquus* in a 30 L airlift flat-panel PBR using primary wastewater effluents, achieving higher nutrient removal efficiencies with a 6-day HRT compared to maintaining a fixed biomass concentration. de Souza Leite *et al.* (2019) studied the treatment of anaerobic reactor effluents in flat-panel PBRs inoculated with *Chlorella sorokiniana*. They achieved complete nitrogen removal primarily through ammonia stripping, while phosphorus removal varied between 60 and 100%. de Francisci *et al.* (2018) cultivated *Chlorella sorokiniana* in a flat-panel PBR using mixed industrial and municipal wastewater, finding similar COD removal across different dilution rates but higher nutrient removal yields with lower dilution rates (higher HRT).

9.5.2 Tubular PBR systems

Tubular PBR systems, due to their design, typically offer better control over environmental parameters and improved light absorption. di Termini *et al.* (2011) compared indoor and outdoor operations of lab-scale horizontal tubular PBRs treating secondary effluents. The indoor setup achieved higher biomass growth and pollutant removal due to continuous illumination, whereas the outdoor setup achieved sufficient nutrient removal levels to comply with discharge regulations. Binnal and Babu

(2017) optimized the operation conditions for *Chlorella protothecoides* cultivation in a secondary effluent using a lab-scale horizontal tubular PBR. They concluded that cultivation yields and water treatment quality were significantly affected by light and dark cycles and irradiance levels.

9.5.3 Cylinder PBR systems

Cylinder PBR systems have also been explored for urban wastewater treatment. Sydney *et al.* (2011) identified *Botryococcus braunii* as the best-performing strain in a BioFlo reactor, achieving complete phosphorus removal and substantial nitrogen removal while accumulating high lipid content in biomass. Ruiz-Martinez *et al.* (2012) used a cylinder bubble column reactor to remediate anaerobic membrane bioreactor effluents, achieving high nutrient removal due to microalgae uptake at controlled pH. Su *et al.* (2012) tested a mixed culture of microalgae and activated sludge for primary wastewater treatment. They found that an algae:sludge ratio of 5:1 yielded the highest nitrogen and phosphorus removals, with significant recovery of these nutrients in the biomass.

9.5.4 Open systems and other configurations

Open systems, including high-rate algal mesocosms (HRAMs) and rotating algal biofilm reactors, have been studied for their simplicity and low energy requirements. Sutherland *et al.* (2015) found that nitrogen and phosphorus removal was most effective in HRAMs without pH control due to volatilization and precipitation rather than microalgae uptake. Zhao *et al.* (2018) modified a pilot-scale open pond raceway to include a revolving mechanism for algae biofilm attachment, which improved nutrient removal compared to conventional raceways. Iman Shayan *et al.* (2016) tested a rotating algae biofilm reactor treating secondary effluents, achieving high nutrient removal yields with different HRTs. Christenson and Sims (2012) also used a rotating algae biofilm reactor, though they achieved lower removal efficiencies, possibly due to the larger system scale and shorter HRTs.

9.6 MICROALGAL CONSORTIA FOR ENHANCED TREATMENT

The integration of microalgae consortia in wastewater treatment represents a sophisticated approach that leverages the synergistic interactions between different microalgal species and other microorganisms such as bacteria. This consortium approach not only enhances the efficiency of contaminant removal but also contributes to the stability and resilience of the treatment process under varying environmental conditions (Bhatt *et al.*, 2022).

9.6.1 Advantages of microalgae consortia

The use of microalgae consortia offers several advantages over monocultures. These systems can achieve higher pollutant removal rates, better withstand environmental fluctuations, and optimize resource use through complementary metabolic activities. For instance, certain microalgae can secrete extracellular substances that support the growth of symbiotic bacteria, which in turn can degrade organic compounds that are difficult for microalgae to process alone. This cooperative interaction enhances the overall degradation capacity of the treatment system, making it more effective at reducing levels of both organic and inorganic pollutants (Gonçalves *et al.*, 2017).

9.6.2 Types of consortia systems

9.6.2.1 Algae–bacteria consortia

These systems are particularly effective in removing nutrients such as nitrogen and phosphorus from wastewater. The bacteria break down organic matter, releasing CO_2 which in turn is utilized by microalgae for photosynthesis. The oxygen produced by microalgae aids bacterial activity, thus creating a self-sustaining cycle that enhances pollutant removal efficiency (Hernández *et al.*, 2013).

9.6.2.2 *Algae–fungi consortia*

Fungal partners in consortia can degrade complex organic pollutants, which are then assimilated by algae as nutrients. This type of consortium is especially beneficial in the treatment of wastewater containing recalcitrant organic compounds. The fungi break down complex molecules into simpler forms that algae can readily absorb and use for growth (Leng *et al.*, 2021).

9.6.2.3 *Integrated PBRs*

Employing consortia within controlled environments such as PBRs allows for optimized light and nutrient conditions, further enhancing the treatment efficiency. These systems can be specifically tailored to target contaminants found in industrial effluents, including heavy metals and synthetic chemicals (Zhang *et al.*, 2020a).

9.6.3 Challenges and optimization

Although the use of consortia offers numerous benefits, it also presents challenges such as the need for careful management of species composition to avoid dominance by a single organism, which can reduce system efficiency. Research into the dynamics of microbial interactions within consortia is essential to optimize these systems for practical applications. Advanced molecular tools and bioinformatics approaches are increasingly being used to understand and manipulate consortia dynamics, leading to more robust and efficient systems (Zhang *et al.*, 2020a).

9.7 TREATMENT EFFICIENCY OF MICROALGAE

Microalgae have emerged as a potent solution for the treatment of wastewater, showing significant capability in the bioremediation of various contaminants such as pharmaceuticals, heavy metals, and industrial chemicals. Their efficiency in treating wastewater not only lies in their ability to degrade harmful substances but also in their capacity to convert these pollutants into valuable biomass, which can be utilized for further applications in a circular economy framework.

Microalgae implement several mechanisms to treat wastewater, primarily through bioadsorption, bioaccumulation, and biodegradation. Bioadsorption involves the attachment of pollutants to the cell surface of microalgae, which is particularly effective for heavy metals and organic pollutants. For instance, strains of *Scenedesmus* and *Chlorella* have shown high adsorption capacities for metals such as lead and arsenic, effectively reducing their concentrations in contaminated waters (Carles *et al.*, 2021).

Bioaccumulation refers to the uptake and internal storage of pollutants by microalgae. This process is more selective and depends significantly on the metabolic activity of the algae. Microalgae such as *Chlorella vulgaris* and *Spirulina* are noted for their ability to bioaccumulate high levels of nutrients and toxic substances from their environment, thereby purifying the water while growing biomass (Nikokherad *et al.*, 2022).

Biodegradation by microalgae involves the metabolic breakdown of pollutants into less harmful substances that can be assimilated into natural biogeochemical cycles. This mechanism is crucial for the degradation of complex organic compounds including pharmaceuticals and pesticides. Microalgae such as *Chlamydomonas* and *Dunaliella* have demonstrated capabilities to degrade and utilize these compounds as carbon sources for their growth (Ofrydopoulou *et al.*, 2022).

9.8 GENETIC ENGINEERING FOR ENHANCED POLLUTANT REMOVAL

Genetic engineering of microalgae offers transformative potential for optimizing their use in wastewater treatment, particularly by enhancing their ability to remove and process a broad spectrum of pollutants. Through genetic modifications, microalgae can be tailored to improve their pollutant uptake, degradation capacities, and tolerance to toxic substances, making them more efficient agents in circular wastewater treatment systems.

9.8.1 Enhancing microalgae capability

Recent advancements in genetic engineering techniques have allowed researchers to manipulate microalgal genomes, thereby enabling these organisms to perform specific tasks more effectively. For instance, the introduction of genes responsible for the synthesis of specific enzymes can enhance the metabolic pathways involved in the degradation of complex pollutants such as pharmaceuticals and POPs. Genetically engineered microalgae have shown increased rates of bioaccumulation and biodegradation, leading to more effective removal of contaminants from wastewater (Zhang *et al.*, 2020a).

9.8.2 Key genetic modifications

One of the notable genetic modifications is the enhancement of photosynthetic efficiency and tolerance to high concentrations of toxic metals. For example, by altering the photosystem-II genes, scientists have improved the light utilization efficiency of microalgae, which is crucial for their growth and pollutant processing capabilities under various light conditions (Nguyen *et al.*, 2020). Additionally, genetic engineering has been used to modify cellular transport mechanisms, enabling microalgae to more effectively sequester and metabolize heavy metals from industrial effluents.

9.8.3 Case studies and applications

In a pioneering study, researchers have successfully introduced genes encoding for specific transporters and enzymes that confer resistance to mercury and arsenic, significantly increasing the bioaccumulation capacity of *Chlorella* strains (Ortiz-Hernández *et al.*, 2013). Another breakthrough involves the engineering of lipid metabolism pathways in microalgae, which not only boosts their ability to degrade organic pollutants but also enhances biomass production for biofuel synthesis, thereby contributing to a circular economy model (Khatiwada *et al.*, 2020).

9.8.4 Challenges and ethical considerations

Although genetic engineering holds promise for revolutionizing microalgal wastewater treatment, it also poses certain challenges and ethical concerns. The release of genetically modified microalgae into natural water bodies could have unforeseen ecological impacts. Thus, stringent biocontainment measures and thorough ecological assessments are essential before deploying these organisms in environmental applications (Bhatt *et al.*, 2022).

9.8.5 Future directions

Continued research is necessary to explore the full potential of genetically engineered microalgae. Future studies should focus on creating robust strains that can withstand variable environmental conditions in wastewater treatment facilities. Moreover, the integration of Clustered Regularly Interspaced Short Palindromic Repeats-CRISPR-associated proteins (CRISPR-Cas) systems offers a promising avenue for precise genetic edits that could tailor microalgae for specific pollutants, further enhancing their utility in sustainable wastewater management (Nomura *et al.*, 2019).

9.9 TECHNO-ECONOMIC ANALYSIS OF MICROALGAL BIOREFINERIES IN CIRCULAR ECONOMY

The circular economy presents a transformative approach to environmental management by integrating CO_2 utilization and wastewater treatment with microalgae cultivation. This integration not only addresses environmental issues but also drives economic gains, making it a compelling model for sustainable development.

9.9.1 Economic advantages of microalgae in circular processes

Microalgae cultivation in a circular economy leverages CO_2 and nutrient-rich wastewater as inputs, reducing the need for synthetic fertilizers and minimizing operational costs. By converting these wastes into valuable biomass, microalgae operations can generate diverse bioproducts such as biofuels, animal feed, and biochemicals, which are economically beneficial. The reduced resource input and the value-added output from microalgae significantly enhance the economic viability of microalgae systems (Chen *et al.*, 2022).

9.9.2 Environmental impacts

The environmental benefits of using microalgae in circular wastewater treatment systems are substantial. Microalgae efficiently sequester CO_2, helping mitigate greenhouse gas emissions. Additionally, their ability to remove pollutants from wastewater contributes to water purification, which can prevent eutrophication and reduce the chemical load entering aquatic ecosystems. These processes inherently decrease the environmental footprint associated with traditional waste disposal and chemical treatment methods, illustrating a significant shift towards sustainability (Chen *et al.*, 2022).

9.9.3 Technological challenges and strategies

Despite these benefits, the implementation of circular microalgae systems faces technological challenges, including the need for optimization of cultivation conditions and enhancement of biomass productivity. Advances in biotechnology, including genetic engineering and bioreactor design, are crucial for overcoming these challenges. Enhancing microalgae's tolerance to high pollutant loads and varying environmental conditions will be key to maximizing their efficiency and reliability in circular treatment systems (Chen *et al.*, 2022).

9.9.4 Policy framework and future directions

The expansion of microalgae-based circular systems requires supportive regulatory frameworks and incentives to encourage investment and innovation. Policies aimed at promoting the use of renewable resources and supporting the commercialization of biobased products can accelerate the adoption of microalgae technologies. Future research should focus on integrating microalgae cultivation with industrial processes that produce CO_2 and wastewater, creating a closed-loop system that benefits both the economy and the environment (Chen *et al.*, 2022).

9.10 ECONOMIC AND ENVIRONMENTAL BENEFITS OF ALGAL-BASED WASTEWATER TREATMENT

Although the integration of microalgae-based systems into circular wastewater treatment and CO_2 utilization offers significant environmental and economic benefits, it also presents unique challenges that can create conflicts between economic viability and environmental sustainability.

9.10.1 Economic challenges

The primary economic challenge lies in the cost associated with microalgae cultivation and processing. Microalgae require specific conditions for optimal growth, including controlled light, temperature, and nutrient availability, which can lead to high operational costs. Additionally, the harvesting and downstream processing of microalgae biomass are energy-intensive, involving steps such as dewatering, drying, and extraction, which significantly contribute to the overall costs. These processes require substantial energy inputs, which, if sourced from non-renewable energy, could offset the environmental benefits gained from microalgae's CO_2 sequestration capabilities (Chen *et al.*, 2022).

9.10.2 Environmental conflicts

From an environmental perspective, the intensive use of resources such as water and energy in microalgae cultivation can lead to significant environmental footprints. Although microalgae mitigate CO_2 emissions and remove pollutants from wastewater, the environmental impact of their cultivation can be substantial if not managed with sustainable practices. For instance, the use of synthetic fertilizers to boost microalgae growth can lead to nutrient runoff, which may cause eutrophication in nearby water bodies, ironically contributing to the environmental issues it aims to solve (Chen *et al.*, 2022).

9.10.3 Balancing economic and environmental aspects

The challenge in microalgae-based production systems lies in balancing economic efficiency with environmental sustainability. Economic pressures may drive the adoption of less environmentally friendly practices, such as the use of cheaper but non-renewable energy sources to reduce operational costs. Conversely, strict environmental regulations can increase production costs, making microalgae products less competitive with traditional alternatives. This tension necessitates innovative solutions to reduce costs while enhancing environmental benefits, such as the development of more efficient bioreactor designs and the use of waste streams as nutrient sources (Chen *et al.*, 2022).

9.10.4 Future directions

Addressing these conflicts requires a multidisciplinary approach, integrating advances in biological sciences, engineering, and economics. Future research should focus on enhancing the energy efficiency of microalgae processing techniques and developing closed-loop systems that minimize waste and maximize resource reuse. Additionally, policies and incentives that encourage the use of renewable energy sources in microalgae cultivation and processing can help align economic activities with environmental goals, promoting a truly sustainable circular economy model (Chen *et al.*, 2022).

9.11 KEY ISSUES AND FUTURE PERSPECTIVES IN MICROALGAL USE FOR CIRCULAR WASTEWATER TREATMENT

9.11.1 Economic viability

One of the primary challenges in employing microalgae for circular wastewater treatment lies in the economic feasibility of the process. The high costs associated with cultivating, harvesting, and processing microalgal biomass can be a significant barrier. Although using wastewater as a nutrient source can reduce some costs, the overall expense remains high compared to conventional wastewater treatment methods (Koyande *et al.*, 2019).

9.11.2 Technological challenges

Efficiently integrating microalgal cultivation with existing wastewater treatment infrastructure is another major issue. Developing cost-effective and scalable technologies for the cultivation and harvesting of microalgae is essential. Innovations in PBR design, optimizing light and nutrient supply, and enhancing mass transfer are critical areas that need attention (Cheirsilp & Torpee, 2012).

9.11.3 Variable wastewater composition

The composition of wastewater can vary significantly depending on the source, which affects the growth and productivity of microalgae. Consistency in nutrient supply and managing inhibitory substances in wastewater are ongoing challenges. Microalgal species selection and genetic engineering may offer solutions to improve resilience and adaptability (Yeesang & Cheirsilp, 2014).

9.11.4 Environmental factors

External factors such as temperature, light intensity, and photoperiod significantly impact microalgal growth and biochemical composition. Achieving optimal growth conditions consistently in outdoor settings or large-scale operations can be difficult. This variability can affect the efficiency and reliability of wastewater treatment processes (Potijun *et al.*, 2021).

9.11.5 Regulatory and public acceptance

Regulatory frameworks for using treated wastewater and microalgal products are still evolving. Public acceptance of wastewater-derived products, especially in food and feed applications, poses additional challenges. Ensuring safety, quality, and compliance with regulatory standards is essential for broader adoption (Bonnet *et al.*, 2010).

9.11.6 Future perspectives

Integration with bio-circular-green economy: Microalgal biorefineries can play a pivotal role in the bio-circular-green economy by converting wastewater into valuable products while reducing environmental impact. Developing integrated systems that combine wastewater treatment with the production of biofuels, biochemicals, and other high-value products will enhance sustainability and economic viability (Cheirsilp & Maneechote, 2022).

9.11.7 Advancements in genetic engineering

Genetic engineering and synthetic biology offer promising avenues for improving microalgal strains. Enhancements in photosynthetic efficiency, nutrient uptake, and stress tolerance can significantly boost productivity. Additionally, tailoring microalgae to produce specific biochemicals or biofuels can add value to wastewater treatment processes (Cheirsilp *et al.*, 2017).

9.11.8 Innovative cultivation systems

Developing novel cultivation systems such as immobilized cell technologies, hybrid PBRs, and co-culture systems with other microorganisms can improve efficiency. For instance, immobilizing microalgae in fungal pellets or using co-cultures with oleaginous yeasts can enhance nutrient uptake and lipid production, making the process more cost-effective and scalable (Maneechote *et al.*, 2023).

9.11.9 Techno-economic and life cycle analyses

Conducting comprehensive techno-economic analyses and life cycle assessments will help identify the most sustainable and cost-effective strategies for microalgal wastewater treatment. These analyses can guide investment decisions, policy-making, and the development of best practices for industry stakeholders (Maneechote *et al.*, 2023).

9.11.10 Policy and incentives

Governments and regulatory bodies can play a crucial role by providing incentives for sustainable wastewater treatment technologies. Policies promoting the use of renewable resources, carbon credits for CO_2 mitigation, and subsidies for green technologies can accelerate the adoption of microalgal systems (Bonnet *et al.*, 2010).

9.11.11 Public engagement and education

Raising awareness about the benefits of microalgal wastewater treatment and its role in the circular economy is essential. Public engagement initiatives and educational programmes can help build acceptance and support for this innovative approach. Highlighting the environmental and economic advantages can drive consumer demand for microalgal products (Chew *et al.*, 2017).

9.12 REGULATORY FRAMEWORKS SUPPORTING ALGAL-BASED CIRCULAR ECONOMY

The integration of microalgae into circular economy strategies for CO_2 utilization and wastewater treatment is a promising approach for addressing climate change and resource recovery. However, the effectiveness of this integration is heavily dependent on supportive governmental regulations that can drive innovation, ensure sustainable practices, and facilitate the economic viability of these technologies.

9.12.1 Need for supportive policy framework

Governmental regulations play a pivotal role in the successful deployment of microalgae-based systems for environmental management. These regulations can provide the necessary framework for ensuring that operations are both environmentally sustainable and economically feasible. For instance, policies that mandate the reduction of industrial CO_2 emissions can create a market for CO_2-utilizing technologies, such as microalgae cultivation, which can transform this waste product into valuable biomass (Chen *et al.*, 2022).

9.12.2 Economic incentives

Economic incentives such as tax breaks, subsidies, and grants are crucial for reducing the financial risks associated with investing in new technologies such as microalgae-based systems. These incentives can help offset the high initial costs associated with setting up advanced bioreactors and processing facilities, making it more attractive for businesses to invest in these green technologies. Furthermore, governments can facilitate market creation for microalgae-based products by setting standards that encourage the use of bio-based materials in various industries (Chen *et al.*, 2022).

9.12.3 Regulatory support for circular economy practices

Regulations that promote the circular economy can drive the adoption of microalgae-based systems by framing guidelines that encourage the reuse of waste materials. For example, regulations that support the use of wastewater as a nutrient source for microalgae cultivation can not only reduce the operational costs associated with microalgae growth but also help in the treatment of wastewater, thereby solving two environmental issues simultaneously. Policies aimed at circular resource use, such as the European Union's Circular Economy Action Plan, serve as models for how regulations can support the integration of biological processes such as microalgae cultivation into wider economic systems (Chen *et al.*, 2022).

9.12.4 Challenges and future directions

Although supportive policies are essential, they must be carefully crafted to avoid unintended consequences. For example, although incentivizing wastewater use for algal growth, regulations must also ensure that these practices do not lead to environmental degradation through nutrient runoff or other forms of pollution. Additionally, as microalgae-based technologies evolve, so too must the regulatory frameworks that govern them, ensuring they remain relevant and effective in promoting sustainable practices. Future regulatory frameworks should aim to integrate more comprehensive, science-based assessments of the environmental impacts of microalgae cultivation. This could involve life cycle assessments that help quantify the true environmental costs and benefits of these systems. Moreover, international cooperation on standards and regulations can help scale the impact of microalgae technologies, making them a viable solution globally (Chen *et al.*, 2022).

9.13 APPLICATIONS OF MICROALGAE IN AQUACULTURE

The integration of microalgae into aquaculture systems offers numerous environmental and economic benefits. Microalgae, known for their ability to efficiently assimilate nutrients from eutrophic water bodies, provide a sustainable solution for wastewater remediation and biomass production. This section explores the applications of algal biomass in aquaculture, emphasizing its role in improving water quality, producing value-added biomass, and enhancing the overall sustainability of aquaculture practices.

9.13.1 Nutrient assimilation and biomass production

Microalgae play a crucial role in nutrient assimilation in aquaculture systems. They can effectively uptake nitrogen, phosphorus, and other nutrients from wastewater, converting these nutrients into valuable biomass. This process not only helps in mitigating water pollution but also results in the production of protein-rich biomass, which can be used as a feed ingredient for aquatic animals (Ansari *et al.*, 2017). The biomass produced from microalgae is rich in essential nutrients, including proteins, lipids, and natural pigments, making it a highly suitable and nutritious feed for aquaculture species (Lu *et al.*, 2019).

9.13.2 Enhancing water quality

The incorporation of microalgae into aquaculture systems significantly improves water quality by reducing the concentration of harmful nutrients and organic matter. Microalgae function as biopumps, generating oxygen through photosynthesis and thus preventing oxygen depletion in the water. This oxygenation helps maintain a healthy aquatic environment, reducing the incidence of harmful algal blooms and promoting the overall health of aquatic ecosystems (Roy & Pal, 2015).

9.13.3 Reduction of antibiotic dependency

One of the major problems with conventional aquaculture is that antibiotics have been overused to stave off diseases, leading to antibiotic resistance and environmental contamination. The use of microalgae-based feeds has been found to enhance the natural immunity of aquatic organisms and prevents the need for antibiotics or medicines. It not only enhances the safety of aquaculture products but also reduces the concomitant environmental pollution from widespread antibiotic usage (Lu *et al.*, 2019).

9.13.4 Economic benefits

There are a lot of economic reasons for employing microalgae for aquaculture. As the wastewater nutrients are effectively converted into desirable biomass, the production system consumes less of the commercial feeds, and they could become independent of the external inputs, thus making it cheaper. Moreover, pollution-free culture products became marketplace values and eventually got higher returns for the producer. In addition, with the development of advanced microalgae biotechnology, the cost of wastewater treatment with aeration ponds was substantially reduced, thus adding to the economic sustainability in aquaculture (Lu *et al.*, 2019).

9.13.5 Substitution of traditional fishfeed

Fish meal is nutritionally considered indispensable for feed in the aquaculture industry because of its high protein content, balanced essential amino acids, palatability, digestibility, minerals, and omega-3 fatty acids. Currently, alternative plant and animal protein feedstuffs are available in such a way that many would be used, but they exist with considerable limitations. Plant proteins are often composed of antinutritional compounds and some essential amino acids are deficient whereas animal proteins are also imbalanced and contain high levels of saturated fats which impact feed preservation. In this regard, they can only be partially replaced by fish meal at levels between 1 and 30%. *Spirulina*, *Chlorella*, and *Nannochloropsis* are microalgal sources with 50–70% protein (Ma & Hu, 2024).

One study reported on the benefits derived from microalgae when used as feeds for fish in relatively different fish species (Ma & Hu, 2024). Feeding aquaculture animals with diets containing, relatively low-to-medium levels (2–10%), microalgae, when compared to control diets showed similar or even better weight gains. The Atlantic salmon effectiveness is on the increment by 31% of the conventional treatment with 5% *Schizochytrium* sp. whereas Pacific white shrimp postlarvae showed 30% higher gain of weight with a lipid diet. Juvenile tilapia gained 69% more weight on a diet containing 15% *Chlorella* sp., whereas 14% defatted *Nannochloropsis oculata* and *Schizochytrium* sp. resulted in 58%, and 10% *N. oculata* caused 46%. Nile tilapia also had significantly higher weight gain on microalgae-based diets (84–97%). All in all, juvenile Senegalese sole, despite some variability among species, benefits can be deduced when microalgae are incorporated into diets.

9.14 APPLICATIONS OF MICROALGAE IN AGRICULTURE

The use of microalgal biomass as biofertilizer has garnered significant attention due to its potential to enhance soil fertility, increase crop productivity, and contribute to sustainable agricultural practices. Microalgae, which include both green algae and cyanobacteria, are versatile microorganisms capable of synthesizing numerous biologically active compounds that can be beneficial for plant growth and soil health. The cultivation of microalgae can be integrated with wastewater treatment processes, thus providing a dual benefit of bioremediation and biofertilizer production (Alvarez *et al.*, 2021).

9.14.1 Nutrient content and benefits

Microalgae are rich in essential nutrients such as nitrogen, phosphorus, and potassium, which are crucial for plant growth. Additionally, they contain a variety of bioactive compounds, including amino acids, phytohormones (such as cytokinins and auxins), vitamins, and antioxidants. These compounds can enhance plant growth by improving root development, nutrient uptake, and stress tolerance (Michalak & Chojnacka, 2015).

For instance, studies have shown that the application of microalgal biofertilizers can increase the nitrogen, potassium, and phosphorus content in plant roots, shoots, and grains. This not only boosts plant growth but also enhances soil microbial activity, which is vital for maintaining soil health (Renuka *et al.*, 2016).

9.14.2 Successful examples

One study documented several successful examples of using algal biomass as biofertilizer (Dagnaisser *et al.*, 2022):

(1) *Wheat cultivation*: A study demonstrated that biofertilizers obtained from two microalgal consortia significantly increased the nitrogen, potassium, and phosphorus content in wheat crops. This treatment also enhanced the microbial biomass carbon in the soil, indicating improved soil health.
(2) *Tomato plants*: The application of biofertilizer from the green algae *Acutodesmus dimorphus* in tomato cultivation resulted in better plant growth, more branches, and flowers compared to control groups. This effect was particularly pronounced when the biofertilizer was applied 22 days before seedling transplant.
(3) *Spinach and chives*: Another study found that treating spinach and Chinese chives with microalgal biofertilizers improved their yield and quality. The treated spinach had higher mineral content, and the yield of Chinese chives was 18.3% higher compared to untreated plants.
(4) *Rice cultivation*: Combining microalgal biofertilizer with chemical fertilizers improved the cost-effectiveness of rice cultivation, suggesting that microalgal biofertilizers can be a suitable organic fertilizer, enhancing crop yields and reducing reliance on chemical fertilizers.

9.14.3 Advantages over conventional fertilizers

Dagnaisser *et al.* (2022) discussed that microalgal biofertilizers offer several advantages over conventional chemical fertilizers (Dagnaisser *et al.*, 2022):

- *Sustainability*: Microalgae can be grown using wastewater, which not only recycles nutrients but also reduces the environmental impact of wastewater disposal.
- *Nutrient efficiency*: Algal biofertilizers provide a slow-release source of nutrients, improving nutrient-use efficiency and reducing the risk of nutrient leaching into waterways.
- *Environmental benefits*: The use of microalgal biofertilizers can decrease greenhouse gas emissions associated with conventional fertilizer production and application.

9.15 APPLICATIONS OF MICROALGAE IN BIOENERGY

The conversion of algal biomass into biofuel represents a promising approach for sustainable energy production. Microalgae, with their high growth rates and ability to thrive in wastewater, offer a renewable and efficient feedstock for biofuel production. This section explores various methods of converting algal biomass into energy, emphasizing successful examples and the underlying processes.

9.15.1 Biodiesel production

Biodiesel, a prominent form of biofuel, is produced from the lipids extracted from microalgae through a process known as transesterification. This process converts triglycerides into fatty acid methyl esters and glycerol. Microalgae are particularly suitable for biodiesel production due to their high lipid content, which can exceed 50% of their dry biomass (Ali *et al.*, 2022). Species such as *Chlorella vulgaris* and *Nannochloropsis* have demonstrated lipid contents ranging from 67 to 98%, making them excellent candidates for biodiesel feedstock (Nguyen *et al.*, 2019; Reddy *et al.*, 2014).

Transesterification techniques vary, with traditional methods using chemical catalysts such as sodium methoxide, and more advanced methods employing nano-catalysts to improve efficiency and reduce costs. The application of mesoporous nanoparticles, for example, has shown potential in continuous biodiesel synthesis without the need for cell lysis, thereby simplifying the process and enhancing lipid recovery (Bhuyan *et al.*, 2021).

9.15.2 Hydrothermal liquefaction

Hydrothermal liquefaction (HTL) is a thermochemical process that converts wet algal biomass into bio-oil, simulating the natural geological processes that form crude oil. This method operates at high temperatures (200–350°C) and pressures (10–25 MPa), effectively breaking down biomass into liquid hydrocarbons (Xia *et al.*, 2021). One of the main advantages of HTL is its ability to process wet biomass directly, eliminating the need for costly drying steps. Studies have shown that HTL can achieve high bio-oil yields, with the aqueous phase containing valuable nutrients that can be recycled for further algal cultivation (Xia *et al.*, 2021).

9.15.3 Pyrolysis

Pyrolysis is another thermochemical conversion process that decomposes organic materials at high temperatures in the absence of oxygen, producing bio-oil, syngas, and biochar. Microalgal biomass is particularly suitable for pyrolysis due to its high lipid and carbohydrate content. The process can be optimized by selecting specific algae strains and adjusting operational parameters such as temperature and catalyst presence (Ali *et al.*, 2022; Xia *et al.*, 2021). The bio-oil produced through pyrolysis has a high-energy density, comparable to fossil fuels, making it a viable alternative for energy production (Xia *et al.*, 2021).

9.15.4 Fermentation

Fermentation is a biochemical process used to convert the carbohydrate content of microalgae into bioethanol. This involves saccharification to break down complex sugars into simple sugars, followed by fermentation using yeast or other microorganisms. Microalgae such as *Chlorella vulgaris* and *Chlorococcum* species have shown high bioethanol yields under optimized conditions. Pretreatment methods, including acid hydrolysis and enzymatic saccharification, are crucial for enhancing sugar release and fermentation efficiency (Cheng *et al.*, 2013; Ho *et al.*, 2013).

9.15.5 Anaerobic digestion

AD is a biological process that breaks down organic matter in the absence of oxygen, producing biogas (a mixture of methane and carbon dioxide) and digestate. Microalgal biomass is an excellent substrate for AD due to its high protein and carbohydrate content. The integration of microalgae cultivation with wastewater treatment enhances the sustainability of this process, providing a low-cost feedstock while simultaneously treating wastewater (Ali *et al.*, 2022; Hossain *et al.*, 2019). Studies have demonstrated that co-digestion of microalgae with other organic wastes can significantly improve biogas yield and process stability (Ali *et al.*, 2022; Reddy *et al.*, 2014).

9.16 CONCLUSION

Microalgal wastewater treatment presents a promising approach within the circular economy framework, offering significant environmental and economic benefits. The ability of microalgae to effectively remove pollutants and produce valuable biomass aligns with the principles of sustainability and resource efficiency. Technological advancements in microalgal cultivation and processing have the potential to overcome existing challenges, such as high operational costs and variable wastewater compositions. The integration of microalgae into wastewater treatment systems not only enhances pollutant removal but also contributes to the production of biofuels, fertilizers, and animal feed, thus supporting various sectors. However, the successful commercialization and widespread adoption of this technology require supportive policies, economic incentives, and public acceptance. Future research should focus on optimizing cultivation systems, advancing genetic engineering techniques, and conducting comprehensive techno-economic analyses to further enhance the viability and sustainability of microalgal wastewater treatment.

REFERENCES

Ali S. S., Al-Tohamy R., Mahmoud Y. A.-G., Kornaros M., Sun S. and Sun J. (2022). Recent advances in the life cycle assessment of biodiesel production linked to azo dye degradation using yeast symbionts of termite guts: a critical review. *Energy Reports*, **8**, 7557–7581, https://doi.org/10.1016/j.egyr.2022.05.240

Alvarez A. L., Weyers S. L., Goemann H. M., Peyton B. M. and Gardner R. D. (2021). Microalgae, soil and plants: a critical review of microalgae as renewable resources for agriculture. *Algal Research*, **54**, 102200, https://doi.org/10.1016/j.algal.2021.102200

Ansari F. A., Singh P., Guldhe A. and Bux F. (2017). Microalgal cultivation using aquaculture wastewater: integrated biomass generation and nutrient remediation. *Algal Research*, **21**, 169–177, https://doi.org/10.1016/j.algal.2016.11.015

Bhatt P., Bhandari G., Bhatt K. and Simsek H. (2022). Microalgae-based removal of pollutants from wastewaters: occurrence, toxicity and circular economy. *Chemosphere*, **306**, 135576, https://doi.org/10.1016/j.chemosphere.2022.135576

Bhuyan N., Dutta A., Mohan R., Bora N. and Kataki R. (2021). Advances in nanotechnology for biofuel production. In: Nanomaterials, 533–562. Elsevier.

Binnal P. and Babu P. N. (2017). Optimization of environmental factors affecting tertiary treatment of municipal wastewater by *Chlorella protothecoides* in a lab scale photobioreactor. *Journal of Water Process Engineering*, **17**, 290–298, https://doi.org/10.1016/j.jwpe.2017.05.003

Bonnet S., Webb E. A., Panzeca C., Karl D. M., Capone D. G. and Wilhelmy S. A. S. (2010). Vitamin B12 excretion by cultures of the marine cyanobacteria *Crocosphaera* and *Synechococcus*. *Limnology and Oceanography*, **55**, 1959–1964, https://doi.org/10.4319/lo.2010.55.5.1959

Carles L., Wullschleger S., Joss A., Eggen R. I. L., Schirmer K., Schuwirth N., Stamm C. and Tlili A. (2021). Impact of wastewater on the microbial diversity of periphyton and its tolerance to micropollutants in an engineered flow-through channel system. *Water Research*, **203**, 117486, https://doi.org/10.1016/j.watres.2021.117486

Cheah W. Y., Show P. L., Juan J. C., Chang J.-S. and Ling T. C. (2018). Microalgae cultivation in palm oil mill effluent (POME) for lipid production and pollutants removal. *Energy Conversion and Management*, **174**, 430–438, https://doi.org/10.1016/j.enconman.2018.08.057

Cheirsilp B. and Maneechote W. (2022). Insight on zero waste approach for sustainable microalgae biorefinery: sequential fractionation, conversion and applications for high-to-low value-added products. *Bioresource Technology Reports*, **18**, 101003, https://doi.org/10.1016/j.biteb.2022.101003

Cheirsilp B. and Torpee S. (2012). Enhanced growth and lipid production of microalgae under mixotrophic culture condition: effect of light intensity, glucose concentration and fed-batch cultivation. *Bioresource Technology*, **110**, 510–516, https://doi.org/10.1016/j.biortech.2012.01.125

Cheirsilp B., Thawechai T. and Prasertsan P. (2017). Immobilized oleaginous microalgae for production of lipid and phytoremediation of secondary effluent from palm oil mill in fluidized bed photobioreactor. *Bioresource Technology*, **241**, 787–794, https://doi.org/10.1016/j.biortech.2017.06.016

Chen J., Dai L., Mataya D., Cobb K., Chen P. and Ruan R. (2022). Enhanced sustainable integration of CO_2 utilization and wastewater treatment using microalgae in circular economy concept. *Bioresource Technology*, **366**, 128188, https://doi.org/10.1016/j.biortech.2022.128188

Cheng Y.-S., Zheng Y., Labavitch J. M. and Vandergheynst J. S. (2013). Virus infection of *Chlorella variabilis* and enzymatic saccharification of algal biomass for bioethanol production. *Bioresource Technology*, **137**, 326–331, https://doi.org/10.1016/j.biortech.2013.03.055

Chew K. W., Yap J. Y., Show P. L., Suan N. H., Juan J. C., Ling T. C., Lee D.-J. and Chang J.-S. (2017). Microalgae biorefinery: high value products perspectives. *Bioresource Technology*, **229**, 53–62, https://doi.org/10.1016/j.biortech.2017.01.006

Christenson L. B. and Sims R. C. (2012). Rotating algal biofilm reactor and spool harvester for wastewater treatment with biofuels by-products. *Biotechnology and Bioengineering*, **109**, 1674–1684, https://doi.org/10.1002/bit.24451

Dagnaisser L. S., Dos Santos M. G. B., Rita A. V. S., Chaves Cardoso J., de Carvalho D. F. and de Mendonça H. V. (2022). Microalgae as bio-fertilizer: a new strategy for advancing modern agriculture, wastewater bioremediation, and atmospheric carbon mitigation. *Water, Air, & Soil Pollution*, **233**, 477, https://doi.org/10.1007/s11270-022-05917-x

de Francisci D., Su Y., Iital A. and Angelidaki I. (2018). Evaluation of microalgae production coupled with wastewater treatment. *Environmental Technology*, **39**, 581–592, https://doi.org/10.1080/09593330.2017.1308441

de Souza Leite L., Hoffmann M. T. and Daniel L. A. (2019). Microalgae cultivation for municipal and piggery wastewater treatment in Brazil. *Journal of Water Process Engineering*, **31**, 100821, https://doi.org/10.1016/j.jwpe.2019.100821

Díaz V., Leyva-Díaz J. C., Almécija M. C., Poyatos J. M., del Mar Muñío M. and Martín-Pascual J. (2022). Microalgae bioreactor for nutrient removal and resource recovery from wastewater in the paradigm of circular economy. *Bioresource Technology*, **363**, 127968, https://doi.org/10.1016/j.biortech.2022.127968

di Termini I., Prassone A., Cattaneo C. and Rovatti M. (2011). On the nitrogen and phosphorus removal in algal photobioreactors. *Ecological Engineering*, **37**, 976–980, https://doi.org/10.1016/j.ecoleng.2011.01.006

Fernandes T. V., Trebuch L. M. and Wijffels R. H. (2022). Chapter 4 – Microalgae-based technologies for circular wastewater treatment. In: Integrated Wastewater Management and Valorization Using Algal Cultures, G. N. Demirer and S. Uludag-Demirer (eds.), 81–112, Elsevier.

Foladori P., Petrini S., Nessenzia M. and Andreottola G. (2018). Enhanced nitrogen removal and energy saving in a microalgal–bacterial consortium treating real municipal wastewater. *Water Science and Technology*, **78**, 174–182, https://doi.org/10.2166/wst.2018.094

Gerardi M. H. (2006). Wastewater bacteria. John Wiley & Sons.

Gonçalves A. L., Pires J. C. M. and Simões M. (2017). A review on the use of microalgal consortia for wastewater treatment. *Algal Research*, **24**, 403–415, https://doi.org/10.1016/j.algal.2016.11.008

Harun H., Anuar A. N., Halim M. H. A., Othman I., Rosman N. H., Hamid N. H. A., Ujang Z. and Van Loosdrecht M. (2020). Granulation and biodegradation by microbial species in granular sequencing batch reactor for soy sauce wastewater treatment. In: Valorisation of Agro-Industrial Residues – Volume I: Biological Approaches, Z. A. Zakaria, R. Boopathy and J. R. Dib (eds.), 287–308, Springer International Publishing, Cham.

Hernández D., Riaño B., Coca M. and García-González M. C. (2013). Treatment of agro-industrial wastewater using microalgae–bacteria consortium combined with anaerobic digestion of the produced biomass. *Bioresource Technology*, **135**, 598–603, https://doi.org/10.1016/j.biortech.2012.09.029

Ho S.-H., Huang S.-W., Chen C.-Y., Hasunuma T., Kondo A. and Chang J.-S. (2013). Bioethanol production using carbohydrate-rich microalgae biomass as feedstock. *Bioresource Technology*, **135**, 191–198, https://doi.org/10.1016/j.biortech.2012.10.015

Hossain N., Mahlia T. and Saidur R. (2019). Latest development in microalgae-biofuel production with nano-additives. *Biotechnology for Biofuels*, **12**, 1–16, https://doi.org/10.1186/s13068-018-1346-y

Iman Shayan S., Agblevor F. A., Bertin L. and Sims R. C. (2016). Hydraulic retention time effects on wastewater nutrient removal and bioproduct production via rotating algal biofilm reactor. *Bioresource Technology*, **211**, 527–533, https://doi.org/10.1016/j.biortech.2016.03.104

Khatiwada B., Sunna A. and Nevalainen H. (2020). Molecular tools and applications of *Euglena gracilis*: from biorefineries to bioremediation. *Biotechnology and Bioengineering*, **117**, 3952–3967, https://doi.org/10.1002/bit.27516

Koyande A. K., Show P.-L., Guo R., Tang B., Ogino C. and Chang J.-S. (2019). Bio-processing of algal bio-refinery: a review on current advances and future perspectives. *Bioengineered*, **10**, 574–592, https://doi.org/10.1080/21655979.2019.1679697

Kumar N., Banerjee C., Chang J.-S. and Shukla P. (2022). Valorization of wastewater through microalgae as a prospect for generation of biofuel and high-value products. *Journal of Cleaner Production*, **362**, 132114, https://doi.org/10.1016/j.jclepro.2022.132114

Leng L., Li W., Chen J., Leng S., Chen J., Wei L., Peng H., Li J., Zhou W. and Huang H. (2021). Co-culture of fungi–microalgae consortium for wastewater treatment: a review. *Bioresource Technology*, **330**, 125008, https://doi.org/10.1016/j.biortech.2021.125008

Lim H. R., Khoo K. S., Chew K. W., Chang C.-K., Munawaroh H. S. H., Kumar P. S., Huy N. D. and Show P. L. (2021). Perspective of *Spirulina* culture with wastewater into a sustainable circular bioeconomy. *Environmental Pollution*, **284**, 117492, https://doi.org/10.1016/j.envpol.2021.117492

Ling Y., Sun L.-P., Wang S.-Y., Lin C. S. K., Sun Z. and Zhou Z.-G. (2019). Cultivation of oleaginous microalga *Scenedesmus obliquus* coupled with wastewater treatment for enhanced biomass and lipid production. *Biochemical Engineering Journal*, **148**, 162–169, https://doi.org/10.1016/j.bej.2019.05.012

Lu Q., Ji C., Yan Y., Xiao Y., Li J., Leng L. and Zhou W. (2019). Application of a novel microalgae-film based air purifier to improve air quality through oxygen production and fine particulates removal. *Journal of Chemical Technology & Biotechnology*, **94**, 1057–1063, https://doi.org/10.1002/jctb.5852

Ma M. and Hu Q. (2024). Microalgae as feed sources and feed additives for sustainable aquaculture: prospects and challenges. *Reviews in Aquaculture*, **16**, 818–835, https://doi.org/10.1111/raq.12869

Maaroff R. M., Md Jahim J., Azahar A. M., Abdul P. M., Masdar M. S., Nordin D. and Abd Nasir M. A. (2019). Biohydrogen production from palm oil mill effluent (POME) by two stage anaerobic sequencing batch reactor (ASBR) system for better utilization of carbon sources in POME. *International Journal of Hydrogen Energy*, **44**, 3395–3406, https://doi.org/10.1016/j.ijhydene.2018.06.013

Maneechote W., Cheirsilp B., Angelidaki I., Suyotha W. and Boonsawang P. (2023). Chitosan-coated oleaginous microalgae–fungal pellets for improved bioremediation of non-sterile secondary effluent and application in carbon dioxide sequestration in bubble column photobioreactors. *Bioresource Technology*, **372**, 128675, https://doi.org/10.1016/j.biortech.2023.128675

Michalak I. and Chojnacka K. (2015). Algae as production systems of bioactive compounds. *Engineering in Life Sciences*, **15**, 160–176, https://doi.org/10.1002/elsc.201400191

Milhazes-Cunha H. and Otero A. (2017). Valorisation of aquaculture effluents with microalgae: the integrated multi-trophic aquaculture concept. *Algal Research*, **24**, 416–424, https://doi.org/10.1016/j.algal.2016.12.011

Nguyen T. D. P., Nguyen D. H., Lim J. W., Chang C.-K., Leong H. Y., Tran T. N. T., Vu T. B. H., Nguyen T. T. C. and Show P. L. (2019). Investigation of the relationship between bacteria growth and lipid production cultivating of microalgae *Chlorella vulgaris* in seafood wastewater. *Energies*, **12**, 2282, https://doi.org/10.3390/en12122282

Nguyen T.-T.-D., Nguyen T.-T., An Binh Q., Bui X.-T., Ngo H. H., Vo H. N. P., Andrew Lin K.-Y., Vo T.-D.-H., Guo W., Lin C. and Breider F. (2020). Co-culture of microalgae-activated sludge for wastewater treatment and biomass production: exploring their role under different inoculation ratios. *Bioresource Technology*, **314**, 123754, https://doi.org/10.1016/j.biortech.2020.123754

Nikokherad H., Esmaili-Sari A., Moradi A. M., Bahramifar N. and Mostafavi P. G. (2022). Bioaccumulation capacity of *Chlorella vulgaris* and *Spirulina platensis* exposed to silver nanoparticles and silver nitrate: bio- and health risk assessment approach. *Algal Research*, **64**, 102671, https://doi.org/10.1016/j.algal.2022.102671

Nobre M. L. F., Tavares D., Fraga C., Oliveira B., Dias M., Mesquita S., Oliveira C. M. and Pires J. C. M. (2024). Techno-economic analysis of a circular microalgal approach for enhanced wastewater treatment and resource recovery in Northern Portugal. *Journal of Cleaner Production*, **434**, 140389, https://doi.org/10.1016/j.jclepro.2023.140389

Nomura T., Inoue K., Uehara-Yamaguchi Y., Yamada K., Iwata O., Suzuki K. and Mochida K. (2019). Highly efficient transgene-free targeted mutagenesis and single-stranded oligodeoxynucleotide-mediated precise knock-in in the industrial microalga *Euglena gracilis* using Cas9 ribonucleoproteins. *Plant Biotechnology Journal*, **17**, 2032–2034, https://doi.org/10.1111/pbi.13174

Novoveská L., Zapata A. K., Zabolotney J. B., Atwood M. C. and Sundstrom E. R. (2016). Optimizing microalgae cultivation and wastewater treatment in large-scale offshore photobioreactors. *Algal Research*, **18**, 86–94, https://doi.org/10.1016/j.algal.2016.05.033

Ofrydopoulou A., Nannou C., Evgenidou E., Christodoulou A. and Lambropoulou D. (2022). Assessment of a wide array of organic micropollutants of emerging concern in wastewater treatment plants in Greece: occurrence, removals, mass loading and potential risks. *Science of the Total Environment*, **802**, 149860, https://doi.org/10.1016/j.scitotenv.2021.149860

Ortiz-Hernández M. L., Sánchez-Salinas E., Dantán-González E. and Castrejón-Godínez M. L. (2013). Pesticide biodegradation: mechanisms, genetics and strategies to enhance the process. *Biodegradation-Life of Science*, **10**, 251–287.

Othman I., Anuar A. N., Ujang Z., Rosman N. H., Harun H. and Chelliapan S. (2013). Livestock wastewater treatment using aerobic granular sludge. *Bioresource Technology*, **133**, 630–634, https://doi.org/10.1016/j.biortech.2013.01.149

Pahunang R. R., Buonerba A., Senatore V., Oliva G., Ouda M., Zarra T., Muñoz R., Puig S., Ballesteros F. C., Li C.-W., Hasan S. W., Belgiorno V. and Naddeo V. (2021). Advances in technological control of greenhouse gas emissions from wastewater in the context of circular economy. *Science of the Total Environment*, **792**, 148479, https://doi.org/10.1016/j.scitotenv.2021.148479

Potijun S., Yaisamlee C. and Sirikhachornkit A. (2021). Pigment production under cold stress in the green microalga *Chlamydomonas reinhardtii*. *Agriculture*, **11**, 564, https://doi.org/10.3390/agriculture11060564

Purba L. D., Abdullah N., Yuzir A., Zamyadi A., Shimizu K. and Hermana J. (2021). Rapid development of microalgae–bacteria granular sludge using low-strength domestic wastewater. *Journal of Water and Environment Technology*, **19**, 96–107, https://doi.org/10.2965/jwet.20-132

Rao P., Kumar R. R., Raghavan B., Subramanian V. and Sivasubramanian V. (2011). Application of phycoremediation technology in the treatment of wastewater from a leather-processing chemical manufacturing facility. *Water SA*, **37**, 7–14.

Razzak S. A., Hossain M. M., Lucky R. A., Bassi A. S. and de Lasa H. (2013). Integrated CO_2 capture, wastewater treatment and biofuel production by microalgae culturing – a review. *Renewable and Sustainable Energy Reviews*, **27**, 622–653, https://doi.org/10.1016/j.rser.2013.05.063

Razzak S. A., Ali S. A. M., Hossain M. M. and Delasa H. (2017). Biological CO_2 fixation with production of microalgae in wastewater – a review. *Renewable and Sustainable Energy Reviews*, **76**, 379–390, https://doi.org/10.1016/j.rser.2017.02.038

Reddy H. K., Muppaneni T., Sun Y., Li Y., Ponnusamy S., Patil P. D., Dailey P., Schaub T., Holguin F. O. and Dungan B. (2014). Subcritical water extraction of lipids from wet algae for biodiesel production. *Fuel*, **133**, 73–81, https://doi.org/10.1016/j.fuel.2014.04.081

Reno U., Regaldo L. and Gagneten A. M. (2020). Circular economy and agro-industrial wastewater: potential of microalgae in bioremediation processes. In: Valorisation of Agro-Industrial Residues – Volume I: Biological Approaches, Z. A. Zakaria, R. Boopathy and J. R. Dib (eds.), 111–129, Springer International Publishing, Cham.

Renuka N., Prasanna R., Sood A., Ahluwalia A. S., Bansal R., Babu S., Singh R., Shivay Y. S. and Nain L. (2016). Exploring the efficacy of wastewater-grown microalgal biomass as a biofertilizer for wheat. *Environmental Science and Pollution Research*, **23**, 6608–6620, https://doi.org/10.1007/s11356-015-5884-6

Roy S. S. and Pal R. (2015). Microalgae in aquaculture: a review with special references to nutritional value and fish dietetics. Proceedings of the Zoological Society, Springer, 1–8.

Ruiz J., Álvarez-Díaz P., Arbib Z., Garrido-Pérez C., Barragán J. and Perales J. (2013). Performance of a flat panel reactor in the continuous culture of microalgae in urban wastewater: prediction from a batch experiment. *Bioresource Technology*, **127**, 456–463, https://doi.org/10.1016/j.biortech.2012.09.103

Ruiz-Martinez A., Garcia N. M., Romero I., Seco A. and Ferrer J. (2012). Microalgae cultivation in wastewater: nutrient removal from anaerobic membrane bioreactor effluent. *Bioresource Technology*, **126**, 247–253, https://doi.org/10.1016/j.biortech.2012.09.022

Sheldon R. A. and Norton M. (2020). Green chemistry and the plastic pollution challenge: towards a circular economy. *Green Chemistry*, **22**, 6310–6322, https://doi.org/10.1039/D0GC02630A

Su Y., Mennerich A. and Urban B. (2012). Synergistic cooperation between wastewater-born algae and activated sludge for wastewater treatment: influence of algae and sludge inoculation ratios. *Bioresource Technology*, **105**, 67–73, https://doi.org/10.1016/j.biortech.2011.11.113

Sutherland D. L., Howard-Williams C., Turnbull M. H., Broady P. A. and Craggs R. J. (2015). The effects of CO_2 addition along a pH gradient on wastewater microalgal photo-physiology, biomass production and nutrient removal. *Water Research*, **70**, 9–26, https://doi.org/10.1016/j.watres.2014.10.064

Sydney E. D., da Silva T., Tokarski A., Novak A. D., de Carvalho J., Woiciecohwski A., Larroche C. and Soccol C. (2011). Screening of microalgae with potential for biodiesel production and nutrient removal from treated domestic sewage. *Applied Energy*, **88**, 3291–3294, https://doi.org/10.1016/j.apenergy.2010.11.024

Urbina-Suarez N. A., Machuca-Martínez F. and Barajas-Solano A. F. (2021). Advanced oxidation processes and biotechnological alternatives for the treatment of tannery wastewater. *Molecules*, **26**, 3222, https://doi.org/10.3390/molecules26113222

Wiley P. E. (2013). Microalgae Cultivation Using Offshore Membrane Enclosures for Growing Algae (OMEGA). University of California, Merced.

Xia C., Van Le Q., Chinnathambi A., Salmen S. H., Alharbi S. A. and Tola S. (2021). Role of ZnO and Fe_2O_3 nanoparticle on synthetic saline wastewater on growth, nutrient removal and lipid content of *Chlorella vulgaris* for sustainable production of biofuel. *Fuel*, **300**, 120924, https://doi.org/10.1016/j.fuel.2021.120924

Yeesang C. and Cheirsilp B. (2014). Low-cost production of green microalga *Botryococcus braunii* biomass with high lipid content through mixotrophic and photoautotrophic cultivation. *Applied Biochemistry and Biotechnology*, **174**, 116–129, https://doi.org/10.1007/s12010-014-1041-9

Zhang B., Li W., Guo Y., Zhang Z., Shi W., Cui F., Lens P. N. L. and Tay J. H. (2020a). Microalgal–bacterial consortia: from interspecies interactions to biotechnological applications. *Renewable and Sustainable Energy Reviews*, **118**, 109563, https://doi.org/10.1016/j.rser.2019.109563

Zhang Z., Pan S.-Y., Li H., Cai J., Olabi A. G., Anthony E. J. and Manovic V. (2020b). Recent advances in carbon dioxide utilization. *Renewable and Sustainable Energy Reviews*, **125**, 109799, https://doi.org/10.1016/j.rser.2020.109799

Zhao X., Kumar K., Gross M. A., Kunetz T. E. and Wen Z. (2018). Evaluation of revolving algae biofilm reactors for nutrients and metals removal from sludge thickening supernatant in a municipal wastewater treatment facility. *Water Research*, **143**, 467–478, https://doi.org/10.1016/j.watres.2018.07.001

doi: 10.2166/9781789064100_0173

Chapter 10

Introduction to bioenergy production: concepts, opportunities, and challenges

Abeer I. M. El-Sayed[1], *Mostafa M. El-Sheekh*[2] *and Wafaa F. Zohir*[1]*

[1]*Botany and Microbiology Department, Faculty of Science, Damanhour University, Damanhour 22511, El-Beheira, Egypt*
[2]*Botany and Microbiology Department, Faculty of Science, Tanta University, 31527 Tanta, Egypt*
**Corresponding author: E-mail: wafaa_phyco_3@yahoo.com or Abeer.elsayed@sci.dmu.edu.eg*

ABSTRACT

Fossil fuels currently yield 88% of the globe's power, and energy consumption is rising dramatically. Using limited resources as an energy generator for an extended length of time is risky because of their finite quantity and non-renewable character. Because bioenergy is more affordable than fossil fuels, it will become the main sustainable energy source in the ensuing decades. The availability of affordable biofuel and a variety of beneficial biofuels that have the ability to produce heat, steam, electricity, and gas, as well as power for transportation, are essential for the growth of bioenergy. The bioenergy formation procedure can be carried out on a range of scales and can employ various materials, comprising different plant parts, bacteria, fungi, manure, algae, or other biomass of lignocellulose origin. This opens the door for the manufacture of biofuel on a worldwide scale. This chapter will discuss major microbial groups, such as cyanobacteria, microalgae bacteria, and fungi, and their roles in the synthesis of bioenergy.

Keywords: Bioenergy, biofuel, biogas, biomass, environmental impacts, microbial conversion, renewable energy, waste bioconversion

10.1 INTRODUCTION

Durable bioenergy microbial metabolism's enormous diversity and adaptability present a chance to create bioenergy and bioproducts that are sustainable and independent of fossil fuels. To fully realize this potential, a fundamental understanding of the behavior of biological systems is needed. This chapter is especially concentrated on the generation of bioproducts and 'advanced' biofuels, which are biologically produced substances that have the potential to function as energy-dense transportation fuels such as aviation, diesel, and gasoline, and that work with the fuel delivery systems and engines that are in use today aiming to create a promising bioenergy industries (Figure 10.1).

Large amounts of organic material are found amid today's societal streams of waste. This material is transported to landfills where it undergoes biological transformations and is discharged into the surroundings, mostly like the powerful greenhouse gas, methane. Generating biofuels out of the energy extracted from this garbage would be an option, which would have a positive economic effect

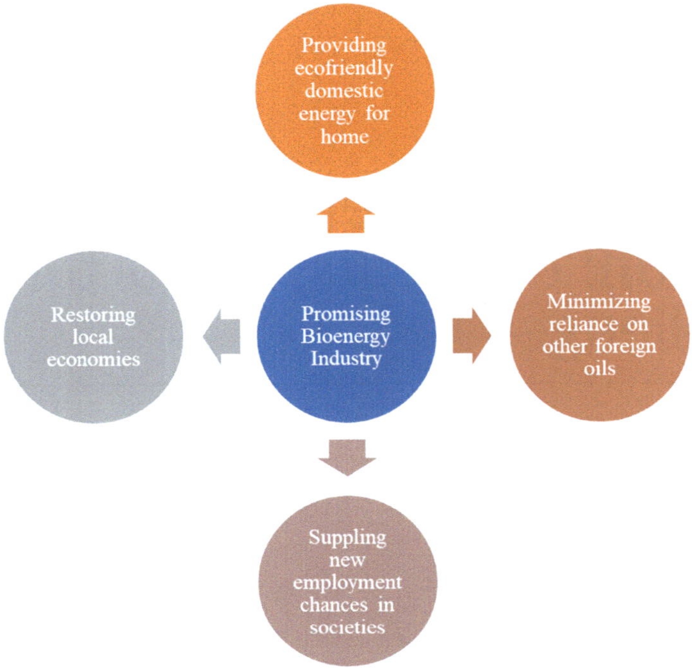

Figure 10.1 Promising bioenergy industry.

and reduce the released greenhouse gas. Abolition of the need for landfills and replacement of fossil fuels with waste producing energy are further advantages.

For the purposes of this chapter, liquid biofuels are defined as bioenergy materials generated from organic elements of local, farming, commercial, and industrial trash that are produced by means of microbial biotechnologies such as algal oils (biodiesel, bio-gasoline, etc.), alcohols (ethanol and butanol), and biogases (methane and hydrogen).

Following that, these forms of bioenergy can be used instead of fossil fuels in the formation of electrical power, heat, and transportation, thereby reducing greenhouse gas emissions and supplying renewable power (Hnain *et al.*, 2011).

Because waste is transformed into more useful fuels, biological technologies come with their own set of drawbacks, including high water requirements, potential contamination of the surroundings, and contributing to the worldwide carbon load (Bergsma *et al.*, 2010; Zanchi *et al.*, 2010). However, their overall impact is far less than that of using non-renewable fuel resources.

Cell walls from plants, which are made up of cellulose wrapped in hemicellulose and lignin, make up the majority of waste in contemporary society (Bayer *et al.*, 2007). This needs to be taken into account in the techniques used to convert waste biomass into bioenergy fuels. Numerous microbes, especially those present in naturally occurring communities that produce methane, have evolved cellulose-breaking and cellulose-metabolizing mechanisms.

These colonies will produce methane from organic molecules if given enough time. Production of cellulosic biofuel, or ethanol, is hampered greatly by the fact that, usually, various microbes breakdown and degrade different types of cellulose. Because methanogenic communities are robust, biogas generation is suitable for handling a variety of waste streams, including municipal trash. Wastes from forestry and agriculture as illustrated in (Figure 10.2) are more homogeneous, which makes them superior substrates for the synthesis of alcohol. Wastewater and wastewater sludge can be grown to provide lipids for biodiesel, bio-jet fuel, and bio-gasoline, or they can be fermented to yield methane, hydrogen, and ethanol. Several industrial processes generate carbon dioxide (CO_2), a by-product of fermentation that may be used to encourage the development of algae and cyanobacteria, which will subsequently yield lipids or hydrogen.

The development and application phases of waste conversion technologies that generate ethanol, methane, hydrogen, and algal lipids for fuel are distinct. Presently, ethanol and methane are manufactured industrially; however, only a small portion of ethanol produced right now is cellulosic ethanol, primarily from crops farmed for commercial purposes, such as corn and sugarcane.

Production of hydrogen and lipids from algae is still in its infancy. But in order to develop the most effective technologies understanding microbial biology is essential for any sort of bioenergy.

Therefore, it is crucial to explain how and why these organisms synthesize fuels for every kind of biopower fuel. To survive, all living beings need to maintain a balance between catabolic and anabolic reactions (i.e., form a redox balance), even in less-than-ideal circumstances. To preserve this equilibrium, organisms have developed specific 'back-up systems' or alternate metabolic processes. For massive regenerative forms of energy to be practical, these various metabolisms must be employed while allowing for the preservation of a steady redox balance in the participating species (Hnain *et al.*, 2011).

The Genomic Science program (GSP), which supports fundamental research aimed at elucidating the fundamental concepts underlying biological systems, is ideally positioned to exploit these opportunities. The goal of GSP research is to supply the basic science required to comprehend, forecast, control, and create the microbial and plant systems and processes that serve as the foundation for energy innovations.

- GSP requested suggestions for sustainable bioenergy in two domains:
  Creating novel model microorganisms or microbial communities with improved or distinct properties in order to generate advanced biofuels and bioproducts.
- Being aware of new microbial functionalities and biosynthetic pathways that are important for producing high-tech biofuels and bioproducts.

Figure 10.2 Different biomass wastes.

One of the most important issues facing contemporary society is waste disposal. It is becoming more and more clear that microorganisms can produce biofuels from most organic waste coming from industrial, agricultural, and residential sources. For instance, these fuels provide essential sources of sustainable energy that have a chance to significantly reduce emissions of greenhouse gases when it comes to the slow release of methane from landfills.

Microbial activity produces four different types of biofuels: (i) algal lipids; (ii) alcohols; (iii) methane; and (iv) hydrogen.

Methane production results from comparatively robust microbial communities, in contrast to the other kinds. Moreover, the leftovers of other biofuel production systems can be used to produce methane. Although single-organism systems are typically used to produce the other biofuels, the idea of producing these fuels through the syntrophic cooperation of microbes is gaining popularity. This is particularly true when it comes to using cellulosic resources to produce ethanol and hydrogen, where the cellulose must first break down into glucose.

The only biofuel that algae can produce effectively is algal lipids, which are generated from waste CO_2 from fuel combustion and wastewater and wastewater sludge as nutrients. Algae grown on waste material, however, may potentially be used as a feedstock for more biofuels. The most efficient way to turn garbage into biofuels could organize biomass using two or more of these factories (Hnain *et al.*, 2011).

10.2 MICROORGANISMS' FUNCTION IN THE SYNTHESIS OF BIOFUEL

It is reasonable to anticipate that fossil fuels will eventually run out because their supply is running low and the global population is growing. Biofuels are therefore viewed as an improved energy source that can satisfy future energy needs. Biomass, which includes microbial, algal, and plant cells, is used to generate biofuels. Converting biomass into biofuels involves a number of processes. A variety of microbial species are also employed in the breakdown of lignocellulosic biomass, the management of decomposition, and the generation of hydrolytic enzymes.

As they are catalysts, enzymes produced by microbes are essential for the conversion of biomass (Machado & Atsumi, 2012; Tabatabai *et al.*, 2019). A fantastic source of bioenergy is microbes such as methanogens, cyanobacteria, microalgae, and archaebacteria (Fu *et al.*, 2016). These organisms generate methane gas, bioelectricity, biohydrogen, and biomethanol, among other forms of bioenergy. Microbial fuel cells (MFCs) are a superior model for producing bioelectric power (Jin *et al.*, 2014; Ni & Sun, 2009). Bioethanol and biomethanol, which are produced from microalgae and are currently added to diesel, are considered excellent energy sources that could eventually replace fossil fuels (Hou *et al.*, 2013; Lutke-Eversloh, 2014; Xue *et al.*, 2017).

As biohydrogen burns without producing any waste, it is thought to be eco-friendly and a pure fuel. A variety of microorganisms, including bacteria and microalgae, produce biohydrogen as they grow (Azwar *et al.*, 2014; Saifuddin & Priatharsini, 2016). Algae and other microorganisms generate bio-oil. Additionally, a variety of plant and agricultural biomass is produced by bio-oils. To convert biomasses into bio-oils, microbes such as bacteria and fungi must create enzymes that are extracellular. One fuel that is thought to be environmentally safe is biogas with a wide range of uses. When methanogenic bacteria bio-convert biomass, biogas is created (Gowen & Fong, 2010; Himmel *et al.*, 2007; Senger, 2010). Microorganisms with the ability to carry out a specific function, such as cyanobacteria and microalgae, are the best substitute for that. Certain microalgae possess the exceptional ability to transform sugars into biofuels, while numerous others have a more than 50% natural oil composition (Zhu *et al.*, 2008).

A wide variety of microbes are involved in the biomass transformation process. Bamboo trees' content of lignin can be reduced by 30% in 30 days by *Echinodontium taxodii*. The ideal temperature range for this kind of fungus is 25–35°C (Philbrook *et al.*, 2013). *Ceriporiopsis subvermispora* possesses excellent qualities for breaking down biomass. In 30 days, it can break down 45% of the corn stover's lignin components (Philbrook *et al.*, 2013).

Utilizing fungi in the breakdown process serves as a significant sign of achievement in producing biogas from empty fruit bunch with the co-digestion approach; consequently, the choice of fungi is vital. Fungi that are frequently used in the fermentation process include *Saccharomyces cerevisiae*, *Aspergillus niger*, *Penicillium chrysogenum*, *Trichoderma reesei*, and *Phanerochaete chrysosporium*.

Earlier researchers (Kovács *et al.*, 2022) have revealed that *T. reesei* can boost biogas generation by 20–30% in anaerobic setups using lignocellulosic trash. *A. niger* has been demonstrated to boost biogas production from starchy feedstock by 15–25% (Silva *et al.*, 2024). It has been found that *P. chrysogenum* increases lignocellulose accessibility for fermentation by 10–20% (Kovács *et al.*, 2022). *P. chrysosporium* is a fungus that can accelerate the breakdown of lignin by 25–35% (Tamilselvan & Immanuel Selwynraj, 2024). *S. cerevisiae* can also lower the sugar composition, which prepares the substrate for digestion by methanogens (Tang *et al.*, 2024).

A number of types of bacteria are also employed in the manufacturing of biofuel. A few instances of bacterial genera that produce ethanol are *Escherichia coli* (Romero-Garcia *et al.*, 2016), *T. reesei* (Huang *et al.*, 2014), and *Caldicellulosiruptor bescii* (Chung *et al.*, 2014). There are some examples of biobutanol-producing microorganisms such as *Clostridium acetobutylicum* (Lutke-Eversloh & Bahl, 2011) and *Pseudomonas putida* (Nielsen *et al.*, 2009).

In simple terms, engineered microbes are the machines that integrate with our sustainable energy sources to generate biofuel (Clomburg & Gonzalez, 2010; Fatma *et al.*, 2018). There are a number of ways to extract biofuel from microorganisms, but more investigation and focus from scientists are needed before they are ready for widespread application.

10.2.1 Cyanobacteria

Cyanobacteria are members of the Monera Kingdom. A unique characteristic of cyanobacteria is the presence of a pigment produced by photosynthesis (Zhou & Li, 2010). Because the majority of cyanobacteria's individuals are so identical to that of bacteria, they are members of the exact same kingdom, Monera. Cyanobacteria do not need fertile land to flourish very quickly. Because they are capable of growing their own nutrition and absorb CO_2 from the environment, cyanobacteria are referred to as organisms that are autotrophic (Lu, 2010). Cyanobacteria are utilized for the creation of biofuels because of their ability to photosynthesize and because they are an excellent resource of carbon (Bandyopadhyay *et al.*, 2010).

Despite their genetic limitation, cyanobacteria offer a promising platform for biofuel research. The main obstacles to the manufacturing of cyanobacterial biofuels are genetically engineered changes to fixation of carbon routes, cyanobacterial reactions involving metabolism, the need for nutrients for industrial production, and increasing the photosynthesis capacity of cyanobacteria under light from the sun (Lindblad *et al.*, 2012; Sakurai & Masukawa, 2007).

10.2.2 Microalgae

Depending on the origin, the manufacturing of biofuel yield varies (Greenwell *et al.*, 2010). As a result, choosing crops or microbes that produce biofuel is crucial to the field of biofuel development (Moreno-Garrido, 2008). The geographic location that best supports the development of an organism determines how much biofuel is produced there. Certain biofuel crops, such as soybeans, need plenty of land for development.

Microalgae may develop in limited space with higher yields (Himmel *et al.*, 2007; Olguin, 2012). As a result, microalgae are a material that is appealing for the manufacturing of biofuel. Following are among the crucial characteristics of microalgae: (1) high output when compared to biomasses from soybean plants and other biological origins; (2) these are assets for non-food feedstock used in the manufacturing of biofuels; (3) microalgae could simply grow on non-arable soil; (4) microalgae can grow in fresh, dark-colored, marine, and saline water as well as sewage; (5) they generate a number of useful products, including biofuels; and (6) garbage has great ability for recycling nutrients and CO_2.

The aforementioned considerations make microalgae a viable option for the fabrication of biofuel. Triglycerides (TAGs), which are made by algae and referred to as algal crude oil, are the source of the algal fuel, also referred to as oilgae (Simionato *et al.*, 2013). The same as biofuels of the second generation, TAGs can be processed using various techniques to produce biodiesel. Anaerobic digestion is another way that algae may be converted into biogas. As a result of the biofuel drying procedure being eliminated, this method has many benefits. Both energies and time are greatly expended during the drying procedure of biomass. Under the sunlight, the chemotrophic creatures can grow in phototrophic fermentation tanks and acquire power. Both open pond systems and closed photobioreactors are common places for phototrophic microbes to grow (Anto *et al.*, 2020). The fermentation vessels are an enclosed structure and require transmitting CO_2 and nutrients along the time. Pond is an open operation which captures CO_2 from the outside world. Algae in the pond grow more quickly thanks to the fertilizing effects of CO_2. In the presence of sunlight and atmospheric CO_2, the algal cells produce food for themselves. However, because laboratory photobioreactors are artificial systems, artificial LED lighting and CO_2 supply are used to keep the system in a suitable state. Nevertheless, a sizable photobioreactor is positioned and exposed to direct sunlight. The manufacturing cost of the pond is considerably less, but because of the pollution issue, it is not suitable for the development of just one species (Karemore *et al.*, 2016). Although cyanobacteria and microalgae form the basis for the manufacture of third-generation biofuels, these fuels are not sold commercially. Research and development are currently underway for the manufacture of third-generation biofuels. Via a variety of techniques, scientists are attempting to increase the synthesis of third-generation biofuels. Some difficulties with third-generation biofuels emerged, as improving biofuel production using a variety of techniques (Rogers *et al.*, 2014). Both macro- and microalgal biomasses can be anaerobically digested. There are other approaches as well, such as gasification and thermal breakdown procedures.

10.2.3 Biofuel types

Biofuels come in different varieties, including biodiesel, biohydrogen, bioethanol, and biomethanol. In this chapter, we have covered a few significant biofuels.

10.2.3.1 Biofuel production and bioconversion

Multiple bioconversion techniques are used to transform complex biomaterials into simple biomaterials. Numerous bioconversion methods yield different types of biofuels, comprising bioethanol, biomethanol, biodiesel, biohydrogen, and biogas (Figure 10.3). The process of producing biofuel and bioconversion involves many microbial catalytic processes.

10.2.3.2 Biohydrogen

Being clean, safe for the environment, and inexpensive, biohydrogen is a fuel source. In contrast to other fuels, it offers a lot more benefits. Fuel efficiency is improved by the high energy produced by biohydrogen. In the biofuel research, the synthesis of biohydrogen is still in its infancy. Nowadays, conventional technologies are typically used to manufacture biohydrogen (Shaishav *et al.*, 2013). Several traditional methods, such as coal gasification and electrolysis, have drawbacks. These technologies' main drawbacks are their high thermal energy requirements and the production of various toxic by-products, such as gasses and sewage (Hsia & Chou, 2014). Although the process of electrolysis of water is benign for the environment, a significant quantity of electricity is needed to generate hydrogen. Therefore, this technique can only be carried out in established areas with regular access to electricity (McKinlay & Harwood, 2010). Therefore, it is imperative to synthesize a biohydrogen generation method that is both low cost and environmentally safe. The process of producing biohydrogen from microbial life is incredibly low cost and environmentally safe (Dincer, 2012). In plants, algae, or cyanobacteria, photosynthetic reactions result in the formation of biohydrogen. Anaerobic or aerobic fermentation processes can also be used to manufacture biohydrogen. There are numerous microbial species that can be employed to produce biohydrogen (Manish & Banerjee, 2008). Some well-known

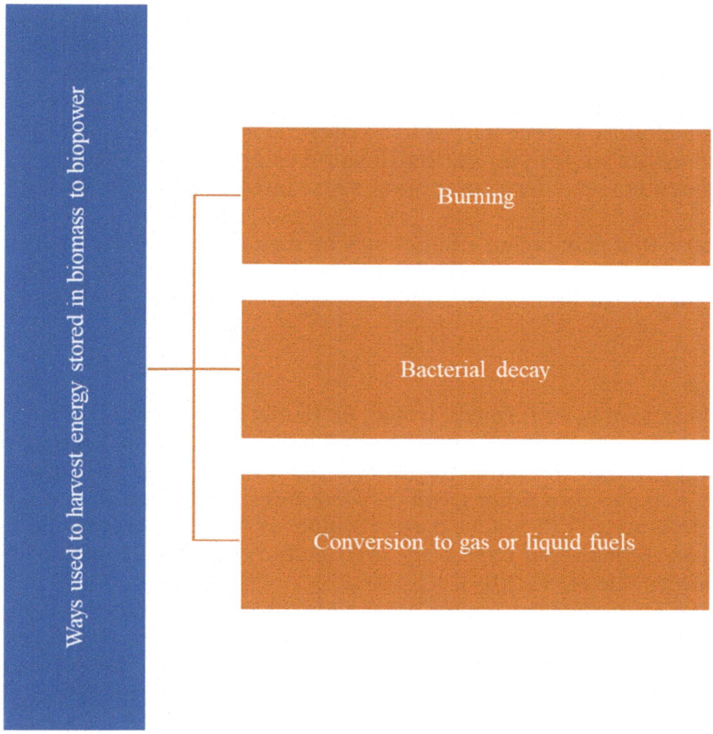

Figure 10.3 Ways to convert biomass to biopower.

examples of biohydrogen-producing microbes are *Chlamydomonas moewusii*, *Scenedesmus obliquus*, *Enterobacter aerogenes*, and *Rhodobacter sphaeroides*.

10.2.3.3 Bioethanol

These days, bioethanol is added to gasoline and diesel as an additive. As such, it is regarded as a substitute for traditional fuels and can eventually substitute the usage of petroleum goods (Guo *et al.*, 2015; Littlewood *et al.*, 2014). It comes from a variety of readily available biomass types found all over the planet. As a result, in comparison to other fuels, bioethanol may have a poor value of production. It is an environmentally safe fuel with a high octane number (Limayen & Ricke, 2012).

Nowadays, a variety of biomasses, including plant, bacterial, fungal, and algae wastes, can be used to produce it. Additionally, a variety of culinary and non-food oils, including mustard, soybean, and corn oil, are used in its production (Gonzalez-Garcia & Bacenetti, 2019; Whitaker *et al.*, 2018). Primary bioethanol source product is microalgae. Bioethanol and other biofuels are produced in vast quantities by microalgae (Porth & El-Kassaby, 2015). By genetic modification, it is possible to increase the amount of bioethanol produced by natural microalgal organisms (Balan, 2014).

Essential function for nanotechnology is observed in the industries based on bioethanol. Large surfaces for biochemical reactions are made possible by nanoscale particles; MaO_2 can maximize the formation of bioethanol via sugarcane leaf biomass (Cherian *et al.*, 2015). Ethanol production is increased by the larger surface area and smaller size of MnO_2, which provide many sites for binding enzymes and various chemicals.

Finding the cleanest liquid fuel with the lowest carbon emissions is currently necessary because fossil fuels are the reason of global warming. As a result, with regard to biomass transformation into fuel, ethanol has been gaining more attention, and cutting-edge technology is being created for generating ethanol through biomass fermentation. A class of microbes that generate cellulase used for hydrolyzing lignocellulose is necessary for the conversion of biomass into ethanol. Microbes that have produced enzymes are trapped for saccharification on an immobile substrate, which is the highly efficient and economically viable fermentation of sugar into ethanol, in order to produce significant amounts of energy.

Numerous microbes that may create ethanol may be included in the decomposition of a non-living body. There are about 58 different species of bacteria, 17 different strains of yeast, and 24 different species of mold that may create ethanol and other metavolatiles through a variety of complex metabolic pathways.

10.2.3.4 Biobutanol

Similar to ethanol, butanol may be formed through fermentation of feedstocks such as sugar, starch, or cellulosic materials. Among the advantages of biobutanol over traditional transportation fuels are as follows: *greater energy content*: compared to gasoline, biobutanol has an energy density that is 10–20% lower. Furthermore, compared to other gasoline substitutes, biobutanol has a significant amount of energy. Reid vapor pressure (RVP) is a measure of volatility and evaporative fluxes. Biobutanol has a lower RVP than ethanol. *Increased energy security*: biobutanol may be produced regionally using a variety of raw materials (Xue *et al.*, 2017).

Reduced emissions: Growing feedstocks absorb CO_2, which lowers the total amount of greenhouse gas fluxes. The fermentation process used to produce biobutanol is carried out by *Clostridium* sp. bacteria along with many microbes. Furthermore, sugars that are obtained from treated lignocellulosic biomass are used as fermentation substrates to guarantee that the production method is cost-effective. Many specific issues, such as substrate inhibition, toxicity of butanol, delayed formation, and consequently reduced cell density, typically limit the solventogenic fermentation process (Xue *et al.*, 2017).

In addition to these drawbacks, the synthesis of ethanol and acetone among other end products lowers the yield of biobutanol. To overcome these problems, scientists have created a variety of microbial strains through genetic engineering that enhance their ability to produce more biobutanol from a variety of bacterial species.

10.2.3.5 Biogas

Being eco-friendly and reasonably priced, biogas is a biofuel. The primary component of biogas is methane, which is created when organic materials are broken down. The microbial species that are involved in the production of biogas include methanogens (Aryal *et al.*, 2018; Romero-Guiza *et al.*, 2016). The manufacture of biogas involves multiple procedures. The primary process in the creation of biogas is hydrolysis. When a sufficient digesting system is present, the substrate is broken down during the hydrolysis process. A big molecule, such as proteins or carbs, is broken down into amino acids and simple sugar, respectively, during the hydrolysis process (Romero Victorica *et al.*, 2020). Acidogenesis method is the second most crucial one. Acetogenesis is the third major stage. Acetic acid participates in a variety of microbiological processes during the acetogenesis. Methanogenesis is the fourth and most significant phase. Methane gas is a result of the biogas manufacturing system through a process called methanogenesis. Many methanogenic bacteria create the methane gas (Arias *et al.*, 2020; Mao *et al.*, 2015; Sekoai *et al.*, 2016; Waqas *et al.*, 2020). The breakdown of sugar and amino acids as well as the synthesis of CO_2, hydrogen, and alcohol are all carried out by the acidogenesis process. The process of acidogenesis involves multiple mechanisms. Low hydrogen in the culture media was necessary for the growth of methanogenic bacteria, which produce biogas through a variety of microbiological processes (Hankamer *et al.*, 2007). Fuel that produces biogas from various biological wastes is regarded as clean, safe, and emits no waste.

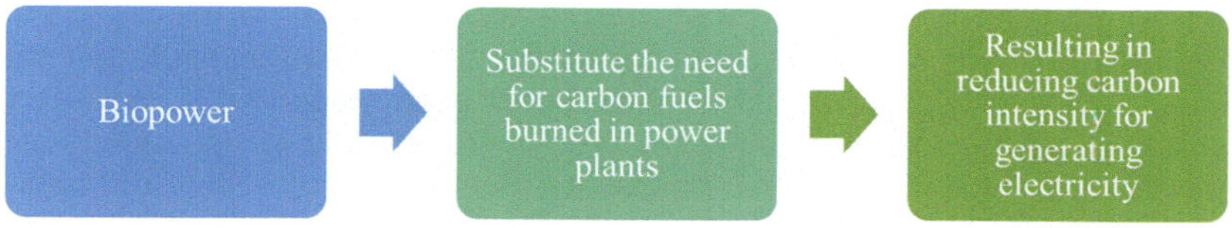

Figure 10.4 Biopower used to generate electricity.

10.2.3.6 Biodiesel

It is produced using a variety of biomass and vegetable oils. In the future, biodiesel can displace fossil fuels as a resource of electricity (Figure 10.4). It is also regarded as an inexpensive, environmentally safe energy source (De Araujo *et al.*, 2013; El-Sheekh *et al.*, 2020; Mohammadshirazi *et al.*, 2014). There are several ways to convert non-food vegetable oils into biofuels. As a significant portion of non-food vegetable oils ends up as waste globally, producing biodiesel from them is a useful procedure. Nanomaterials are crucial to the manufacturing of biodiesel. Numerous studies revealed that employing nanomaterials to modify the transesterification reaction could increase the generation of biofuels (Chen *et al.*, 2018; Lee *et al.*, 2015). It has been found that ZnMg(Al)O and Fe_3O_4 nanoparticles can improve biodiesel production. Many biofuel crops, including jatropha, can be used to generate biofuels such as biodiesel. The plant species Jatropha is a member of the Euphorbiaceae family of flowering plants. For the manufacturing of biodiesel, the oil extracted from jatropha seeds is quite beneficial. Fish and other animals have been fed the leftover material (cake) from the oil extraction process. As a result of the resultant materials' high protein content, they are much more nutritious than other feeds (Peralta-Yahya & Keasling, 2010).

10.3 LIPID PRODUCTION IN ALGAE

Third-generation biofuels as well obtained from the fatty acids of algae, may prove to be the greatest viable and sustainable source of renewable energy in the years to come as algae can be grown in barren landscapes and coastal zones, develop faster than terrestrial vegetation, and generate more biomass per unit area than agricultural crops.

The American Society for Testing Materials claims that algal biodiesels have qualities similar to those of our contemporary diesel fuels, with viscosity, heating value, and stability being the more significant attributes due to their comparatively high temperatures (Huang *et al.*, 2010). Lipids are additionally processed to produce high-grade hydrocarbon-based goods such as bio-jet fuel and green gasoline, alongside biodiesel (El-Sheekh *et al.*, 2020; Marsh, 2009; Pienkos & Darzins, 2009). In living things, lipids have a variety of functions. Polar lipids can function as cellular signaling molecules and are essential parts of plant waxes and cell membranes (Guschina & Harwood, 2009). Non-polar lipids, such as esterified waxes and triacylglycerols, often known as triglycerides (TAGs), can function as metabolic energy storage (Guschina & Harwood, 2009). Free-fatty acids are the main ingredients of both polar and non-polar lipids, but they are not often seen in nature in their free form. Various types of lipids are displayed by organisms under varying stresses from the environment. Algae may reroute metabolic power into the production of TAGs in order to live under less-than-ideal circumstances. Chloroplast membrane constituents may potentially be stored in fatty acid reservoirs by polyunsaturated fatty acid-rich TAGs (Guschina & Harwood, 2009). These TAGs are synthesized under the influence of light most of the time, stored in cytosolic lipid bodies, and then repurposed for polar lipid creation at nighttime.

In their native state, suitable algae have total lipid levels ranging from 5 to 20% by dry weight (Huang *et al.*, 2010). When conditions are nutrient-rich, algae typically have low TAG lipid contents. TAG levels have a great range, from approximately 2 to 77% of total lipid content, and are dependent on the species and other environmental conditions. TAG synthesis can generally be enhanced by persistently high light intensities, low CO_2 concentrations, rising salinities, falling temperatures, and raising pH (Guschina & Harwood, 2009). Different types of algae respond differently to these changed circumstances. It is necessary to comprehend the type of climate and environmental conditions that give rise to soil-borne, freshwater, or saltwater algae in order to properly utilize and anticipate changes in their lipid profiles. For this reason, it is very important that TAGs form and extract efficiently in order to produce lipid biofuels (Huang *et al.*, 2010).

10.3.1 Natural gaseous fuel bioconversion to liquid fuel

Biogas is generated when organic molecules break down. The generation of biogas is caused by a variety of bacteria, including various methanogen species. The primary component of biogas is methane (Haibach *et al.*, 2012). The new strategy in the bioenergy industry is the transformation of biogas to liquid fuel. Because it is safer and easier to carry out, liquid fuel is in higher demand than gaseous fuel. Therefore, it is crucial to convert gaseous fuels into liquid fuels (Fabbri & Torri, 2016). In both anaerobic and aerobic environments, methanogenic bacteria can use methane as a source of carbon and create methane gas (Fei *et al.*, 2014; Hu *et al.*, 2016).

10.3.2 Utilizing wastewater treatment for biofuel production

Enough carbon-containing compounds are present in wastewater, including both industrial and home effluents (Zhang *et al.*, 2014). These carbon compounds promote the microbial growth that is thought to be reliable producers of bioenergy. The primary step in the treatment of wastewater is sludge activation, during which organic materials convert to CO_2 and are involved in a variety of microbial metabolic processes (Abdelaziz *et al.*, 2013). Lipid concentration in household sewage ranges from 30 to 40% of total organic matter. The primary lipid content found in municipal sludge is triacylglycerol (Shreve & Brennan, 2019). Numerous bacteria can either create additional carbon sources or absorb lipid from urban sewage. Lipids can be stored by these bacteria within the cell's intracellular region. Bacteria retain oils in the intracellular space in the form of wax esters, polyhydroxyalkanoates, and triacylglycerols (Chinnasamy *et al.*, 2010; Singh *et al.*, 2016).

10.3.3 Microbial fuel cells

MFC is the process by which microorganisms use their metabolic processes to bioconvert chemical energy to electrical energy (Yu *et al.*, 2012). Plant MFCs are devices that use plant cells as their primary power source. Potential uses of MFCs include wastewater treatment, biowaste conversion, biosensors, bioremediation of contaminants, and power generation. The primary process used to produce energy is hydrogenesis. Microbial cells produce hydrogen molecules during their metabolism. The most well-known instance of biohydrogen synthesis occurs in the microbial electron transport chain. MFC devices collect these hydrogen or proton species and use them to produce electricity (Balasubramaniam *et al.*, 2020; Mani *et al.*, 2020; Mathuriya, 2020; Yadav *et al.*, 2019; Zhang *et al.*, 2020).

10.3.3.1 Challenges of bioenergy formation

Utilizing ensembles of different species or solitary individuals can help overcome the difficulties associated with converting trash into biofuel. This chapter covers the topic of organism upgrades, including mixed species production in communities, metabolic and genetic engineering, strain selection, and evolutionary issues, which are being used to improve present production systems. A comparison of the environmental factors needed to transform by-products into the various biopower fuels and gasses discussed next. The several ways that single and mixed species, as well as autotrophic and heterotrophic, can produce bioenergy, are also taken into account. Furthermore, we examine the possible connections between systems that produce alcohol, methane, hydrogen, and algal lipids in order to maximize waste conversion to bioenergy fuels (Hnain *et al.*, 2011).

To improve access to cellulose by dissolving lignin and cracking open cell walls, waste feedstocks are subjected to a variety of chemical, physical, and biological pretreatments (Stephanopoulos, 2007; Sun & Cheng, 2002). Recent developments in microbial prior treatments include the aerobic degradation of lignin in wood by the white-rot fungus *P. chrysosporium*, even if the majority of pretreatments only attempt to remove lignin in order to release cellulose (Shi *et al.*, 2009). Prospective developments in the usage of lignin for the generation of bioenergy could result from studies conducted on the aerobic bacterial genera *Streptomyces*, *Arthrobacter*, *Nocardia*, *Pseudomonas*, and *Alcaligenes* (Li & Fang, 2009) and the unexplored anaerobic conversion of aromatic substances by bacteria (Fuchs, 2008). It has long been known that pretreatment of hemicellulose and cellulose can hydrolyze them into 5- and 6-carbon sugars using extracellular enzymes that come primarily from the *Trichoderma* fungus (Wackett, 2008).

Enzymatic hydrolysis is finished ahead of and apart from fermentation in separate hydrolysis and fermentation (SHF). Simultaneous saccharification and fermentation (SSF) is the term used to describe the process of combining cellulases with anaerobic fermentation, typically using yeast. The SSF mechanism effectively eliminates glucose, which inhibit cellulase action. This boosts the pace and yield of cellulose hydrolysis, but it also creates a new problem: ethanol inhibits cellulases (Sun & Cheng, 2002). In an experiment conducted in a laboratory contrasting SHF and SSF, SSF was able to achieve greater ethanol quantities and levels (Öhgren *et al.*, 2007). Nevertheless, SHF is still widely used in the fermentation sector. The goal of consolidated bioprocessing, or CBP, is to synthesize saccharolytic enzymes, hydrolyze polysaccharides, and ferment both hexose and pentose sugars in one step (Demain *et al.*, 2005; Lynd *et al.*, 2002, 2005), ideally in one organism (Van Zyl *et al.*, 2007). In comparison to sequential techniques, CBP has an added advantage of decreasing the price of producing biofuels, reducing the requirement for prior treatments, improving the power balance by reducing energy inputs, and boost transformation rates (Carere *et al.*, 2008). A possible method to achieve CBP is by modifying the genome of a creature with an elevated ethanol yield, such yeast, to incorporate cellulose-utilization routes, or by strain-selecting natural cellulolytic microorganisms (thermophiles such as *Clostridium*) to increase their ethanol output (Lynd *et al.*, 2005). In this case, it is notable how difficult it is to include the required enzymes at the right quantities so that the species is not negatively impacted (Van Zyl *et al.*, 2007).

10.4 CONCLUSION

Biofuels are regarded as an economical and sustainable source of energy. They are generated from many substances, comprising bacterial biomass, plants, fungi, and algae. Agricultural wastes and edible or inedible oils can also be used to generate them. The finite supplies of fossil fuels, which include gasoline, diesel, and LPG, are found in the crust of the Earth and may run out eventually. Therefore, the development of an appropriate and renewable energy source that may take the place of fossil fuels is imperative. As they originate from biological sources that are sustainable, biofuels are regarded as renewable resources. Biofuels have various types, including biohydrogen, biogas, bioethanol, and biodiesel. They are generated from a variety of materials, including microalgae, fungal biomass, and multiple bacterial species. Scientists have determined that microorganisms have a potential use in the generation of biofuels based on their present study. Nevertheless, the yield of the material produced by native strains is not cost-effective, necessitating their development and improvement through the application of genetic and metabolic engineering techniques. Our research has focused on explaining the major microbial groups as cyanobacteria, microalgae bacteria, and fungi, and their role for the synthesis of bioenergy.

REFERENCES

Abdelaziz A. E., Leite G. B. and Hallenbeck P. C. (2013). Addressing the challenges for sustainable production of algal biofuels: I. algal strains and nutrient supply. *Environmental Technology*, **34**, 1783–1805, https://doi.org/10.1080/09593330.2013.827748

Arias A., Behera C. R., Feijoo G., Sin G. and Moreira M. T. (2020). Unravelling the environmental and economic impacts of innovative technologies for the enhancement of biogas production and sludge management in wastewater systems. *Journal of Environmental Management*, **270**, 110965, https://doi.org/10.1016/j.jenvman.2020.110965

Aryal N., Kvist T., Ammam F., Pnat D. and Ottosen L. D. M. (2018). An overview of microbial biogas enrichment. *Bioresource Technology*, **264**, 359–369, https://doi.org/10.1016/j.biortech.2018.06.013

Anto S., Mukherjee S. S., Muthappa R., Mathimani T., Deviram G., Kumar S. S., Verma T. N. and Pugazhendhi A. (2020). Algae as green energy reserve: technological outlook on biofuel production. *Chemosphere*, **242**, 125079, https://doi.org/10.1016/j.chemosphere.2019.125079

Azwar M. Y., Hussain M. A. and Abdul-Wahab A. K. (2014). Development of biohydrogen production by photobiological: fermentation and electrochemical processes: a review. *Renewable and Sustainable Energy Reviews*, **31**, 158–173, https://doi.org/10.1016/j.rser.2013.11.022

Balan V. (2014). Current challenges in commercially producing biofuels from lignocellulosic biomass. *ISRN Biotechnology*, **2014**(2014), 1–31, https://doi.org/10.1155/2014/463074

Balasubramaniam A., Adi P., Do Thi T. M., Yang J. H., Labibah A. S. and Huang C. M. (2020). Skin bacteria mediate glycerol fermentation to produce electricity and resist UV-B. *Microorganisms*, **8**, 1092, https://doi.org/10.3390/microorganisms8071092

Bandyopadhyay A., Stockel J., Min H., Sherman L. A. and Pakrasi H. B. (2010). High rates of photobiological H_2 production by a cyanobacterium under aerobic conditions. *Nature Communications*, **1**, 139, https://doi.org/10.1038/ncomms1139

Bayer E. A., Lamed R. and Himmel M. E. (2007). The potential of cellulases and cellulosomes for cellulosic waste management. *Current Opinion in Biotechnology*, **18**(3), 237–245, https://doi.org/10.1016/j.copbio.2007.04.004. PMID: 17462879

Bergsma G. C., Croezen H. J., Otten M. B. J. and van Valkengoed M. P. J. (2010). Biofuels: Indirect Land use Change and Climate Impact. CE Delft, Delft, Netherlands, http://www.birdlife.org/eu/pdfs/Biofuels_indirect_land_use_change_and_climate_impact_- CE_Del.pdf (accessed 23 March 2011)

Boyle N. R. and Morgan J. A. (2009). Flux balance analysis of primary metabolism in *Chlamydomonas reinhardtii*. *BMC Systems Biology*, **3**(4), 1–14, https://doi.org/10.1186/1752-0509-3-4. PMID: 19118495

Carere C. R., Sparling R., Cicek N. and Levin D. B. (2008). Third generation biofuels via direct cellulose fermentation. *International Journal of Molecular Sciences*, **9**(7), 1342–1360, https://doi.org/10.3390/ijms9071342. PMID: 19325807

Cherian E., Dharmendirakumar M. and Baskar G. (2015). Immobilization of cellulose onto MnO_2 nanoparticles for bioethanol production by enhanced hydrolysis of agricultural waste. *Chinese Journal of Catalysis*, **36**(8), 1223–1229, https://doi.org/10.1016/S1872-2067(15)60906-8

Chen Y., Liu T., He H. and Liang H. (2018). Fe_3O_4/ZnMg(Al)O magnetic nanoparticles for efficient biodiesel production. *Applied Organometallic Chemistry*, **32**, 1–10.

Chinnasamy S., Bhatnagar A., Hunt R. W. and Das K. C. (2010). Microalgae cultivation in a wastewater dominated by carpet mill effluents for biofuel applications. *Bioresource Technology*, **101**, 3097–3105, https://doi.org/10.1016/j.biortech.2009.12.026

Chung D., Cha M., Guss A. M. and Westpheling J. (2014). Direct conversion of plant biomass to ethanol by engineered *Caldicellulosiruptor bescii*. *Proceedings of the National Academy of Sciences of the USA*, **111**(24), 8931–8936, https://doi.org/10.1073/pnas.1402210111

Clomburg J. M. and Gonzalez R. (2010). Biofuel production in *Escherichia coli*: the role of metabolic engineering and synthetic biology. *Applied Microbiology and Biotechnology*, **86**(2), 419–434, https://doi.org/10.1007/s00253-010-2446-1

De Araujo C. D. M., De Andade C. C., De Souza E., Silva E. and Dupas F. A. (2013). Biodiesel production from used cooking oil: a review. *Renewable and Sustainable Energy Reviews*, **27**, 445–452, https://doi.org/10.1016/j.rser.2013.06.014

Demain A. L., Newcomb M. and Wu J. H. D. (2005). Cellulase, clostridia, and ethanol. *Microbiology and Molecular Biology Reviews*, **69**(1), 124–154, https://doi.org/10.1128/MMBR.69.1.124-154.2005. PMID: 15755956

Dincer I. (2012). Green methods for hydrogen production. *International Journal of Hydrogen Energy*, **37**, 1954–1971, https://doi.org/10.1016/j.ijhydene.2011.03.173

El-Sheekh M. M., El-Mohsnawy E., Mabrouk M. E. and Zoheir W. F. (2020). Enhancement of biodiesel production from the green microalga *Micractinium reisseri* via optimization of cultivation regimes. *Journal of Taibah University for Science*, **14**(1), 437–444, https://doi.org/10.1080/16583655.2020.1745505

Fabbri D. and Torri C. (2016). Linking pyrolysis and anaerobic digestion (Py-AD) for the conversion of lignocellulosic biomass. *Current Opinion in Biotechnology*, **38**, 167–173, https://doi.org/10.1016/j.copbio.2016.02.004

Fatma S., Hameed A., Noman M., Ahmed T., Shahid M., Tariq M., Sohail I. and Tabassum, R. (2018). Lignocellulosic biomass: a sustainable bioenergy source for the future. *Protein & Peptide Letters*, **25**(2), 148–163, https://doi.org/10.2174/0929866525666180122144504

Fei Q., Guarnieri M. T., Tao L., Laurens L. M. L., Dowe N. and Pienkos P. T. (2014). Bioconversion of natural gas to liquid fuel: opportunities and challenges. *Biotechnology Advances*, **32**, 596–614, https://doi.org/10.1016/j.biotechadv.2014.03.011

Fu Y., Chen L. and Zhang W. (2016). Regulatory mechanisms related to biofuel tolerance in producing microbes. *Journal of Applied Microbiology*, **121**, 320–332, https://doi.org/10.1111/jam.13162

Fuchs G. (2008). Anaerobic metabolism of aromatic compounds. *Annals of the New York Academy of Sciences*, **1125**(1), 82–99, https://doi.org/10.1196/annals.1419.010. PMID: 18378589

Gonzalez-Garcia S. and Bacenetti J. (2019). Exploring the production of bio-energy from wood biomass. Italian case study. *Science of the Total Environment*, **647**(647), 158–168, https://doi.org/10.1016/j.scitotenv.2018.07.295

Gowen C. M. and Fong S. S. (2010). Exploring biodiversity for cellulosic biofuel production. *Chemistry & Biodiversity*, **7**(5), 1086–1097, https://doi.org/10.1002/cbdv.200900314

Greenwell H. C., Laurens L. M. L., Shields R. J., Lovitt R. W. and Flynn K. J. (2010). Placing microalgae on the biofuels priority list: a review of the technological challenges. *Journal of The Royal Society Interface*, **7**(46), 703–726, https://doi.org/10.1098/rsif.2009.0322

Guo M., Li C., Facciotto G., Bergante S., Bhatia R., Comolli R., Ferré C. and Murphy R. (2015). Bioethanol from poplar clone Imola: an environmentally viable alternative to fossil fuel? *Biotechnology for Biofuels*, **8**, 134, https://doi.org/10.1186/s13068-015-0318-8

Guschina I. A. and Harwood J. L. (2009). Algal lipids and effect of the environment on their biochemistry. In: Lipids in Aquatic Ecosystems, Ch. 1, Springer, New York, pp. 1–24, https://doi.org/10.1007/978-0-387-89366-2_1

Haibach M. C., Kundu S., Brookhart M. and Goldman A. S. (2012). Alkane metathesis by tandem alkane dehydrogenation-olefin-metathesis catalysis and related chemistry. *Accounts of Chemical Research*, **45**, 947–958, https://doi.org/10.1021/ar3000713

Hankamer B., Lehr F., Rupprecht J., Mussgnug J. H., Posten C. and Kruse O. (2007). Photosynthetic biomass and H_2 production by green algae: from bioengineering to bioreactor scale-up. *Physiologia Plantarum*, **131**(1), 10–21, https://doi.org/10.1111/j.1399-3054.2007.00924.x

Himmel M. E., Ding S. Y., Johnson D. K., Adney W. S., Nimlos M. R., Brady J. W. and Foust T. D. (2007). Biomass recalcitrance: engineering plants and enzymes for biofuels production. *Science*, **315**(5813), 804–807, https://doi.org/10.1126/science.1137016

Hnain A. K., Cockburn L. M. and Lefebvre D. D. (2011). Microbiological processes for waste conversion to bioenergy products: approaches and directions. *Environmental Reviews*, **19**, 214–237, https://doi.org/10.1139/A11-007

Hou X., Peng W., Xiong L., Huang C., Chen X., Xinde Chen X. and Zhanget W. (2013). Engineering *Clostridium acetobutylicum* for alcohol production. *Journal of Biotechnology*, **166**(1–2), 25–33, https://doi.org/10.1016/j.jbiotec.2013.04.013

Hsia S. Y. and Chou Y. T. (2014). Optimization of biohydrogen production with biomechatronics. *Journal of Nanomaterials*, **2014**, 1–11, https://doi.org/10.1155/2014/721267

Hu P., Chakraborty S., Kumar A., Woolston B., Liu H., Emerson D. and Stephanopoulos G. (2016). Integrated bioprocess for conversion of gaseous substrates to liquids. *Proceedings of the National Academy of Sciences of the USA*, **113**, 3773–3778, doi: 10.1073/pnas.1516867113

Huang G., Chen F., Wei D., Zhang X. and Chen G. (2010). Biodiesel production by microalgal biotechnology. *Applied Energy*, **87**(1), 38–46, https://doi.org/10.1016/j.apenergy.2009.06.016

Huang J., Chen D., Wei Y., Wang Q., Li Z., Chen Y. and Huang R. (2014). Direct ethanol production from lignocellulosic sugars and sugarcane bagasse by a recombinant *Trichoderma reesei* strain HJ48. *The Scientific World Journal*, **2014**, 1–8, https://doi.org/10.1155/2014/798683

Jin H., Chen L., Wang J. and Zhang W. (2014). Engineering biofuel tolerance in non-native producing microorganisms. *Biotechnology Advances*, **32**, 541–548, https://doi.org/10.1016/j.biotechadv.2014.02.001

Jorquera O., Kiperstok A., Sales E. A., Embiruçu M. and Ghirardi M. L. (2010). Comparative energy life-cycle analyses of microalgal biomass production in open ponds and photobioreactors. *Bioresource Technology*, **101**(4), 1406–1413, https://doi.org/10.1016/j.biortech.2009.09.038. PMID: 19800784

Karemore A., Nayak M. and Sen R. (2016). Recent inventions and trends in algal biofuels research. *Recent Patents on Biotechnology*, **10**(1), 30–42, https://doi.org/10.2174/1872208310666160830164929

Kovács E., Szücs C., Farkas A., Szuhaj M., Maróti G., Bagi Z., Rákhely G. and Kovács K. L. (2022). Pretreatment of lignocellulosic biogas substrates by filamentous fungi. *Journal of Biotechnology*, **360**, 160–170, https://doi.org/10.1016/j.jbiotec.2022.10.013

Lee Y., Lee K. and Oh Y. (2015). Recent nanoparticles engineering advances in microalgal cultivation and harvesting processes of biodiesel production: a review. *Bioresource Technology*, **184**, 63–72, https://doi.org/10.1016/j.biortech.2014.10.145

Li R. Y. and Fang H. H. P. (2009). Heterotrophic photo fermentative hydrogen production. *Critical Reviews in Environmental Science and Technology*, **39**(12), 1081–1108, https://doi.org/10.1080/10643380802009835

Limayen A. and Ricke S. (2012). Lignocellulosic biomass for bioethanol production: current perspectives, potential issues and future prospects. *Progress in Energy and Combustion Science*, **38**, 449–467, https://doi.org/10.1016/j.pecs.2012.03.002

Lindblad P., Lindberg P., Oliveira P., Stensjo K. and Heidorn T. (2012). Design, engineering, and construction of photosynthetic microbial cell factories for renewable solar fuel production. *Ambio*, **41**, 163–168, https://doi.org/10.1007/s13280-012-0274-5

Littlewood J., Guo M., Boerjan W. and Murphy R. J. (2014). Bioethanol from poplar: a commercially viable alternative to fossil fuel in the European Union. *Biotechnology for Biofuels*, **7**, 113, https://doi.org/10.1186/1754-6834-7-113

Lu X. (2010). A perspective: photosynthetic production of fatty acid-based biofuels in genetically engineered cyanobacteria. *Biotechnology Advances*, **28**(6), 742–746, https://doi.org/10.1016/j.biotechadv.2010.05.021

Lutke-Eversloh T. (2014). Application of new metabolic engineering tools for *Clostridium acetobutylicum*. *Applied Microbiology and Biotechnology*, **98**, 5823–5837, https://doi.org/10.1007/s00253-014-5785-5

Lutke-Eversloh T. and Bahl H. (2011). Metabolic engineering of *Clostridium acetobutylicum*: recent advances to improve butanol production. *Current Opinion in Biotechnology*, **22**(5), 634–647, https://doi.org/10.1016/j.copbio.2011.01.011

Lynd L. R., Weimer P. J., van Zyl W. H. and Pretorius I. S. (2002). Microbial cellulose utilization: fundamentals and biotechnology. *Microbiology and Molecular Biology Reviews*, **66**(3), 506–577, https://doi.org/10.1128/MMBR.66.3.506-577.2002. PMID: 12209002

Lynd L. R., van Zyl W. H., McBride J. E. and Laser M. (2005). Consolidated bioprocessing of cellulosic biomass: an update. *Current Opinion in Biotechnology*, **16**(5), 577–583, https://doi.org/10.1016/j.copbio.2005.08.009. PMID: 16154338

Machado I. M. and Atsumi S. (2012). Cyanobacterial biofuel production. *Journal of Biotechnology*, **162**, 50–56, https://doi.org/10.1016/j.jbiotec.2012.03.005

Mani P., Fidal V. T., Bowman K., Chandra T. S., Keshavarz T. and Kyazze G. (2020). Development of an electroactive aerobic biocathode for microbial fuel cell applications. *Environmental Microbiology Reports*, **12**(5), 607–612, https://doi.org/10.1111/1758-2229.12871

Manish, S. and Rangan Banerjee (2008). Comparison of biohydrogen production processes. *International Journal of Hydrogen Energy*, **33**(1), 279–286.

Mao C., Feng Y., Wang X. and Ren G. (2015). Review on research achievement of biogas from anaerobic digestion. *Renewable and Sustainable Energy Reviews*, **45**, 540–555, https://doi.org/10.1016/j.rser.2015.02.032

Marsh G. (2009). Small wonders: biomass from algae. *Renewable Energy Focus*, **9**(7), 74–78, https://doi.org/10.1016/S1755-0084(09)70046-8

Mata T. M., Martins A. A. and Caetano N. S. (2010). Microalgae for biodiesel production and other applications: a review. *Renewable and Sustainable Energy Reviews*, **14**(1), 217–232, https://doi.org/10.1016/j.rser.2009.07.020

Mathuriya A. S. (2020). Development of trickling bio-electrochemical reactor (TrickBER) for large scale energy efficient wastewater treatment. *Environmental Technology*, **2**, 1–35.

McKinlay J. B. and Harwood C. S. (2010). Photobiological production of hydrogen gas as a biofuel. *Current Opinion in Biotechnology*, **21**(3), 244–251, https://doi.org/10.1016/j.copbio.2010.02.012

Miao X. and Wu Q. (2006). Biodiesel production from heterotrophic microalgal oil. *Bioresource Technology*, **97**(6), 841–846, https://doi.org/10.1016/j.biortech.2005.04.008. PMID: 15936938

Mohammadshirazi A., Akram A., Rafiee S. and Kalhor E. B. (2014). Energy and cost analyses of biodiesel production from waste cooking oil. *Renewable and Sustainable Energy Reviews*, **33**, 44–49, https://doi.org/10.1016/j.rser.2014.01.067

Moreno-Garrido I. (2008). Microalgae immobilization: current techniques and uses. *Bioresource Technology*, **99**(10), 3949–3964, https://doi.org/10.1016/j.biortech.2007.05.040

Ni Y. and Sun Z. (2009). Recent progress on industrial fermentative production of acetone-butanolethanol by *Clostridium acetobutylicum* in China. *Applied Microbiology and Biotechnology*, **83**, 415–423, https://doi.org/10.1007/s00253-009-2003-y

Nielsen D. R., Leonard E., Yoon S. H., Tseng H. C., Yuan C. and Prather K. L. J. (2009). Engineering alternative butanol production platforms in heterologous bacteria. *Metabolic Engineering*, **11**(4–5), 262–273, https://doi.org/10.1016/j.ymben.2009.05.003

Öhgren K., Bura R., Lesnicki G., Saddler J. and Zacchi G. (2007). A comparison between simultaneous saccharification and fermentation and separate hydrolysis and fermentation using steam pretreated corn stover. *Process Biochemistry*, **42**(5), 834–839, https://doi.org/10.1016/j.procbio.2007.02.003

Olguin E. J. (2012). Dual purpose microalgae-bacteria-based systems that treat wastewater and produce biodiesel and chemical products within a biorefinery. *Biotechnology Advances*, **30**(5), 1031–1046, https://doi.org/10.1016/j.biotechadv.2012.05.001

Órpez R., Martínez M. E., Hodaifa G., Yousfi F. E., Jbari N. and Sánchez S. (2009). Growth of the microalga *Botryococcus braunii* in secondarily treated sewage. *Desalination*, **246**(1–3), 625–630, https://doi.org/10.1016/j.desal.2008.07.016

Peralta-Yahya P. P. and Keasling J. D. (2010). Advanced biofuel production in microbes. *Biotechnology Journal*, **5**, 147–162, https://doi.org/10.1002/biot.200900220

Philbrook A., Alissandratos A. and Easton C. J. (2013). Biochemical processes for generating fuels and commodity chemicals from lignocellulosic biomass. In: Environmental Biotechnology, 39–63. InTech Open, London, https://doi.org/10.5772/55309

Pienkos P. T. and Darzins A. (2009). The promise and challenges of microalgal-derived biofuels. *Biofuels, Bioproducts and Biorefining*, **3**(4), 431–440, https://doi.org/10.1002/bbb.159

Porth I. and El-Kassaby Y. A. (2015). Using *Populus* as a lignocellulosic feedstock for bioethanol. *Biotechnology Journal*, **10**, 510–524, https://doi.org/10.1002/biot.201400194

Rogers J. N., Rosenberg J. N., Guzman B. J., Oh V. H., Mimbela L. E., Ghassemi A., Betenbaugh M. J., Oyler G. A. and Donohue M. D. (2014). A critical analysis of paddlewheel-driven raceway ponds for algal biofuel production at commercial scales. *Algal Research*, **4**, 76–88, https://doi.org/10.1016/j.algal.2013.11.007

Romero-Garcia J. M., Martínez-Pati C., Ruiz E., Romero I. and Castro E. (2016). Ethanol production from olive stone hydrolysates by xylose fermenting microorganisms. *Bioethanol*, **2**, 51–65.

Romero-Guiza M., Vila J., Mata-Alvarez J., Chimenos J. and Astals S. (2016). The role of additives on anaerobic digestion: a review. *Renewable and Sustainable Energy Reviews*, **58**, 1486–1499, https://doi.org/10.1016/j.rser.2015.12.094

Romero Victorica M., Soria M. A., Batista-García R. A., Ceja-Navarro J. A., Vikram S., Ortiz M., Ontañon O., Ghio S., Martínez-Ávila L., Quintero García O. J. and Etcheverry C. (2020). Neotropical termite microbiomes as sources of novel plant cell wall degrading enzymes. *Scientific Reports*, **10**, 3864, https://doi.org/10.1038/s41598-020-60850-5

Rosenberg J. N., Oyler G. A., Wilkinson L. and Betenbaugh M. J. (2008). A green light for engineered algae: redirecting metabolism to fuel a biotechnology revolution. *Current Opinion in Biotechnology*, **19**(5), 430–436, https://doi.org/10.1016/j.copbio.2008.07.008. PMID: 18725295

Saifuddin N. and Priatharsini P. (2016). Developments in bio-hydrogen production from algae: a review. *Research Journal of Applied Sciences, Engineering and Technology*, **12**(9), 968–982, https://doi.org/10.19026/rjaset.12.2815

Sakurai H. and Masukawa H. (2007). Promoting R&D in photobiological hydrogen production utilizing mariculture-raised cyanobacteria. *Marine Biotechnology*, **9**(2), 128–145, https://doi.org/10.1007/s10126-006-6073-x

Sekoai P. T., Yoro K. and Daramola M. O. (2016). Batch fermentative biohydrogen production process using immobilized anaerobic sludge from organic solid waste. *Environments*, **38**, 1–10.

Senger R. S. (2010). Biofuel production improvement with genome-scale models: the role of cell composition. *Biotechnology Journal*, **5**, 671–685, https://doi.org/10.1002/biot.201000007

Shaishav S., Singh R. N. and Satyendra T. (2013). Biohydrogen from algae: fuel of the future. *International Research Journal of Environmental Sciences*, **2**(4), 44–47.

Shen Y., Yuan W., Pei Z. and Mao E. (2010). Heterotrophic culture of *Chlorella protothecoides* in various nitrogen sources for lipid production. *Applied Biochemistry and Biotechnology*, **160**(6), 1674–1684, https://doi.org/10.1007/s12010-009-8659-z. PMID: 19424668

Shi J., Sharma-Shivappa R. R. and Chinn M. S. (2009). Microbial pretreatment of cotton stalks by submerged cultivation of *Phanerochaete chrysosporium*. *Bioresource Technology*, **100**(19), 4388–4395, https://doi.org/10.1016/j.biortech.2008.10.060. PMID: 19423334

Shreve M. J. and Brennan R. A. (2019). Trace organic contaminant removal in six full-scale integrated fixed-film activated sludge (IFAS) systems treating municipal wastewater. *Water Research*, **151**, 318–331, https://doi.org/10.1016/j.watres.2018.12.042

Silva D. F., Bezerra T. D., Rodrigues L. B., Franco M. and Santana N. B. (2024). Biological pretreatment of cassava husk with *Aspergillus niger* ATCC 1004 to improve biogas production. *Waste and Biomass Valorization*, 1–12, https://doi.org/10.1007/s12649-024-02456-w.

Simionato D., Basso S., Giacometti G. M. and Morosinotto T. (2013). Optimization of light use efficiency for biofuel production in algae. *Biophysical Chemistry*, **182**, 71–78, https://doi.org/10.1016/j.bpc.2013.06.017

Singh V., Singh M. P., Verma V., Singh P., Srivastava R. and Singh A. K. (2016). Characteristics of cold adapted enzyme and its comparison with mesophilic and thermophilic counterpart. *Cellular and Molecular Biology*, **62**, 3, https://doi.org/10.4172/1165-158X.1000144

Stephanopoulos G. (2007). Challenges in engineering microbes for biofuels production. *Science*, **315**(5813), 801–804, https://doi.org/10.1126/science.1139612. PMID: 17289987

Sun Y. and Cheng J. (2002). Hydrolysis of lignocellulosic materials for ethanol production: a review. *Bioresource Technology*, **83**(1), 1–11, https://doi.org/10.1016/S0960-8524(01)00212-7. PMID: 12058826

Tabatabai B., Fathabad S. G., Bonyi E., Rajini S., Aslan K. and Sitther V. (2019). Nanoparticle-mediated impact on growth and fatty acid methyl ester composition in the cyanobacterium *Fremyella diplosiphon*. *BioEnergy Research*, **12**, 409–418, https://doi.org/10.1007/s12155-019-09966-9

Tamilselvan R. and Immanuel Selwynraj A. (2024). Enhancing biogas generation from lignocellulosic biomass through biological pretreatment: exploring the role of ruminant microbes and anaerobic fungi. *Anaerobe*, **85**, 102815, https://doi.org/10.1016/j.anaerobe.2023.102815

Tang X., Liao C., Zhou S., Chen C., Li L., Lu G., Huang X., Zhang M., Chen C. and Li P. (2024). Potential of perennial sorghum for biogas production: pretreatment with yeast-contained inoculants during anaerobic storage. *Fuel*, **359**, 130365, https://doi.org/10.1016/j.fuel.2023.130365

Van Zyl W. H., Lynd L. R., den Haan R. and McBride J. E. (2007). Consolidated bioprocessing for bioethanol production using *Saccharomyces cerevisiae*. *Advances in Biochemical Engineering/Biotechnology*, **108**, 205–235, https://doi.org/10.1007/10_2007_061. PMID: 17846725

Wackett L. (2008). Biomass to fuels via microbial transformations. *Current Opinion in Chemical Biology*, **12**(2), 187–193, https://doi.org/10.1016/j.cbpa.2008.01.025. PMID: 18275861

Waqas S., Bilad M. R., Man Z., Wibisono Y., Jaafar J., Mahlia T. M. I., Khan A. L. and Aslam M. (2020). Recent progress in integrated fixed-film activated sludge process for wastewater treatment: a review. *Journal of Environmental Management*, **268**, 110718, https://doi.org/10.1016/j.jenvman.2020.110718

Whitaker J., Field J. L., Bernacchi C. J., Cerri C. E., Ceulemans R., Davies C. A., DeLucia E. H., Donnison I. S., McCalmont J. P., Paustian K. and Rowe R. L. (2018). Consensus, uncertainties and challenges for perennial bioenergy crops and land use. *GCB Bioenergy*, 10:150–164. **10**(3), 150–164, https://doi.org/10.1111/gcbb.12488.

Xu L., Weathers P. J., Xiong X.-R. and Liu C.-Z. (2009). Microalgal bioreactors: challenges and opportunities. *Engineering in Life Sciences*, **9**(3), 178–189, https://doi.org/10.1002/elsc.200800111

Xue C., Zhao J., Chen L., Yang S. T. and Bai F. (2017). Recent advances and state-of-the-art strategies in strain and process engineering for biobutanol production by *Clostridium acetobutylicum*. *Biotechnology Advances*, **35**, 310–322, https://doi.org/10.1016/j.biotechadv.2017.01.007

Yadav V. K., Singh V. and Mishra V. (2019). Alkaline protease: a tool to manage solid waste and its utility in detergent industry. In: Microbial Genomics in Sustainable Agroecosystems, V. Tripathi, P. Kumar, P. Tripathi, A. Kishore and M. Kamle (eds), Springer, Singapore, https://doi.org/10.1007/978-981-32-9860-6_14

Yu D., Wang G., Xu F. and Chen L. (2012). Constitution and optimization on the performance of microbial fuel cell based on sulfate-reducing bacteria. *Energy Procedia*, **16**(Part C), 1664–1670, https://doi.org/10.1016/j.egypro.2012.01.258

Zanchi G., Pena N. and Bird N. (2010). The Upfront Carbon Debt of Bioenergy. Joanneum Research, Graz, Austria, www.transportenevironment.org/Publications/prep_hand_out/lid/591 (accessed 21 March 2011)

Zhang T. Y., Wu Y. H. and Hu H. Y. (2014). Domestic wastewater treatment and biofuel production by using microalga *Scenedesmus* sp. ZTY1. *Water Science and Technology*, **69**, 2492–2496, https://doi.org/10.2166/wst.2014.160

Zhang J., Liu Y., Sun Y., Wang H., Cao X. and Li X. (2020). Effect of soil type on heavy metals removal in bioelectrochemical system. *Bioelectrochemistry*, **136**, 107596, https://doi.org/10.1016/j.bioelechem.2020.107596

Zhou J. and Li Y. (2010). Engineering cyanobacteria for fuels and chemicals production. *Protein & Cell*, **1**(3), 207–210, https://doi.org/10.1007/s13238-010-0043-9

Zhu L. Y., Zong M. H. and Wu H. (2008). Efficient lipid production with *Trichsporon fermentas* and its use for biodiesel preparation. *Bioresource Technology*, **99**(16), 7881–7885, https://doi.org/10.1016/j.biortech.2008.02.033

doi: 10.2166/9781789064100_0191

Chapter 11

Sustainable routes in the production of biodiesel, bioethanol, and biomethane with subsequent bioenergy conversion

Umer Hayyat[1], Muhammad Shahid[2], Jonathan Lee Tian En[3,4], Nalok Dutta[5,6] and Muhammad Usman Khan[2,7]

[1]Department of Thermal and Fluid Engineering, Faculty of Engineering Technology, University of Twente, Drienerlolaan 5, 7522 NB Enschede, Netherlands
[2]Department of Energy Systems Engineering, Faculty of Agricultural Engineering and Technology, University of Agriculture Faisalabad, 38000 Faisalabad, Pakistan
[3]NUS Environmental Research Institute, National University of Singapore, T-Lab Building 5A Engineering Drive 1, Singapore 117411, Singapore
[4]Energy and Environmental Sustainability for Megacities (E2S2) Phase II, Campus for Research Excellence and Technological Enterprise (CREATE), 1 Create Way, Singapore 138602, Singapore
[5]Department of Biochemical Engineering, University College London, London, UK
[6]Department of Anthropology, University of Washington, Seattle, WA, USA
[7]Bioproducts Sciences and Engineering Laboratory (BSEL), Washington State University Tricities Campus, Richland, WA 99354, USA
**Correspondence: u.hayyat@utwente.nl, mshahid9220@outlook.com, eriltej@nus.edu.sg, nalok.dutta@ucl.ac.uk, nalok.dutta@wsu.edu, muhammadusman.khan@wsu.edu, usman.khan@uaf.edu.pk*

ABSTRACT

Given the decline of fossil fuels and growing concern over environmental pollution, sustainable bioenergy has emerged as a promising avenue for future energy development. The production of bioenergy from various waste materials offers a promising and sustainable solution to meet future energy goals. It helps reduce waste and minimize environmental impact. This chapter explores the sustainable routes for producing biofuels such as biodiesel, biomethane, and bioethanol from biomass. The focus is on the key conversion technologies and processes involved in these production methods. It is seen that different methods are used for producing biodiesel and biomethane, including both conventional and advanced technologies, whereas bioethanol is primarily produced through fermentation and distillation processes. Applications of these biofuels are also discussed in this chapter, including their potential use in transportation, power generation, and other sectors. This chapter critically analyses these production methods to provide a comprehensive understanding of the current state and prospects of bioenergy production.

Keywords: Biodiesel, biomethane, bioethanol, biofuel production, biofuel applications.

11.1 INTRODUCTION

The depletion of fossil fuels has underscored the urgent need for the human race to transition toward alternative energy sources. Currently, fossil fuels contribute to 86% of global energy consumption, a figure that continues to increase over time (Abas *et al.*, 2015). This indicates that the sustainable energy alternative sources are needed to meet the growing energy demands of the world. Based on the report published by the International Energy Agency (IEA), it has been projected that the reserves of fossil fuels, including oil and natural gas, are estimated to be depleted within a span of 50–53 years (IEA, 2021). In recent years, researchers have shown a notable surge in interest in alternative fuels, spurred largely by factors such as domestic availability, cost-effectiveness, and environmental benefits. Among the array of alternative energy resources, biofuels have emerged as a promising solution to counteract the challenges associated with fossil fuel depletion. Their potential to achieve both environmental and economic sustainability makes them a focal point in the quest for sustainable energy solutions (Joshi *et al.*, 2017). Biofuels are a type of fuels that are produced from biomass using either direct or indirect methods, resulting in the formation of bioethanol, biodiesel, or biomethane (Saladini *et al.*, 2016). Biofuels, which can be solid, liquid, or gaseous, are derived from sources such as wood, agricultural residue, animal waste, and sludge. They are primarily used for producing heat, light, and power.

Biofuels are mainly used in their liquid or gaseous states within the global transportation and energy sectors. This alternative fuel demonstrates the potential to effectively contribute to sustainable energy goals by with reduced emissions during combustion, as well as mitigating carbon dioxide (CO_2) and greenhouse gas emissions. Consequently, it helps reduce the overall pollution burden and environmental footprint (Khan *et al.*, 2022b). Gaseous and liquid biofuels have gained widespread acknowledgement for their positive impact on society and the economy. These benefits include creating employment opportunities in rural areas, reducing reliance on foreign currency, minimizing CO_2 emissions, providing localized energy supplies, and strengthening energy security (Manikandan *et al.*, 2021).

Biofuels are produced from organic matter, which has recently been categorized as abundant and renewable sources. In addition to their advantageous impact on fossil fuel usage, biofuels hold great potential for promoting sustainability and economic growth in industrialized nations by utilizing locally available renewable resources. The classification of biofuels into four generations is based on the feedstock utilized. First-generation biofuels are derived from feedstocks traditionally used for food production. However, concerns regarding the competition between fuel and food production have raised questions about the sustainability of first-generation biofuels, as they place pressure on food commodities. Second-generation biofuels prioritize the use of non-edible plant materials, particularly those rich in lignocellulosic substances. Algal biomasses are primarily associated with third-generation biofuels, and they can also be partially linked to the utilization of carbon dioxide as a raw material (Lee *et al.*, 2013).

Biodiesel, bioethanol, and biomethane are pivotal contemporary biofuels derived from biomass, holding significant importance in the transportation sector. Carbohydrates present in biomass, including cellulose and hemicelluloses, can be effectively converted into sugars, which then undergo anaerobic fermentation to yield bioethanol (Vickram *et al.*, 2023). Currently, biodiesel and bioethanol stand as the primary biomass-derived transportation biofuels globally. Biodiesel is sourced from plant-based oils and animal fats, whereas bioethanol is produced from starch and sugar (Vickram *et al.*, 2023). In contrast, biomethane is primarily generated through the process of biochemical anaerobic digestion using agricultural, livestock, and municipal waste materials (Kadam & Panwar, 2017; Ware & Power, 2016). This chapter provides a detailed description of routes for the production of biofuels such as biodiesel, biomethane, and bioethanol. Section 11.2 describes the feedstock and conventional and advanced methods for biodiesel production. Section 11.3 discusses the production process of biogas and biogas upgrading for biomethane production. Section 11.4 describes the production process of bioethanol from various feedstocks through the fermentation process.

11.2 PRODUCTION OF BIODIESEL

Biodiesel is a sustainable and environmentally friendly form of bioenergy obtained from sources such as vegetable oils, animal fats, and microalgal oil through various conversion technologies, as shown in Figure 11.1. Its characteristics are very similar to those of diesel fuel, making it a highly promising

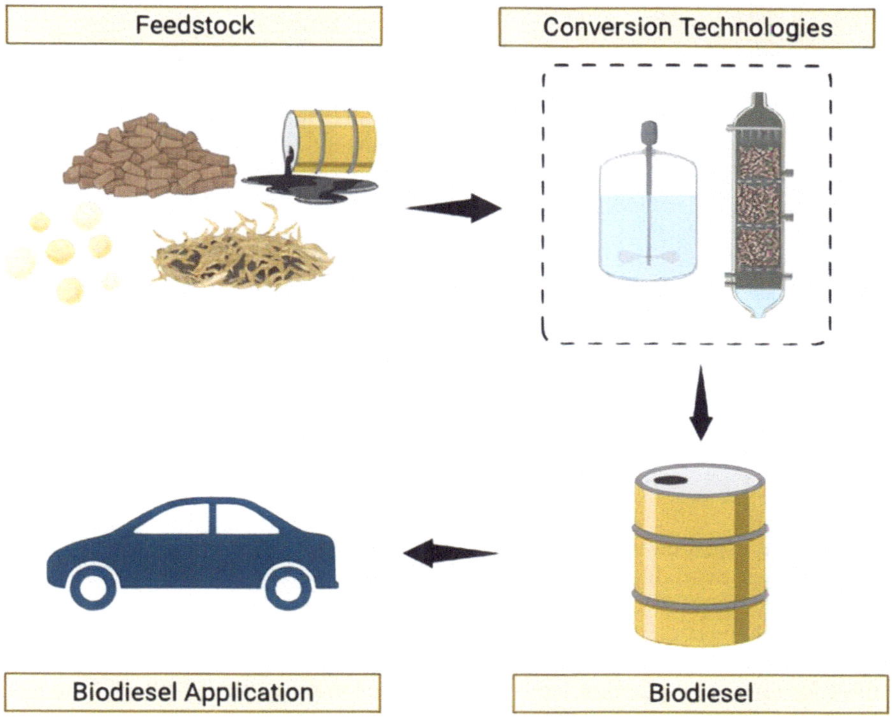

Figure 11.1 Biodiesel production and utilization flowchart.

substitute for traditional diesel (Fukuda *et al.*, 2001). Biodiesel offers numerous advantages, including its biodegradability, non-toxic nature, low emission profile including potentially harmful substances, and its status as a renewable resource (Ma & Hanna, 1999). Moreover, it does not contribute to the rise of carbon dioxide concentrations in the atmosphere, therefore mitigating the impacts associated with the greenhouse effect.

11.2.1 Feedstock for biodiesel

In the past few years, biodiesel production has seen a wide variety of feedstock being utilized. The choice of feedstock is primarily determined by factors such as availability, desired properties, and economic viability. Different types of vegetable oils, waste oils, animal fats, unconventional oils, and non-edible oils have been employed in the production of biodiesel. A detailed list outlining the various feedstocks used can be found in Table 11.1.

11.2.1.1 Animal fats

Animal fats are widely utilized as a primary feedstock in the production of biodiesel. In industrial settings, choices such as chicken fat, tallow, and lard frequently serve as popular options for commercial-scale biodiesel production (Gürü *et al.*, 2010; Lu *et al.*, 2007; Öner & Altun, 2009). Using animal fats as a raw material for producing biodiesel not only eliminates the need to dispose of them but also helps increase the availability of biodiesel (Janaun & Ellis, 2010). In terms of economic value, animal fats often offer a cost advantage compared to commonly utilized edible vegetable oils (Gürü *et al.*, 2010).

Table 11.1 Merits and demerits of feedstock for biodiesel.

Feedstock	Merits	Demerits	References
Animal fat	Economically more viable Substantially low cost Easily available	Contain contaminants High cost point Different to process	Toldrá-Reig *et al.* (2020), Barua *et al.* (2022)
Vegetable oils	Reduce dependency on imported oils Can easily blend Biodegradable and less toxic than fossil fuels	High cost Competition with food supply Low energy content	Abdul Hakim Shaah *et al.* (2021)
Waste oils	Environmentally friendly Cost effective Reduce dependency on fossil fuels	Presence of unwanted impurities	Linganiso *et al.* (2022), Asaad *et al.* (2023)
Non-traditional oils	High energy yield Desire properties	High cost, Environmental impact	Behera *et al.* (2015)

11.2.1.2 Vegetable oils

The use of non-edible vegetable oils is highly favored to produce biodiesel. These non-consumable vegetable oils are widely available in almost all regions of the world (Pramanik, 2003). Certain types of edible vegetable oils have the potential to be used in the production of biodiesel. However, it is important to note that these oils tend to be relatively costly due to their common usage as cooking oils in households. Furthermore, it is generally recommended to segregate oils used for consumption from those used for energy production feedstocks in order to mitigate the risk of food scarcity (Buyx & Tait, 2011; Lin *et al.*, 2011).

11.2.1.3 Waste oils

Yellow grease, also referred to as waste oil, holds a significant promise as an affordable source material for the production of biodiesel. Utilizing waste cooking oil in biodiesel production greatly improves its economic feasibility by addressing one of the major concerns regarding feedstock prices (Chhetri *et al.*, 2008). One possible solution for the collection of waste oil from restaurants and houses is to implement specialized containers in each establishment or residence. However, it is important to conduct public awareness campaigns prior to initiating the collection process (Yaakob *et al.*, 2013).

11.2.1.4 Non-traditional oils

The interest in using alternative oils, including microalgae oil, camelina oil, and castor oil, for biodiesel production is on the rise. Microalgae oil, in particular, has garnered attention as a promising option due to its potential to substitute traditional feedstocks such as vegetable oil while exhibiting properties comparable to diesel fuel (Francisco *et al.*, 2010). Camelina possesses numerous advantageous agronomic traits that have the potential to greatly improve the production of biofuels domestically (Moser, 2010). Castor oil is a highly valuable commodity due to its affordable price and superior quality. Of particular importance is the fact that this non-consumable plant-based oil does not pose any concerns or endanger food security (Osorio-González *et al.*, 2020).

11.2.1.5 Alternative and lesser-known feedstocks

There are feedstocks that are less known or less research has been conducted on them but has great potential to be used as feedstock for biodiesel production. These feedstocks include salt-tolerant plants such as halophytes which can be grown on marginal lands and do not affect the food consumption issue (Abideen *et al.*, 2012). Pongamia a less known feedstock that can add benefits to sustainable biodiesel production (Scott *et al.*, 2008). Pennycress is another feedstock option as it contains high properties which are required for biodiesel production (Moser *et al.*, 2009). Table 11.1 shows the merits and demerits of these feedstocks available for biodiesel production.

11.2.2 Biodiesel production technologies

The production technologies of biodiesel have experienced substantial development and can be categorized into two main groups: conventional and advanced methods. Traditional methods, including blending, microemulsion, pyrolysis, and transesterification, have long been established as the primary techniques in this field. The process of blending involves the direct combination of biodiesel and regular diesel, whereas microemulsion entails the formation of stable and finely dispersed mixtures of oil and water, often facilitated by the presence of surfactants. Pyrolysis is a thermal decomposition process characterized by the breakdown of feedstocks in the absence of oxygen. In contrast, transesterification, the predominant technique, entails the chemical conversion of triglycerides into biodiesel, facilitated by the use of catalysts. On the contrary, the domain of advanced technologies presents novel methodologies, such as the use of the supercritical fluid (SCF) technique and the superheated vapor (SHV) technique. The utilization of the SCF technique is characterized by its operation under high pressure and temperature conditions, thereby obviating the requirement for conventional catalysts and potentially expediting the reaction process. In contrast, the SHV technique uses heated vapors to enhance and streamline the process of biodiesel conversion. As scientific advancements continue, these technologies hold the potential to augment the efficacy and ecological viability of biodiesel manufacturing.

11.2.2.1 Conventional biodiesel production technologies

Biodiesel is produced via conventional physical and chemical processing techniques. The objective of these methods is to alter the intrinsic properties of feedstock oils, with a primary focus on reducing viscosity and improving volatility. Various techniques, such as the formation of petroleum-derived diesel through blending, the formation of microemulsions, the utilization of pyrolysis, and the application of transesterification, are essential methodologies in the initial processing of biodiesel. These preliminary measures guarantee that the resulting fuel satisfies the necessary criteria for optimal combustion and engine functionality (Binhweel *et al.*, 2023).

11.2.2.1.1 Blending

The blending method involves the combination of raw oil and petroleum fuel in accordance with specified percentages and proportions. In specific situations, the achievement of optimal blending necessitates the implementation of purification and preheat treatment procedures. The application of this treatment results in a reduction in the viscosity and an increase in the volatility of the resulting fuel. The direct blending of raw oils with conventional diesel fuel represents a cost-effective and uncomplicated approach to the production of biodiesel. However, this methodology frequently leads to a lower standard of fuel quality. The final outcome often demonstrates greater levels of density, viscosity, and undesirable components such as gum and free-fatty acids. In addition, the diminished volatility and sluggish responsiveness of the unsaturated hydrocarbon chains in question may potentially undermine the overall efficiency of the system in which the yield biodiesel is used (Tabatabaei *et al.*, 2019).

It can be rather difficult to find the optimal mixture of conventional diesel and biodiesel when it comes to blending, which is one of the challenges surrounding the process. Success lies in understanding the optimal combination of biodiesel and traditional diesel, and carefully following the correct proportions as outlined by regulations. Another issue to consider is the compatibility between various types of biodiesel and the base diesel. The issues discussed earlier are linked to the separation of phases, plugging of filters, and decreased engine performance (Kamaraj *et al.*, 2022). The key to successfully navigating these challenges lies in ensuring that biodiesel blends meet regulatory requirements and consistently perform well. It is crucial to maintain and monitor quality throughout the blending process.

11.2.2.1.2 Microemulsion

Micro-emulsification is a widely used method for the production of biodiesel. In essence, the formation of an emulsion occurs when two or more immiscible liquids are blended together. A distinguishing characteristic of a microemulsion is its stability as a clear diffusion of fluid microstructures, typically ranging in size from 1 to 150 nm. The state under consideration is attained in a spontaneous manner when two liquids that are not compatible with each other are combined, facilitated by one or more amphiphiles. These amphiphiles can possess either an ionic or non-ionic character. The amphiphiles function as intermediary molecules, enabling the homogenization of otherwise distinct liquids (Karmakar & Halder, 2019; Ma & Hanna, 1999; Schwab *et al.*, 1987).

There are numerous challenges that must be addressed throughout the entire production process, starting from microemulsion to biodiesel. One of the main difficulties that individuals may encounter is achieving and sustaining the stability of their microemulsion. Several factors, such as the type and concentration of surfactant, temperature, and agitation, commonly influence the formation of a stable microemulsion. Thus, achieving a smaller droplet size and better control over droplet size distribution in a microemulsion is crucial for meeting the optimization criterion in biodiesel production (Senusi *et al.*, 2024).

11.2.2.1.3 Pyrolysis

Biodiesel is manufactured using a thermochemical process called pyrolysis. In this method, various raw materials such as vegetable oils and animal-based fats undergo exposure to elevated temperatures in the absence of oxygen, leading to a substantial modification in their chemical compositions. Pyrolysis can be categorized into three specific types based on the applied temperature range: standard pyrolysis (550–900°C), fast pyrolysis (850–1250°C), and flash pyrolysis (1050–1300°C). This thermal procedure facilitates the decomposition of complicated molecular bonds present in the oils, leading to the conversion of complex molecules into simpler components, ultimately yielding a liquid fuel. The efficiency of pyrolysis can be improved through the addition of specific catalysts during the process (Karmakar & Halder, 2019).

One of the major challenges is ensuring the accuracy of the temperature and reaction conditions, as they are crucial in the pyrolysis biodiesel process. The temperature plays a crucial role in controlling the desired temperature for converting biomass into biodiesel. Temperature fluctuations can lead to incomplete conversions, resulting in lower yield and the formation of unwanted by-products. Managing the by-products and wastes that come from the pyrolysis process can be quite challenging (Pawar & Panwar, 2022).

11.2.2.1.4 Transesterification

Transesterification is widely recognized as the primary and highly effective method used for the synthesis of biodiesel. The procedure mainly involves a substitution reaction in which the alkyl group of esters is substituted with an alkyl group from an alcohol, resulting in the formation of alkyl esters, commonly referred to as biodiesel, along with glycerol as a by-product. In this process, the stoichiometric ratio commonly includes a three-to-one molar ratio of alcohol to triglyceride. To achieve maximum conversion, the quantity of alcohol required is doubled to 6 mol, with any excess being eliminated after the reaction. To improve the rate of the chemical reaction under consideration, catalysts such as acids, alkalis, or enzymes are introduced, thereby frequently leading to increased yields of biodiesel within shorter time periods (Karmakar & Halder, 2019).

One of the key challenges for consideration in the transesterification process for biodiesel production is achieving full conversion of triglycerides into biodiesel, while also minimizing the formation of undesirable by-products such as soap and glycerol. Accomplishing this task necessitates meticulous management of reaction parameters such as temperature, pressure, and catalyst concentration (Farouk *et al.*, 2024).

11.2.2.2 Advanced biodiesel production technologies

Advanced technologies are increasingly being favored as the preferred approach in the realm of industrial biodiesel production. The catalyst separation phase poses significant challenges for conventional transesterification methods. In response, recent developments in the field of biodiesel production have witnessed a shift toward the adoption of non-catalyzed transesterification techniques. The use of these novel techniques offers several advantages, such as the removal of initial treatment procedures, enhanced reaction kinetics, and a simplified approach to the separation of alkyl esters and glycerol. This chapter explores two advanced technologies, namely SCF technique and SHV technique.

11.2.2.2.1 SCF technique

The transesterification process using SCF technique was initially proposed by Saka and Kusdiana (2001) as a catalyst-free method. In this process, the alcohol is subjected to higher temperatures exceeding the critical temperature, resulting in improved diffusivity. The supercritical state induces modifications to the hydrogen bonding within the alcohol, leading to a fluid exhibiting greater viscosity and diminished polarity. As a result, the formation of a singular phase consisting of triglycerides and methanol occurs. With the elevation of temperature and pressure, there is a concurrent augmentation in the abundance of reactive radicals as well as the solubility of triglycerides and methanol. It is worth mentioning that the SCF transesterification method has demonstrated rates of conversion that surpass 95% (Binhweel *et al.*, 2023).

There are several challenges that arise in the process of biodiesel in the SCF technique. There is a significant risk of equipment failure and safety hazards associated with the use of high pressure and high temperature in this process. In addition, it can be quite challenging to maintain the fluid in a supercritical state throughout the process. This may involve the need for regular cleaning and strict control. Separation and recovery of the produced biodiesel from the SCF mixture can be a complex and costly process. It is crucial to find innovative ways of separation that are both cost-effective and environmentally friendly.

11.2.2.2.2 SHV technique

The SHV technique, as proposed by Yamazaki *et al.* (2007), presents a novel and innovative method for the production of biodiesel. When alcohol is subjected to temperatures exceeding its normal boiling point, specifically within the range of 220–290°C, it undergoes a change into a state known as superheating. The alcohol vapor, which possesses a significant amount of energy, has the ability to engage with triglycerides present in the feedstock. This interaction serves as the catalyst for an esterification process, ultimately resulting in the production of biodiesel. One notable benefit of this technique lies in its ability to produce biodiesel of exceptional quality and improved performance characteristics. This technique requires the conversion of reaction products into gaseous forms through the use of superheated vapor. Significantly, it is crucial to note that the temperatures used for this particular procedure are maintained at levels below the autoignition temperature of the fatty acid methyl ester and the boiling point of the glycerol that is obtained as a result. This precautionary measure is implemented to guarantee the safe extraction of these substances (Karmakar & Halder, 2019).

The common issues surrounding the superheated vapor process in biodiesel production apply to any other related processes. First and foremost, achieving and sustaining the superheated state of vapor is essential for a successful conversion of feedstock to biodiesel. Another possible outcome is the occurrence of prolonged superheating, which can be influenced by various factors such as the type of material being used, the pressure applied, and the rate of heating. Another excellent example is how this technique allows for safe handling at high temperatures and pressures using the SHV technique. For a safe and efficient thermal transfer, it is essential to ensure the proper functionality of the confinement system.

11.2.3 Applications of biodiesel

Biodiesel has various applications in the automobile sector and sustainable industries. In the automobile sector, biodiesel can be used as an alternative fuel source for diesel engines. It is a clean and biodegradable fuel that reduces engine emissions, making it an environmentally friendly option. Biodiesel can be blended with diesel fuel, allowing it to be used in existing diesel engines without any modifications. Apart from the automobile sector, biodiesel has a wide range of sustainable applications. It can be used as heating oil, providing a renewable and cleaner fuel source for heating systems (Xing *et al.*, 2021). Additionally, biodiesel can be used as a plasticizer, power production fuel, and high-boiling absorbent for cleaning gaseous industrial emissions. Moreover, biodiesel can be utilized as a lubricant and in various solvent applications.

11.3 PRODUCTION OF BIOMETHANE

Biomethane, usually regarded as an environmentally friendly substitute for conventional natural gas, can be generated in a sustainable manner via the conversion of organic matter, including agricultural residues, sewage, and discarded waste deposited in landfills (D'Adamo *et al.*, 2023). By using these wastes, not only is the emission of methane, a potent greenhouse gas, into the atmosphere diminished, but also a valuable source of renewable energy is produced. The concept of sustainable production places emphasis on the use of advanced and optimized conversion processes, the reduction of environmental harm, and the use of effective biomass feedstock. The biomethane production process flowchart is shown in Figure 11.2. This approach ensures that biomethane plays a significant role in facilitating the shift toward a circular and low-carbon economy.

11.3.1 Feedstock for biomethane

Different types of biomass feedstocks are used in the process of producing biomethane using conversion technologies. However, the use of sustainable feedstocks ensures the production of environmentally friendly biofuels, while also promoting waste-to-energy solutions and adhering to the principles of a circular bioeconomy. Organic municipal waste, agriculture waste (manure and crop residues), sludge and organic materials, and dedicated energy crops (non-food crops) are the main source of biomethane production. Recent studies have shown that biomethane derived from such feedstocks possesses the capacity to substitute compressed natural gas, thereby being regarded as a sustainable solution from an environmental standpoint (Ammenberg & Feiz, 2017; Collet *et al.*, 2017; Valli *et al.*, 2017; Vo *et al.*, 2018).

11.3.2 Biogas production technologies

Biogas production technologies encompass both thermochemical and biochemical processes. Thermochemical technologies involve the use of high temperatures to convert organic waste into biogas. Biogas production technologies can be broadly classified into two categories: thermochemical and biochemical technologies. Thermochemical technologies involve the conversion of biomass into biogas through high-temperature processes such as gasification or pyrolysis. These processes break down the organic materials in the biomass into synthesis gas, which can then be further processed to yield biogas. On the contrary, biochemical technologies rely on anaerobic digestion, a process that involves the breakdown of organic materials by microorganisms in the absence of oxygen. This process produces biogas as a by-product, which can be collected and used as a renewable energy source.

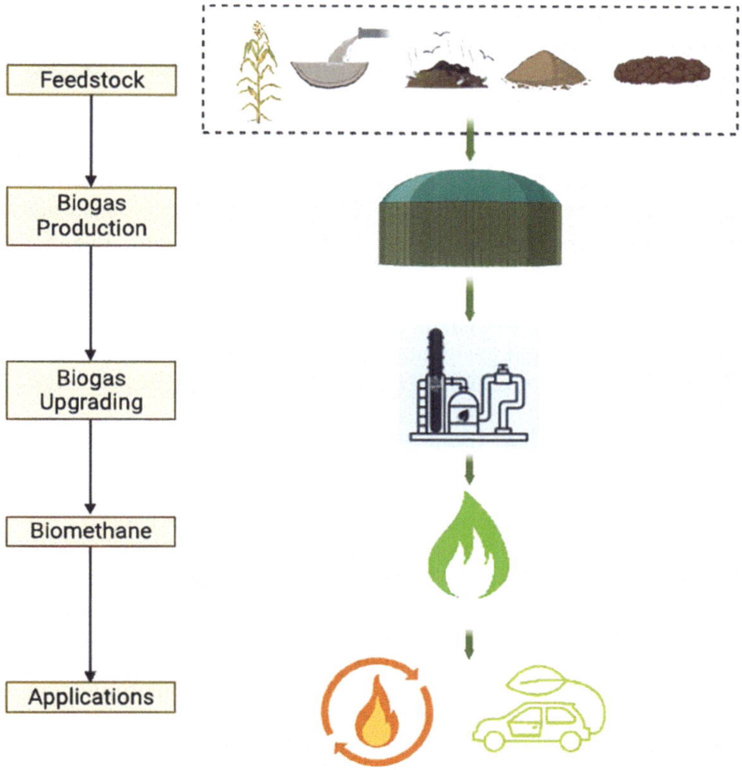

Figure 11.2 Biomethane production process flowchart.

11.3.2.1 Conventional biogas production technologies

11.3.2.1.1 Anaerobic digestion

Anaerobic digestion is a biochemical process characterized by the oxygen-free decomposition of biodegradable organic matter, including food waste. This process is carried out by a diverse group of microorganisms called methanogens, which convert the organic material into biogas. Multiple stages constitute the anaerobic digestion process. Organic waste is initially gathered and introduced into an anaerobic digester. A digester is a sealed container that fosters optimal conditions for the proliferation of microorganisms. The moisture content, temperature, and pH level are all under control under these conditions. The organic matter is decomposed by microorganisms in the digester via a succession of biochemical reactions. Most of the biogas produced consists of simpler molecules, such as carbon dioxide and methane, which are produced by these reactions from complex organic compounds (Awasthi *et al.*, 2023; Monlau *et al.*, 2013). Following collection, the biogas can be upgraded by means of biomethane separation.

There are certain challenges associated with anaerobic digestion for biogas production. The primary challenge is ensuring the optimal conditions for the microbes involved in the digestion process. Temperature, pH condition, and nature of substrate are crucial factors that significantly impact the performance of the anaerobic process and its energy production outcomes. Even the slightest change in these factors can lead to decreased efficiency and lower biogas production, and in some cases, even total process failure. One additional challenge to overcome is the effective management of by-products and wastes produced during anaerobic digestion processes (Hayyat *et al.*, 2024; Uddin & Wright, 2023).

11.3.2.1.2 Gasification

Gasification is a thermochemical process that involves the conversion of organic waste into a methane-rich syngas. Unlike anaerobic digestion, gasification operates at higher temperatures and in the presence of oxygen or steam. In the gasification process, the organic waste material is first dried and then subjected to high temperatures. This thermal decomposition breaks down the complex organic compounds into a mixture of gases known as syngas. Syngas mainly consists of methane, carbon monoxide, carbon dioxide, and small amounts of other gases (Awasthi *et al.*, 2023). Syngas produced during gasification is then converted into biomethane using biogas upgrading processes.

There are numerous challenges in gasification, such as the characteristics of feedstock, including moisture content, particle size, and nutrient composition, which can vary significantly and play a crucial role in determining the efficiency of the gasification process and the quality of the resulting biogas. Factors to consider are the temperature of gasification, residence time, and flow rate of gas. Fluctuations in these factors may lead to insufficient gasification, decreased gas quality, and higher tar levels, which can have a negative impact on the overall effectiveness of the process (Engvall *et al.*, 2011).

11.3.2.2 Advanced biogas production technologies

11.3.2.2.1 Two-stage anaerobic process

An anaerobic digestion process primarily comprises four conversion steps, namely hydrolysis, acidogenesis, acetogenesis, and methanogenesis. When these processes occur simultaneously, a problem arises within a digester due to the differing operating conditions required for each process (Coats *et al.*, 2013). The acidogenesis and methanogenesis microorganisms necessitate distinct nutrient and environmental conditions, as exemplified. To tackle this problem, a new approach has been devised that consists of a two-step procedure. In this process, the conversion stages are optimized separately in two distinct digesters. The two-stage digestion system includes different sequential processes such as hydrolysis and acidogenesis, followed by acetogenesis and methanogenesis (Lindner *et al.*, 2016). By creating ideal pH and temperature conditions for four different microorganisms, this setup allows for improved process control. The syngas produced in two-stage anaerobic digestion has more biomethane content as compared to conventional anaerobic digestion process. The biogas upgrading of gas produced from two-stage anaerobic digestion yields high calorific value of biomethane which meets the specifications comparable to natural gas (Kougias *et al.*, 2017).

One of the main challenges encountered in two-stage anaerobic processes for biogas production is its sensitivity to changes in pH and temperature. Even the slightest changes in these factors can lead to significant fluctuations in the produced biogas. The wide ranges of each parameter can lead to an imbalance in the microbial community at every stage, which can be challenging for the overall performance of the process. A two-stage anaerobic process presents certain difficulties in biogas production, particularly in Anaerobic Inoculum with High Biomass (AIHB) (Zheng & Li, 2024).

11.3.2.2.2 Thermal gasification with integrated methanation

The process of thermal gasification involves the thermochemical conversion of solid biomass feedstock into syngas through high-temperature transformation. Thermal gasification is conducted within a controlled environment, with a restricted oxygen supply, at an operational temperature of 700°C. The gasification process involves the decomposition of the complex organic constituents of the feedstock, resulting in the production of hydrogen (H_2), carbon monoxide (CO), carbon dioxide (CO_2), and, to a lesser extent, methane (CH_4) (D'Aquino *et al.*, 2022). Subsequently, these components undergo a catalytic process called methanation, which serves to further refine and process them. During the process of methanation, the chemical reaction between carbon monoxide and hydrogen primarily results in the production of methane, resulting in a notable increase in its concentration within the gas mixture. The methane with a high concentration, known as biomethane, can be further purified and enhanced to comply with the criteria set for natural gas (Cucchiella *et al.*, 2017).

Dealing with contaminants such as tars and particulates in the syngas can be a hurdle during the methanation process. This can result in catalyst deactivation and a decrease in methane production. Efficiently eliminating these impurities is crucial for maintaining the seamless functioning of the integrated gasification and methanation process (Calbry-Muzyka & Schildhauer, 2020).

11.3.3 Biogas upgrading

A biogas upgrading process accounts for 90% of the biomethane produced globally. Biogas is comprised of certain undesirable substances, primarily in the form of carbon dioxide (CO_2) and hydrogen sulfide (H_2S). These substances can be effectively eliminated through the process of biogas upgrading (Garlapati *et al.*, 2021).

11.3.3.1 Pressure swing adsorption

A pressure swing adsorption (PSA) method is used to upgrade the quality of biogas by increasing its methane content. During the PSA process, the unprocessed biogas is directed through a surface that contains a particular material known as an adsorbent. This material grabs onto carbon dioxide (CO_2) from the biogas, allowing the methane to pass through (Zhao *et al.*, 2010). Over time, the adsorbent becomes saturated with carbon dioxide (CO_2). In preparation for the further rounds, the surface undergoes a depressurization process, thereby facilitating the release of the confined carbon dioxide (CO_2). To maintain a continuous purification process, multiple surfaces are used. During the process of surface cleaning, another surface is simultaneously engaged in the active purification of biogas. The ultimate output product achieved high methane concentration with concentration ranging between 95 and 99%.

One major drawback is the presence of impurities in the gases used for biogas production. These impurities can vary and ultimately hinder the effectiveness of the adsorption process, resulting in a lower purity of the upgraded biogas. Efficiently selecting the right adsorbents and optimizing operating parameters such as pressure, temperature, and cycle time are crucial for achieving purity in biogas (Kohlheb *et al.*, 2021).

11.3.3.2 Absorption method

An absorption method for biogas upgrading can be differentiated from an adsorption method based on the process involved. In an absorption method, gases are dissolved in a liquid medium, whereas in adsorption, gases adhere to the surface of a solid material. Absorption techniques usually use organic solvents, such as methanol and polyethylene glycol ethers, in addition to more commonly used solvents such as water. Water scrubbing method, a type of absorption method, possesses the capacity to effectively remove carbon dioxide (CO_2) while also serving as a pretreatment process for the removal of hydrogen sulfide (H_2S) (Rotunno *et al.*, 2017). A chemical scrubbing method, which is widely used for the absorption of carbon dioxide (CO_2), uses chemicals such as amines for this purpose (Chen *et al.*, 2015). An absorption method involves the use of an absorber tank, wherein carbon dioxide (CO_2) is effectively captured from biogas. This is subsequently followed by a stripper, which facilitates the release of CO_2 through the application of heat. The concentration level of methane produced by this method is 99%.

An adsorption technique for biogas upgrading presents several challenges that need to be addressed for successful implementation. There is, however, a challenge in selecting an appropriate material for the adsorbent, which must effectively capture most of the impurities from the biogas stream, particularly methane. The adsorbent material needs to have a high selectivity for CO_2 and contaminants with a high adsorption capacity to ensure efficient biogas upgrading. It is crucial to ensure the reliability of the material and its ability to be recovered in multiple adsorption–desorption cycles (Struk *et al.*, 2020).

11.3.3.3 Membrane separation method

A process of membrane separation involves the use of a selective barrier to effectively separate undesirable components, such as carbon dioxide (CO_2). These membranes possess selective permeability, enabling the passage of specific components while blocking the entry of undesired molecules, dependent upon factors such as diffusion rate, size, and solubility. The use of pressure differences allows the transportation of biogas across the membrane, while simultaneously preventing the passage of CO_2 and other contaminants. Presently, there exist three distinct categories of membrane separation techniques: inorganic membrane separation, polymeric membrane separation, and mixed matrix membrane separation. The use of mixed matrix membrane separation technique has gained substantial acceptance in various industries for the purpose of biomethane separation (Ullah Khan *et al.*, 2017). In addition to mixed matrix separation method, an inorganic method offers several advantages compared to the polymeric method. These advantages stem from its high thermal stability, chemical resistance, and strong mechanical properties (Chen *et al.*, 2015). To enhance the capacity of the membrane separation technique, a strategy involving the use of multiple membranes is employed to increase the yield of biomethane production from biogas, achieving a yield of 99.5% (Chen *et al.*, 2015).

A membrane separation technique for biogas upgrading also comes with its own set of unique challenges. The main issue is fouling caused by the contaminants and toxins found in the feed biogas. The separation cycle becomes less viable, resulting in higher working charges due to the need for frequent cleaning and replacement (Muñoz *et al.*, 2015). One important aspect to consider is selecting a suitable film material that has a high affinity for CO_2 and CH_4 and is highly porous.

11.3.3.4 Cryogenic separation method

A cryogenic separation technique used for the process of biogas upgrading is founded on the principle of using low temperature and high pressure to effectively separate biomethane from biogas. The biogas is subjected to a process of cooling to an extremely low temperature of −170°C and is then pressurized to a high pressure of 80 bar. During this process, each component of the biogas condenses into a liquid at different stages. During the separation process, the carbon dioxide (CO_2) along with other components such as water and hydrogen sulfide (H_2S) undergo condensation and can subsequently be separated. The cryogenic technique demonstrates the capability to effectively separate biomethane from biogas, while minimizing biomethane losses during the separation process. The production of biomethane yields significant results; however, achieving the necessary operating conditions, such as low temperature and high pressure, can be challenging (Sahota *et al.*, 2018).

There are certain challenges that come with implementing the cryogenic separation strategy for biogas upgrading. One of the major challenges faced during cryogenic separation is the energy-intensive nature of the process (Spitoni *et al.*, 2019). The need for extremely cold temperatures to separate the different components of biogas necessitates a significant amount of energy, which can greatly impact the overall operational costs.

11.3.3.5 Biological upgrading method

A biological upgrading method refers to a biogas purification process that involves the use of microorganisms, such as bacteria, to convert impurities present in biogas, with a particular focus on carbon dioxide (CO_2) conversion (Kougias *et al.*, 2017; Lee *et al.*, 2012). Microorganisms, specifically bacteria, undergo a reaction with carbon dioxide (CO_2) in the presence of hydrogen (H_2) to facilitate the process of biological methanation. The method's output leads to an increase in biomethane production, consequently resulting in an additional yield of biomethane. This method offers a sustainable and environmentally friendly approach to biogas upgrading, yet with the requirement of a consistent hydrogen supply and a controlled microorganism functioning environment. The merits and demerits of all biogas upgrading technologies are presented in Table 11.2.

Table 11.2 Merits and demerits of biogas upgrading technologies.

Upgrading Technology	Merits	Demerits	References
Pressure swing adsorption	• No chemical addition • Highly flexible • Easy installation	• High-cost process • High methane losses • High installation cost	Ullah Khan *et al.* (2017), Cavenati *et al.* (2006), Persson *et al.* (2006)
Absorption method	• High methane yield • Low methane losses • Regenerative	• High operational and installation cost • High cost of solvent • Foaming issues	López *et al.* (2012), Hoyer *et al.* (2016)
Membrane separation	• High reliability • Easy installation • Low energy input required	• High operation cost as membranes are costly • Methane losses occur • Degradation of membrane with time	Persson *et al.* (2006), Molino *et al.* (2013), Andriani *et al.* (2014), Makaruk *et al.* (2010)
Cryogenic separation	• High separation • No chemical addition • High methane yield	• Expensive as compared to other • Required high energy input • Temperature effective process	Andriani *et al.* (2014), Baena-Moreno *et al.* (2020), Xu *et al.* (2015)
Biological upgrading	• Simple process • Cost effective • No unwanted by-products	• Not adoptable for industrial scale • H_2 addition	Muñoz *et al.* (2015), Sun *et al.* (2015)

11.3.4 Application of biomethane

Biomethane exhibits superior properties in comparison to biogas and demonstrates comparable characteristics to natural gas, making it a highly viable alternative to the natural gas. Biomethane has the potential to be utilized for domestic cooking purposes and can also be stored in cylinders in the form of compressed methane. Biomethane, specifically in the form of Bio CNG, is composed of 97% biomethane and exhibits notable thermal efficiency and reduced emissions (Azam *et al.*, 2021). Consequently, it holds potential for application in gas power plants. The use of biomethane in fuel cells presents a highly promising strategy for the direct and efficient conversion of biogas into electrical energy. This approach presents an ecologically sustainable substitute for conventional power generation methods reliant on combustion (Kabeyi & Olanrewaju, 2022).

11.3.4.1 Challenges in biological upgrading method

One of the challenges is the medium's high sensitivity to changes in environmental factors such as pH, temperature, and nutrient levels. Essentially, the result of trying to maintain conditions that lead to improved efficiency among the microorganisms responsible for converting CO_2 and other pollutants into methane is often costs high (Sahota *et al.*, 2018).

11.4 PRODUCTION OF BIOETHANOL

Bioethanol, also known as ethanol or simply 'biofuel', has emerged as a potentially viable alternative to traditional fossil fuels. It is produced by fermenting various types of agricultural and industrial wastes, such as fermentable sugars, amylaceous feedstocks, and cellulosic substrates such as bagasse and agricultural waste. This sustainable production process makes bioethanol a renewable and environmentally friendly energy source. Additionally, bioethanol offers several advantages over fossil-derived fuels. These advantages include its nearly CO_2-neutral properties, natural availability, and safety in storage and handling. A general overview of feedstock, bioethanol production process, and application are shown in Figure 11.3.

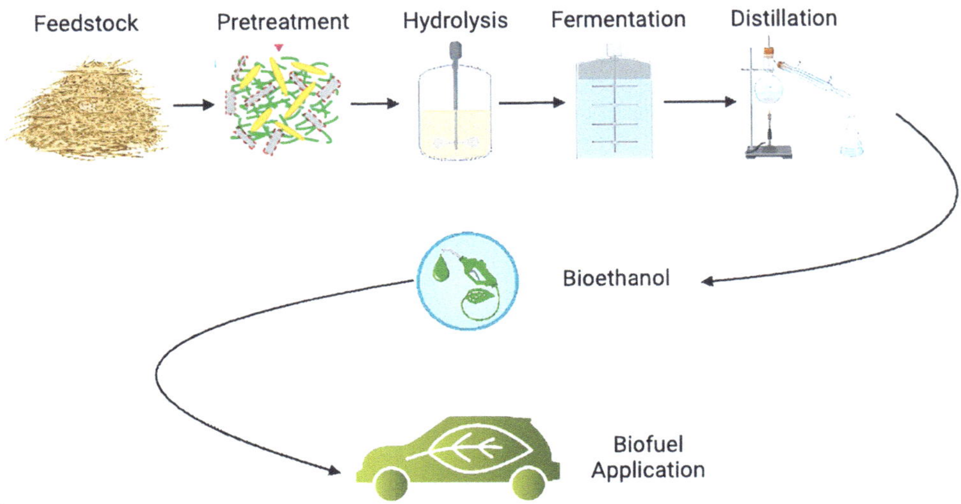

Figure 11.3 Bioethanol production showing feedstock, production processes, and application.

11.4.1 Feedstock for bioethanol

There are several different feedstocks that can be used for the production of bioethanol, depending on factors such as availability, cost, and sustainability. One commonly used feedstock is agro-industrial residues, which are agricultural by-products that are left over after processing. These residues, such as straw, corn stover, and sugarcane bagasse, contain carbohydrates that can be converted into bioethanol through microbial fermentation processes.

Another option for feedstock is starch-containing crops, such as corn, wheat, and cassava. These crops have high levels of starch, which can be easily converted into fermentable sugars and subsequently into bioethanol (Maas *et al.*, 2008). Furthermore, lignocellulosic feedstocks, such as wood chips and sawdust, are also being explored for their potential in bioethanol production (Micic & Jotanovic, 2015). These feedstocks have the advantage of being widely available, relatively inexpensive, and non-competitive with food production. Additionally, these lignocellulosic feedstocks are sustainable in terms of CO_2 emissions, making them an attractive option for large-scale bioethanol production. Moreover, the utilization of these agro-industrial residues and lignocellulosic feedstocks for bioethanol production helps to address the issue of competition with food resources. Unlike first-generation bioethanol, which is produced from edible agricultural crops such as cereals and sugarcane juices, second-generation bioethanol utilizes these non-food feedstocks. This not only reduces the conflict with food and feed production but also allows for the utilization of raw materials with low commercial value that would otherwise go to waste.

11.4.2 Pretreatment process

In the pretreatment stage, the biomass undergoes harsh conditions to disrupt the lignocellulosic matrix and make the cellulose and hemicellulose more accessible for hydrolysis. Various pretreatment methods are available, including physical, chemical, and biological techniques, each with its advantages and limitations. Several commonly used techniques for pretreatment include steam explosion, acid pretreatment, alkaline pretreatment, and microbial pretreatment, whose general overview is presented in Table 11.3. The selection of a specific method is determined by various factors such as the nature of the biomass, desired sugar production output, cost efficiency considerations, and environmental implications.

Table 11.3 Overview of pretreatment process for bioethanol.

	Classification	Remarks	References
Physical pretreatment	Mechanical size reduction	Use of milling, grinding, and chipping machine to reduce the size of feedstock	Sarkar *et al.* (2012)
	Electron beam treatment	Use of thermal and non-thermal effects generated during radiation	Bjerre *et al.* (2000)
Chemical pretreatment	Acid-based	Various acids such as sulfuric acid, nitric acid, and phosphoric acid are used	Cardona *et al.* (2010)
	Alkaline-based	Use of NaOH to reduce the lignin content in feedstock	Pandey *et al.* (2000), Kumar and Wyman (2009)
	Wet oxidization	Treatment of feedstock under 120°C or above temperature in the presence of oxygen	Martín *et al.* (2007)
Biological pretreatment	Microbial pretreatment	Breakdown of complex feedstock using microorganisms such as bacteria and fungi	Talebnia *et al.* (2010)
	Fungal pretreatment	Common fungi such as brown, soft, and white rot are used for pretreatment process to reduce lignin content	Khan *et al.* (2022a)
Physiochemical pretreatment	Wet explosion	Reduce the lignin content with oxygen addition under 170°C or above	Usman Khan and Kiaer Ahring (2020)

11.4.3 Hydrolysis

A hydrolysis process plays a crucial role in bioethanol production from lignocellulosic biomass. It involves breaking down the complex structure of biomass to release fermentable sugars, mainly cellulose and hemicellulose, which can then be converted into ethanol. There are various methods available for hydrolyzing carbohydrates, including both chemical and biological approaches. Biological methods are generally preferred for hydrolyzing cellulose and hemicellulose due to their effectiveness and environmental sustainability. Enzymes, such as cellulases and hemicellulases, are commonly used to catalyze hydrolysis processes. These enzymes break down the glycosidic bonds present in the carbohydrates, converting them into glucose and other sugar molecules. Chemical methods, on the contrary, involve the use of strong acids or bases to break down the carbohydrates. However, chemical methods are less preferred due to their high energy consumption and potential environmental hazards associated with the use and disposal of these chemicals. Furthermore, the choice of hydrolysis method can also influence the overall yield and quality of bioethanol.

A biological hydrolysis method is widely used process in bioethanol production process, one of which is enzymatic hydrolysis. This process is crucial as it converts complex carbohydrates in the biomass into simple sugars that can be fermented to produce bioethanol (Vasić *et al.*, 2021). Efforts are being made to improve the efficiency and cost-effectiveness of enzymatic hydrolysis. Researchers are exploring various effective enzymes for this purpose. Lignocellulosic biomass, such as wood feedstocks, agricultural wastes, and marine algae, is used as substrates to produce bioethanol more efficiently. Among these substrates, algae show promise as a good option for bioethanol production. Studies have focused on assessing the potential of chemical and biological hydrolysis methods using macroalgae and microalgae. These methods aim to pretreat the cell wall and generate sugars from these algal species through hydrolysis. The resulting sugars can then be fermented to produce bioethanol (Hamouda *et al.*, 2018).

A chemical hydrolysis process involves the use of acids or alkalis to break down the complex carbohydrates in biomass into simple sugars (Damayanti *et al.*, 2021). Acids (such as H_2SO_4, HCl, and organic acids) and alkalis (such as NaOH, Na_2CO_3, CaOH, and NH_3) are commonly used in chemical processes. Acid treatment is known for its ability to selectively hydrolyze hemicellulose components, whereas alkalis have a higher selectivity for lignin degradation. Bioethanol production from algae is considered promising. In a recent study, researchers assessed the potential of macrogreen alga *Ulva fasciata* and microgreen alga *Chlorella vulgaris* for pretreatment of cell walls and sugar production through both chemical and biological hydrolysis methods. The findings suggest that chemical hydrolysis proves to be the most effective method for algae pretreatment (Hamouda *et al.*, 2018).

11.4.3.1 Challenges in hydrolysis

One of the significant difficulties in this stage is the productive breakdown of perplexing polysaccharides into straightforward sugars. The efficiency of hydrolysis can be influenced by various factors such as the difficulty in handling the substrate, the activity of enzymes, and the design of the reactor. Enzymatic action plays a crucial role in determining the effectiveness of hydrolysis. The determination and execution of hydrolytic compounds play a significant role in determining the overall yield of sugars from biomass. Factors such as compound explicitness, security, and restraint can pose challenges when trying to achieve successful hydrolysis (Gollakota *et al.*, 2018).

11.4.4 Fermentation

Fermentation is a crucial step in the bioethanol production process. During fermentation, the fermentable sugars obtained from the hydrolysis of biomass are converted into bioethanol by microorganisms such as bacteria, yeast, or fungi (Adeleye *et al.*, 2020). These microorganisms consume the sugars and produce ethanol and carbon dioxide as by-products. This process involves the conversion of sugars, such as glucose, into alcohol (ethanol) and carbon dioxide. The general equation for this process is

$$C_6H_{12}O_6 \text{ (glucose)} \rightarrow 2C_2H_5OH \text{ (ethanol)} + 2CO_2 \text{ (carbon dioxide)}$$

This process, often carried out using *Saccharomyces cerevisiae*, plays a vital role in the overall efficiency and success of bioethanol production (Behzadnia *et al.*, 2020). Without fermentation, the sugars derived from biomass would not be converted into bioethanol, making this step essential in the bioethanol production process. In addition, the choice of the microorganism used for fermentation can greatly impact the quality and quantity of bioethanol produced. Furthermore, the fermentation process in bioethanol production is not limited to the conversion of simple sugars, as it can also involve the breakdown of more complex carbohydrates such as cellulose and hemicellulose found in lignocellulosic biomass (Sahlan *et al.*, 2018). Other microorganisms that have shown potential in bioethanol fermentation include *Pichia stipitis*, *Kluyveromyces fragilis*, *Kluyveromyces marxianus*, *Escherichia coli*, *Klebsiella oxytoca*, and *Zymomonas mobilis*. These microorganisms have distinct characteristics that make them suitable for bioethanol production (Phwan *et al.*, 2018). *P. stipitis*, also known as the xylose-fermenting yeast, is capable of converting xylose, a sugar found in hemicellulose, into ethanol. This makes it an ideal candidate for bioethanol production from lignocellulosic biomass. Furthermore, *K. fragilis* and *K. marxianus* have shown high tolerance to inhibitory compounds present in lignocellulosic biomass, allowing for more efficient and productive fermentation processes.

11.4.4.1 Challenges in fermentation

The main concern is to prevent microbial contamination during the process, as it can lead to competition for nutrients and substrates in forming the final results. The primary challenge encountered throughout the process is its impact on the cycle's efficiency and the yield of bioethanol. The negative aspects include the influence of substances, such as acidic acid and furfural, on microorganism fermentation (Yang *et al.*, 2016).

11.4.5 Distillation

After the fermentation process, the next step in bioethanol production is distillation. Distillation is a crucial step that allows for the separation and concentration of ethanol from the fermentation mixture. The fermentation mixture also known as 'fermentation broth' contains ethanol, water, and other by-products. This mixture heats up the ethanol, which has a lower boiling point than water (78.37°C compared to 100°C for water), starts to vaporize first. The ethanol vapor rises up the distillation column and is collected at the top, where it is condensed back into a liquid. This liquid is a concentrated form of ethanol (Bušić *et al.*, 2018).

There are several techniques that can be employed for the distillation of ethanol from water–ethanol mixtures. Azeotropic distillation, extractive distillation, reactive distillation, and adsorptive distillation are commonly used methods for the fractionation and enrichment of ethanol. In azeotropic distillation, a third component is added to the mixture to break the azeotrope and enhance the separation of ethanol and water. Extractive distillation involves the addition of a solvent that forms a separate liquid phase with ethanol, allowing for easier separation. Reactive distillation combines the processes of distillation and chemical reactions to achieve simultaneous separation and conversion of components. Adsorptive distillation involves use of adsorbents to selectively adsorb ethanol, allowing for the separation from water. These various distillation techniques are used to overcome the challenge of effective and economical water removal in bioethanol production. The use of an appropriate distillation technique is crucial in the bioethanol production process to effectively separate and concentrate ethanol from water–ethanol mixtures.. Ultimately, the choice of distillation technique depends on factors such as the desired level of purity, cost-effectiveness, and specific characteristics of the bioethanol mixture. The distilled ethanol still contains some water, so it undergoes a dehydration process to remove the remaining water. This results in anhydrous (water-free) ethanol, which is suitable for use as a biofuel (Cutzu & Bardi, 2017).

11.4.5.1 Challenges in distillation

One of the major challenges is the significant amount of energy required for refining, especially when dealing with diluted ethanol solutions obtained from fermentation. Another significant challenge involves the formation of azeotropes, particularly the ethanol–water azeotrope, which complicates the process of obtaining highly pure ethanol using traditional distilling methods (Furzer, 2000).

11.4.6 Application of bioethanol

The application of bioethanol extends beyond just being an engine fuel. It has a wide range of uses and potential benefits in various industries. One of the key applications of bioethanol is in the production of ethylene, a crucial component in the manufacturing of plastics, solvents, and other chemical products. Another important application of bioethanol is its use as a solvent in industries such as pharmaceuticals, cosmetics, and cleaning products. Furthermore, bioethanol can be utilized as a renewable and sustainable source of heat and power generation in industries and residential settings (Joshi *et al.*, 2019).

11.5 FUTURE DIRECTIONS

The future of biofuels, including biodiesel, biomethane, and bioethanol, holds great promise and are expected to play a significant role in meeting global energy demands. Advancements in feedstock expansion, process improvement, and blending techniques are anticipated to tackle existing production challenges and enhance biodiesel yield. Exploring the use of waste materials and non-food crops as feedstocks shows great potential for making biodiesel production more viable and cost-effective. Biomethane creation can benefit from advancements in anaerobic digestion and gasification technologies. There will likely be a focus on improving the efficiency of these processes and enhancing the quality of the produced biomethane. Significant advancements are anticipated in the field of bioethanol production, particularly in expanding the range of feedstock, enhancing yeast strains through genetic engineering, and optimizing the fermentation process. The use of lignocellulosic biomass as a feedstock is a particularly promising area for bioethanol production.

11.6 CONCLUSION

The production of biofuels, including biodiesel, biomethane, and bioethanol, necessitates the use of a variety of processes and the encountering of a wide range of challenges. Despite the challenges that are inherent to every technology, biofuels are contributing to the development of energy solutions that are both more efficient and more environmentally friendly all over the world. These biofuels have discovered a variety of applications in a variety of industries, including transportation, power generation, and industrial processes. Incorporating these biofuels into the existing fuel infrastructure, improving fuel quality, and reducing production costs are all areas that will require significant research and development in the future.

REFERENCES

Abas N., Kalair A. and Khan N. (2015). Review of fossil fuels and future energy technologies. *Futures*, **69**, 31–49, https://doi.org/10.1016/j.futures.2015.03.003

Abdul Hakim Shaah M., Hossain M. S., Salem Allafi F. A., Alsaedi A., Ismail N., AbKadir M. O. and Ahmad M. I. (2021). A review on non-edible oil as a potential feedstock for biodiesel: physicochemical properties and production technologies. *RSC Advances*, **11**(40), 25018–25037, https://doi.org/10.1039/D1RA04311K

Abideen Z., Ansari R., Gul B. and Khan M. A. (2012). The place of halophytes in Pakistan's biofuel industry. *Biofuels*, **3**(2), 211–220, https://doi.org/10.4155/bfs.11.158

Adeleye T. M., Kareem S. O., Bankole M. O., Atanda O. and Adeogun A. I. (2020). Ethanol production from cassava starch by protoplast fusants of *Wickerhamomyces anomalus* and *Galactomyces candidum*. *Egyptian Journal of Basic and Applied Sciences*, **7**(1), 67–81, https://doi.org/10.1080/2314808X.2020.1746884

Ammenberg J. and Feiz R. (2017). Assessment of feedstocks for biogas production, part II – results for strategic decision making. *Resources, Conservation and Recycling*, **122**, 388–404, https://doi.org/10.1016/j.resconrec.2017.01.020

Andriani D., Wresta A., Atmaja T. D. and Saepudin A. (2014). A review on optimization production and upgrading biogas through CO_2 removal using various techniques. *Applied Biochemistry and Biotechnology*, **172**(4), 1909–1928, https://doi.org/10.1007/s12010-013-0652-x

Asaad S. M., Inayat A., Jamil F., Ghenai C. and Shanableh A. (2023). Optimization of biodiesel production from waste cooking oil using a green catalyst prepared from glass waste and animal bones. *Energies (Basel)*, **16**(5), 2322, https://doi.org/10.3390/en16052322

Awasthi M. K., Sar T., Gowd S. C., Rajendran K., Kumar V., Sarsaiya S., Li Y., Sindhu R., Binod P., Zhang Z., Pandey A. and Taherzadeh M. J. (2023). A comprehensive review on thermochemical, and biochemical conversion methods of lignocellulosic biomass into valuable end product. *Fuel*, **342**, 127790, https://doi.org/10.1016/j.fuel.2023.127790

Azam A., Rafiq M., Shafique M. and Yuan J. (2021). Renewable electricity generation and economic growth nexus in developing countries: An ARDL approach. *Economic Research-Ekonomska Istraživanja*, **34**(1), 2423–2446, https://doi.org/10.1080/1331677X.2020.1865180

Baena-Moreno F. M., le Saché E., Pastor-Pérez L. and Reina T. R. (2020). Membrane-based technologies for biogas upgrading: a review. *Environmental Chemistry Letters*, **18**(5), 1649–1658, https://doi.org/10.1007/s10311-020-01036-3

Barua P., Dhana Raju V., Soudagar M. E. M. and Hossain N. (2022). Animal fat-derived biodiesel and nano-technology applications. pp. 335–357, https://doi.org/10.1007/978-981-19-3582-4_11

Behera S., Singh R., Arora R., Sharma N. K., Shukla M. and Kumar S. (2015). Scope of algae as third generation biofuels. *Frontiers in Bioengineering and Biotechnology*, **2**, 1–13, https://doi.org/10.3389/fbioe.2014.00090

Behzadnia A., Moosavi-Nasab M., Ojha S. and Tiwari B. K. (2020). Exploitation of ultrasound technique for enhancement of microbial metabolites production. *Molecules*, **25**(22), 5473, https://doi.org/10.3390/molecules25225473

Binhweel F., Hossain M. S. and Ahmad M. I. (2023). Recent trends, potentials, and challenges of biodiesel production from discarded animal fats: a comprehensive review. *BioEnergy Research*, **16**(2), 778–800, https://doi.org/10.1007/s12155-022-10527-w

Bjerre A. B., Olesen A. B., Fernqvist T., Plöger A. and Schmidt A. S. (2000). Pretreatment of wheat straw using combined wet oxidation and alkaline hydrolysis resulting in convertible cellulose and hemicellulose. *Biotechnology and Bioengineering*, **49**(5), 568–577, https://doi.org/10.1002/(SICI)1097-0290(19960305)49:5<568::AID-BIT10>3.0.CO;2-6

Bušić A., Mardetko N., Kundas S., Morzak G., Belskaya H., Šantek M. I., Komes D., Novak S. and Šantek B. (2018). Bioethanol production from renewable raw materials and its separation and purification: a review. *Food Technology and Biotechnology*, **56**(3), 289–311, https://doi.org/10.17113/ftb.56.03.18.5546

Buyx A. and Tait J. (2011). Ethical framework for biofuels. *Science*, **332**(6029), 540–541, https://doi.org/10.1126/science.1206064

Calbry-Muzyka A. S. and Schildhauer T. J. (2020). Direct methanation of biogas – technical challenges and recent progress. *Frontiers in Energy Research*, **8**, 114, https://doi.org/10.3389/fenrg.2020.570887

Cardona C. A., Quintero J. A. and Paz I. C. (2010). Production of bioethanol from sugarcane bagasse: status and perspectives. *Bioresource Technology*, **101**(13), 4754–4766, https://doi.org/10.1016/j.biortech.2009.10.097

Cavenati S., Grande C. A. and Rodrigues A. E. (2006). Removal of carbon dioxide from natural gas by vacuum pressure swing adsorption. *Energy & Fuels*, **20**(6), 2648–2659, https://doi.org/10.1021/ef060119e

Chen X. Y., Vinh-Thang H., Ramirez A. A., Rodrigue D. and Kaliaguine S. (2015). Membrane gas separation technologies for biogas upgrading. *RSC Advances*, **5**(31), 24399–24448, https://doi.org/10.1039/C5RA00666J

Chhetri A., Watts K. and Islam M. (2008). Waste cooking oil as an alternate feedstock for biodiesel production. *Energies (Basel)*, **1**(1), 3–18, https://doi.org/10.3390/en1010003

Coats E. R., Searcy E., Feris K., Shrestha D., Mcdonald A. G., Briones A., Magnuson T. and Prior M. (2013). An integrated two-stage anaerobic digestion and biofuel production process to reduce life cycle GHG emissions from US dairies. *Biofuels, Bioproducts and Biorefining*, **7**(4), 459–473, https://doi.org/10.1002/bbb.1408

Collet P., Flottes E., Favre A., Raynal L., Pierre H., Capela S., Peregrina C. (2017). Techno-economic and life cycle assessment of methane production via biogas upgrading and power to gas technology. *Applied Energy*, **192**, 282–295, https://doi.org/10.1016/j.apenergy.2016.08.181

Cucchiella F., D'Adamo I. and Gastaldi M. (2017). Biomethane: a renewable resource as vehicle fuel. *Resources*, **6**(4), 58, https://doi.org/10.3390/resources6040058

Cutzu R. and Bardi L. (2017). Production of bioethanol from agricultural wastes using residual thermal energy of a cogeneration plant in the distillation phase. *Fermentation*, **3**(2), 24, https://doi.org/10.3390/fermentation3020024

D'Adamo I., Ribichini M. and Tsagarakis K. P. (2023). Biomethane as an energy resource for achieving sustainable production: economic assessments and policy implications. *Sustainable Production and Consumption*, **35**, 13–27, https://doi.org/10.1016/j.spc.2022.10.014

Damayanti D., Supriyadi D., Amelia D., Saputri D. R., Devi Y. L. L., Auriyani W. A. and Wu H. S. (2021). Conversion of lignocellulose for bioethanol production, applied in bio-polyethylene terephthalate. *Polymers (Basel)*, **13**(17), 2886, https://doi.org/10.3390/polym13172886

D'Aquino C. A., Santos S. C. and Sauer I. L. (2022). Biogas as an alternative source of decentralized bioelectricity for large waste producers: an assessment framework at the University of São Paulo. *Energy*, **239**, 122326, https://doi.org/10.1016/j.energy.2021.122326

Engvall K., Kusar H., Sjöström K. and Pettersson L. J. (2011). Upgrading of raw gas from biomass and waste gasification: challenges and opportunities. *Topics in Catalysis*, **54**(13–15), 949–959, https://doi.org/10.1007/s11244-011-9714-x

Farouk S. M., Tayeb A. M., Abdel-Hamid S. M. S. and Osman R. M. (2024). Recent advances in transesterification for sustainable biodiesel production, challenges, and prospects: a comprehensive review. *Environmental Science and Pollution Research*, **31**, 12722–12747, https://doi.org/10.1007/s11356-024-32027-4

Francisco É. C, Neves D. B., Jacob-Lopes E. and Franco T. T. (2010). Microalgae as feedstock for biodiesel production: carbon dioxide sequestration, lipid production and biofuel quality. *Journal of Chemical Technology & Biotechnology*, **85**(3), 395–403, https://doi.org/10.1002/jctb.2338

Fukuda H., Kondo A. and Noda H. (2001). Biodiesel fuel production by transesterification of oils. *Journal of Bioscience and Bioengineering*, **92**(5), 405–416, https://doi.org/10.1016/S1389-1723(01)80288-7

Furzer I. A. (2000). Optimization of a wastewater system containing the ternary homogeneous azeotropic system ethyl acetate–ethanol–water. *Industrial & Engineering Chemistry Research*, **39**(6), 1539–1545, https://doi.org/10.1021/ie990133x

Garlapati V. K., Sharma S. and Sevda S. (2021). Photosynthetic biogas upgrading: an attractive biological technology for biogas upgrading. In: Emerging Technologies and Biological Systems for Biogas Upgrading, Elsevier, Academic Press, London, pp. 383–409, https://doi.org/10.1016/B978-0-12-822808-1.00014-3

Gollakota A. R. K., Kishore N. and Gu S. (2018). A review on hydrothermal liquefaction of biomass. *Renewable and Sustainable Energy Reviews*, **81**, 1378–1392, https://doi.org/10.1016/j.rser.2017.05.178

Gürü M., Koca A., Can Ö, Çınar C. and Şahin F. (2010). Biodiesel production from waste chicken fat based sources and evaluation with Mg based additive in a diesel engine. *Renewable Energy*, **35**(3), 637–643, https://doi.org/10.1016/j.renene.2009.08.011

Hamouda R. A., Sherif S. A. and Ghareeb M. M. (2018). Bioethanol production by various hydrolysis and fermentation processes with micro and macro green algae. *Waste and Biomass Valorization*, **9**(9), 1495–1501, https://doi.org/10.1007/s12649-017-9936-7

Hayyat U., Khan M. U., Sultan M., Zahid U., Bhat S. A. and Muzamil M. (2024). A review on dry anaerobic digestion: existing technologies, performance factors, challenges, and recommendations. *Methane*, **3**(1), 33–52, https://doi.org/10.3390/methane3010003

Hoyer K., Hulteberg C., Svensson M., Jernberg J. and Nörregård Ö. (2016). Biogas upgrading – technical review.

IEA (2021). World Energy Outlook 2021. IEA, Paris. https://www.iea.org/reports/world-energy-outlook-2021, License: CC BY 4.0.

Janaun J. and Ellis N. (2010). Perspectives on biodiesel as a sustainable fuel. *Renewable and Sustainable Energy Reviews*, **14**(4), 1312–1320, https://doi.org/10.1016/j.rser.2009.12.011

Joshi G., Pandey J. K., Rana S. and Rawat D. S. (2017). Challenges and opportunities for the application of biofuel. *Renewable and Sustainable Energy Reviews*, **79**, 850–866, https://doi.org/10.1016/j.rser.2017.05.185

Joshi J., Dhungana P., Prajapati B., Maharjan R., Poudyal P., Yadav M., Bhattarai T. and Sreerama L. (2019). Enhancement of ethanol production in electrochemical cell by *Saccharomyces cerevisiae* (CDBT2) and *Wickerhamomyces anomalus* (CDBT7). *Frontiers in Energy Research*, **7**, 474–887. https://doi.org/10.3389/fenrg.2019.00070

Kabeyi M. J. B. and Olanrewaju O. A. (2022). Biogas production and applications in the sustainable energy transition. *Journal of Energy*, **2022**, 1–43, https://doi.org/10.1155/2022/8750221

Kadam R. and Panwar N. L. (2017). Recent advancement in biogas enrichment and its applications. *Renewable and Sustainable Energy Reviews*, **73**, 892–903, https://doi.org/10.1016/j.rser.2017.01.167

Kamaraj R., Rao Y. K. S. S. and Balakrishna B. (2022). Biodiesel blends: a comprehensive systematic review on various constraints. *Environmental Science and Pollution Research*, **29**(29), 43770–43785, https://doi.org/10.1007/s11356-021-13316-8

Karmakar B. and Halder G. (2019). Progress and future of biodiesel synthesis: advancements in oil extraction and conversion technologies. *Energy Conversion and Management*, **182**, 307–339, https://doi.org/10.1016/j.enconman.2018.12.066

Khan M. U., Usman M., Ashraf M. A., Dutta N., Luo G. and Zhang S. (2022a). A review of recent advancements in pretreatment techniques of lignocellulosic materials for biogas production: opportunities and limitations. *Chemical Engineering Journal Advances*, **10**, 100263, https://doi.org/10.1016/j.ceja.2022.100263

Khan S., Naushad M., Iqbal J., Bathula C. and Sharma G. (2022b). Production and harvesting of microalgae and an efficient operational approach to biofuel production for a sustainable environment. *Fuel*, **311**, 122543, https://doi.org/10.1016/j.fuel.2021.122543

Kohlheb N., Wluka M., Bezama A., Thrän D., Aurich A. and Müller R. A. (2021). Environmental-economic assessment of the pressure swing adsorption biogas upgrading technology. *BioEnergy Research*, **14**(3), 901–909, https://doi.org/10.1007/s12155-020-10205-9

Kougias P. G., Treu L., Benavente D. P., Boe K., Campanaro S. and Angelidaki I. (2017). Ex-situ biogas upgrading and enhancement in different reactor systems. *Bioresource Technology*, **225**, 429–437, https://doi.org/10.1016/j.biortech.2016.11.124

Kumar R. and Wyman C. E. (2009). Effects of cellulase and xylanase enzymes on the deconstruction of solids from pretreatment of poplar by leading technologies. *Biotechnology Progress*, **25**(2), 302–314, https://doi.org/10.1002/btpr.102

Lee J. C., Kim J. H., Chang W. S. and Pak D. (2012). Biological conversion of CO_2 to CH_4 using hydrogenotrophic methanogen in a fixed bed reactor. *Journal of Chemical Technology & Biotechnology*, **87**(6), 844–847, https://doi.org/10.1002/jctb.3787

Lee R.-J., Tamm T., Temmer R., Aabloo A. and Kiefer R. (2013). Two formation mechanisms and renewable antioxidant properties of suspensible chitosan–PPy and chitosan–PPy–BTDA composites. *Synthetic Metals*, **164**, 6–11, https://doi.org/10.1016/j.synthmet.2012.12.032

Lin L., Cunshan Z., Vittayapadung S., Xiangqian S. and Mingdong D. (2011). Opportunities and challenges for biodiesel fuel. *Applied Energy*, **88**(4), 1020–1031, https://doi.org/10.1016/j.apenergy.2010.09.029

Lindner J., Zielonka S., Oechsner H. and Lemmer A. (2016). Is the continuous two-stage anaerobic digestion process well suited for all substrates? *Bioresource Technology*, **200**, 470–476, https://doi.org/10.1016/j.biortech.2015.10.052

Linganiso E. C., Tlhaole B., Magagula L. P., Dziike S., Linganiso L. Z., Motaung T. E., Moloto N. and Tetana Z. N. (2022). Biodiesel production from waste oils: a South African outlook. *Sustainability*, **14**(4), 1983, https://doi.org/10.3390/su14041983

López M. E., Rene E. R., Veiga M. C. and Kennes C. (2012). Biogas technologies and cleaning techniques, pp. 347–377, https://doi.org/10.1007/978-94-007-2439-6_9

Lu J., Nie K., Xie F., Wang F. and Tan T. (2007). Enzymatic synthesis of fatty acid methyl esters from lard with immobilized *Candida* sp. 99–125. *Process Biochemistry*, **42**(9), 1367–1370, https://doi.org/10.1016/j.procbio.2007.06.004

Ma F. and Hanna M. A. (1999). Biodiesel production: a review. *Bioresource Technology*, **70**(1), 1–15, https://doi.org/10.1016/S0960-8524(99)00025-5

Maas R. H., Bakker R. R., Boersma A. R., Bisschops I., Pels J. R., de Jong E., Weusthuis R. A. and Reith H. (2008). Pilot-scale conversion of lime-treated wheat straw into bioethanol: quality assessment of bioethanol and valorization of side streams by anaerobic digestion and combustion. *Biotechnology for Biofuels*, **1**(1), 113, https://doi.org/10.1186/1754-6834-1-14

Makaruk A., Miltner M. and Harasek M. (2010). Membrane biogas upgrading processes for the production of natural gas substitute. *Separation and Purification Technology*, **74**(1), 83–92, https://doi.org/10.1016/j.seppur.2010.05.010

Manikandan S., Karmegam N., Subbaiya R., Karthiga Devi G., Arulvel R., Ravindran B. and Awasthi M. K. (2021). Emerging nano-structured innovative materials as adsorbents in wastewater treatment. *Bioresource Technology*, **320**, 124394, https://doi.org/10.1016/j.biortech.2020.124394

Martín C., Klinke H. B. and Thomsen A. B. (2007). Wet oxidation as a pretreatment method for enhancing the enzymatic convertibility of sugarcane bagasse. *Enzyme and Microbial Technology*, **40**(3), 426–432, https://doi.org/10.1016/j.enzmictec.2006.07.015

Micic V. and Jotanovic M. (2015). Bioethanol as fuel for internal combustion engines. *Zastita Materijala*, **56**(4), 403–408, https://doi.org/10.5937/ZasMat1504403M

Molino A., Migliori M., Ding Y., Bikson B., Giordano G. and Braccio G. (2013). Biogas upgrading via membrane process: modelling of pilot plant scale and the end uses for the grid injection. *Fuel*, **107**, 585–592, https://doi.org/10.1016/j.fuel.2012.10.058

Monlau F., Barakat A., Trably E., Dumas C., Steyer J.-P. and Carrère H. (2013). Lignocellulosic materials into biohydrogen and biomethane: impact of structural features and pretreatment. *Critical Reviews in Environmental Science and Technology*, **43**(3), 260–322, https://doi.org/10.1080/10643389.2011.604258

Moser B. R. (2010). Camelina (*Camelina sativa* L.) oil as a biofuels feedstock: golden opportunity or false hope? *Lipid Technology*, **22**(12), 270–273, https://doi.org/10.1002/lite.201000068

Moser B. R., Knothe G., Vaughn S. F. and Isbell T. A. (2009). Production and evaluation of biodiesel from field pennycress (*Thlaspi arvense* L.) oil. *Energy & Fuels*, **23**(8), 4149–4155, https://doi.org/10.1021/ef900337g

Muñoz R., Meier L., Diaz I. and Jeison D. (2015). A review on the state-of-the-art of physical/chemical and biological technologies for biogas upgrading. *Reviews in Environmental Science and Bio/Technology*, **14**(4), 727–759, https://doi.org/10.1007/s11157-015-9379-1

Öner C. and Altun Ş (2009). Biodiesel production from inedible animal tallow and an experimental investigation of its use as alternative fuel in a direct injection diesel engine. *Applied Energy*, **86**(10), 2114–2120, https://doi.org/10.1016/j.apenergy.2009.01.005

Osorio-González C. S., Gómez-Falcon N., Sandoval-Salas F., Saini R., Brar S. K. and Ramírez A. A. (2020). Production of biodiesel from castor oil: a review. *Energies (Basel)*, **13**(10), 2467, https://doi.org/10.3390/en13102467

Pandey A., Soccol C. R., Nigam P. and Soccol V. T. (2000). Biotechnological potential of agro-industrial residues. I: sugarcane bagasse. *Bioresource Technology*, **74**(1), 69–80, https://doi.org/10.1016/S0960-8524(99)00142-X

Pawar A. and Panwar N. L. (2022). Influence of pyrolysis technologies on biofuel production and its physicochemical properties: a review. pp. 569–600, https://doi.org/10.1007/978-981-16-6875-3_47

Persson M., Jönsson O. and Wellinger A. (2006). Biogas upgrading to vehicle fuel standards and grid injection. In: IEA Bioenergy Task, IEA Bioenergy, Paris, France, pp. 1–34.

Phwan C. K., Ong H. C., Chen W.-H., Ling T. C., Ng E. P. and Show P. L. (2018). Overview: comparison of pretreatment technologies and fermentation processes of bioethanol from microalgae. *Energy Conversion and Management*, **173**, 81–94, https://doi.org/10.1016/j.enconman.2018.07.054

Pramanik K. (2003). Properties and use of jatropha curcas oil and diesel fuel blends in compression ignition engine. *Renewable Energy*, **28**(2), 239–248, https://doi.org/10.1016/S0960-1481(02)00027-7

Rotunno P., Lanzini A. and Leone P. (2017). Energy and economic analysis of a water scrubbing based biogas upgrading process for biomethane injection into the gas grid or use as transportation fuel. *Renewable Energy*, **102**, 417–432, https://doi.org/10.1016/j.renene.2016.10.062

Sahlan M., Saefuddin M., Muryanto H., Hermansyah and Wijanarko A. (2018). Extraction and characterization of yeast extract bioethanol byproduct from empty palm oil bunch for raw material of cosmetic products. *E3S Web of Conferences*, **67**, 03040, https://doi.org/10.1051/e3sconf/20186703040

Sahota S., Shah G., Ghosh P., Kapoor R., Sengupta S., Singh P., Vijay V., Sahay A., Vijay V. K. and Thakur I. S. (2018). Review of trends in biogas upgradation technologies and future perspectives. *Bioresource Technology Reports*, **1**, 79–88, https://doi.org/10.1016/j.biteb.2018.01.002

Saka S. and Kusdiana D. (2001). Biodiesel fuel from rapeseed oil as prepared in supercritical methanol. *Fuel*, **80**(2), 225–231, https://doi.org/10.1016/S0016-2361(00)00083-1

Saladini F., Patrizi N., Pulselli F. M., Marchettini N. and Bastianoni S. (2016). Guidelines for emergy evaluation of first, second and third generation biofuels. *Renewable and Sustainable Energy Reviews*, **66**, 221–227, https://doi.org/10.1016/j.rser.2016.07.073

Sarkar N., Ghosh S. K., Bannerjee S. and Aikat K. (2012). Bioethanol production from agricultural wastes: an overview. *Renewable Energy*, **37**(1), 19–27, https://doi.org/10.1016/j.renene.2011.06.045

Schwab A. W., Bagby M. O. and Freedman B. (1987). Preparation and properties of diesel fuels from vegetable oils. *Fuel*, **66**(10), 1372–1378, https://doi.org/10.1016/0016-2361(87)90184-0

Scott P. T., Pregelj L., Chen N., Hadler J. S., Djordjevic M. A. and Gresshoff P. M. (2008). *Pongamia pinnata*: an untapped resource for the biofuels industry of the future. *BioEnergy Research*, **1**(1), 2–11, https://doi.org/10.1007/s12155-008-9003-0

Senusi W., Ahmad M. I., Binhweel F., Shalfoh E., Alsaedi S. and Shakir M. A. (2024). Biodiesel production and characteristics from waste frying oils: sources, challenges, and circular economic perspective. *Environmental Science and Pollution Research*, **31**(23), 33239–33258, https://doi.org/10.1007/s11356-024-33533-1

Spitoni M., Pierantozzi M., Comodi G., Polonara F. and Arteconi A. (2019). Theoretical evaluation and optimization of a cryogenic technology for carbon dioxide separation and methane liquefaction from biogas. *Journal of Natural Gas Science and Engineering*, **62**, 132–143, https://doi.org/10.1016/j.jngse.2018.12.007

Struk M., Kushkevych I. and Vítězová M. (2020). Biogas upgrading methods: recent advancements and emerging technologies. *Reviews in Environmental Science and Bio/Technology*, **19**(3), 651–671, https://doi.org/10.1007/s11157-020-09539-9

Sun Q., Li H., Yan J., Liu L., Yu Z. and Yu X. (2015). Selection of appropriate biogas upgrading technology – a review of biogas cleaning, upgrading and utilisation. *Renewable and Sustainable Energy Reviews*, **51**, 521–532, https://doi.org/10.1016/j.rser.2015.06.029

Tabatabaei M., Aghbashlo M., Dehhaghi M., Panahi H. K. S, Mollahosseini A., Hosseini M. and Soufiyan M. M. (2019). Reactor technologies for biodiesel production and processing: a review. *Progress in Energy and Combustion Science*, **74**, 239–303, https://doi.org/10.1016/j.pecs.2019.06.001

Talebnia F., Karakashev D. and Angelidaki I. (2010). Production of bioethanol from wheat straw: an overview on pretreatment, hydrolysis and fermentation. *Bioresource Technology*, **101**(13), 4744–4753, https://doi.org/10.1016/j.biortech.2009.11.080

Toldrá-Reig F., Mora L. and Toldrá F. (2020). Trends in biodiesel production from animal fat waste. *Applied Sciences*, **10**(10), 3644, https://doi.org/10.3390/app10103644

Uddin M. M. and Wright M. M. (2023). Anaerobic digestion fundamentals, challenges, and technological advances. *Physical Sciences Reviews*, **8**(9), 2819–2837, https://doi.org/10.1515/psr-2021-0068

Ullah Khan I., Hafiz Dzarfan Othman M., Hashim H., Matsuura T., Ismail A. F., Rezaei-DashtArzhandi M. and Azelee W. (2017). Biogas as a renewable energy fuel – a review of biogas upgrading, utilisation and storage. *Energy Conversion and Management*, **150**, 277–294, https://doi.org/10.1016/j.enconman.2017.08.035

Usman Khan M. and Kiaer Ahring B. (2020). Anaerobic digestion of biorefinery lignin: effect of different wet explosion pretreatment conditions. *Bioresource Technology*, **298**, 122537, https://doi.org/10.1016/j.biortech.2019.122537

Valli L., Rossi L., Fabbri C., Sibilla F., Gattoni P., Dale B. E., Kim S., Ong R. G. and Bozzetto S. (2017). Greenhouse gas emissions of electricity and biomethane produced using the Biogasdoneright™ system: four case studies from Italy. *Biofuels, Bioproducts and Biorefining*, **11**(5), 847–860, https://doi.org/10.1002/bbb.1789

Vasić K., Knez Ž and Leitgeb M. (2021). Bioethanol production by enzymatic hydrolysis from different lignocellulosic sources. *Molecules*, **26**(3), 753, https://doi.org/10.3390/molecules26030753

Vickram S., Manikandan S., Deena S. R., Mundike J., Subbaiya R., Karmegam N., Jones S., Yadav K. K., Chang S. W., Ravindran B. and Awasthi M. K. (2023). Advanced biofuel production, policy and technological implementation of nano-additives for sustainable environmental management – a critical review. *Bioresource Technology*, **387**, 129660, https://doi.org/10.1016/j.biortech.2023.129660

Vo T. T. Q., Wall D. M., Ring D., Rajendran K. and Murphy J. D. (2018). Techno-economic analysis of biogas upgrading via amine scrubber, carbon capture and ex-situ methanation. *Applied Energy*, **212**, 1191–1202, https://doi.org/10.1016/j.apenergy.2017.12.099

Ware A. and Power N. (2016). Biogas from cattle slaughterhouse waste: energy recovery towards an energy self-sufficient industry in Ireland. *Renewable Energy*, **97**, 541–549, https://doi.org/10.1016/j.renene.2016.05.068

Xing Y., Zheng Z., Sun Y. and Agha Alikhani M. (2021). A review on machine learning application in biodiesel production studies. *International Journal of Chemical Engineering*, **2021**, 1–12, https://doi.org/10.1155/2021/2154258

Xu Y., Huang Y., Wu B., Zhang X. and Zhang S. (2015). Biogas upgrading technologies: energetic analysis and environmental impact assessment. *Chinese Journal of Chemical Engineering*, **23**(1), 247–254, https://doi.org/10.1016/j.cjche.2014.09.048

Yaakob Z., Mohammad M., Alherbawi M., Alam Z. and Sopian K. (2013). Overview of the production of biodiesel from waste cooking oil. *Renewable and Sustainable Energy Reviews*, **18**, 184–193, https://doi.org/10.1016/j.rser.2012.10.016

Yamazaki R., Iwamoto S., Nabetani H., Osakada K., Miyamaki O. and Sagara Y. (2007). Noncatalytic alcoholysis of oils for biodiesel fuel production by a semi-batch process. *Japan Journal of Food Engineering*, **8**(1), 11–19, https://doi.org/10.11301/jsfe2000.8.11

Yang X., Wang K., Wang H., Zhang J., Tang L. and Mao Z. (2016). Control of pH by acetic acid and its effect on ethanol fermentation in an integrated ethanol–methane fermentation process. *RSC Advances*, **6**(63), 57902–57909, https://doi.org/10.1039/C6RA04129A

Zhao Q., Leonhardt E., MacConnell C., Frear C., Chen S. (2010). Purification technologies for biogas generated by anaerobic digestion. CSANR Research Report 2010-001, Climate Friendly Farming: Improving the Carbon Footprint of Agriculture in the Pacific Northwest, pp. 1–24. Washington State University.

Zheng X. and Li R. (2024). Critical review on two-stage anaerobic digestion with H_2 and CH_4 production from various wastes. *Water (Basel)*, **16**(11), 1608, https://doi.org/10.3390/w16111608

doi: 10.2166/9781789064100_0215

Chapter 12

Bioenergy from forest resources: wood pellets, biochar, and heat generation

Umer Hayyat[1], Muhammad Shahid[2], Jonathan Lee Tian En[3,4], Nalok Dutta[5,6] and Muhammad Usman Khan[2,7]

[1]Department of Thermal and Fluid Engineering, Faculty of Engineering Technology, University of Twente, Drienerlolaan 5, 7522 NB Enschede, Netherlands
[2]Department of Energy Systems Engineering, Faculty of Agricultural Engineering and Technology, University of Agriculture Faisalabad, 38000 Faisalabad, Pakistan
[3]NUS Environmental Research Institute, National University of Singapore, T-Lab Building 5A Engineering Drive 1, Singapore 117411, Singapore
[4]Energy and Environmental Sustainability for Megacities (E2S2) Phase II, Campus for Research Excellence and Technological Enterprise (CREATE), 1 Create Way, Singapore 138602, Singapore
[5]Department of Biochemical Engineering, University College London, London, UK
[6]Department of Anthropology, University of Washington, Seattle, WA, USA
[7]Bioproducts Sciences and Engineering Laboratory (BSEL), Washington State University Tricities Campus, Richland, WA 99354, USA
*Correspondence: u.hayyat@utwente.nl, mshahid9220@outlook.com, eriltej@nus.edu.sg, nalok.dutta@ucl.ac.uk, nalok.dutta@wsu.edu, muhammadusman.khan@wsu.edu, usman.khan@uaf.edu.pk

ABSTRACT

In recent years, there has been a widespread focus on utilizing bioenergy derived from forests as an effective and sustainable renewable energy alternative. The utilization of wood wastes and forest residues to generate heat and produce energy has gained significant importance. However, there are certain challenges associated with using these materials in their original form. Biomass pelletizing presents a promising solution by converting raw biomass into fuel pellets that possess high energy density. Furthermore, the resulting biochar plays a vital role in enhancing soil quality. This chapter delves into the process of manufacturing wood fuel pellets and explores their diverse applications across residential and industrial sectors. It also examines technologies for producing biochar from this waste, highlighting potential benefits such as carbon sequestration and soil enrichment they offer. Moreover, this chapter also discusses various heat generation technologies that can be used in conjunction with biomass pellet production.

Keywords: Wood pellets, pellet production, biochar, heat generation.

12.1 INTRODUCTION

Over the past few years, there has been a burgeoning fascination with the exploration of renewable energy sources. This interest stems from a collective effort to diminish our reliance on fossil fuels and alleviate the adverse environmental effects associated with conventional energy production methods (Boquera *et al.*, 2021). One of these promising sources is bioenergy derived from forest resources, specifically wood-based biofuel. Wood pellets, biochar, and heat generation are some of the key components of this bioenergy sector. Forest residues, specifically the leftover wood from commercial harvesting operations and silvicultural practices, play a crucial role in the production of wood pellets, as well as in generating heat and electricity (Grodsky *et al.*, 2016). These residues are processed and transformed into bio-based products, such as wood pellets, that are highly efficient in terms of heating quality and can replace traditional fossil fuels in various applications.

Wood pellets are a type of biofuel that are obtained from compressed sawdust and wood shavings. They have gained significant popularity in recent years due to their high energy density and clean burning properties (Artemio *et al.*, 2018). Wood pellets are commonly used in residential heating applications, as well as in large-scale industrial facilities. Biochar, on the contrary, is a type of charcoal that is produced from heating biomass, such as wood or agricultural waste, in a low-oxygen environment. It is primarily used as a soil amendment, as it has the ability to enhance soil fertility and sequester carbon dioxide from the atmosphere (Criscuoli *et al.*, 2019). Heat generation, another important aspect of bioenergy from forest resources, involves utilizing wood as a direct heat source.

Utilizing forest-based biomass as a feedstock for bioenergy production presents numerous advantages. Primary among these is the diminished dependence on non-renewable energy sources, contributing significantly to the mitigation of climate change through a notable reduction in greenhouse gas emissions. By utilizing forest resources for bioenergy production, we can not only reduce our reliance on fossil fuels but also promote sustainable forest management practices. Forest-based biomass, such as waste wood and harvest residues, can be effectively utilized to manufacture wood pellets and produce heat (Artemio *et al.*, 2018).

Wood pellets, in particular, offer excellent heating quality and can serve as a viable alternative to natural gas, oil, and coal. They are a renewable energy source that can be used in heated livestock buildings, residential heating systems, and industrial facilities. The global demand for wood pellets is expected to increase significantly in the coming years, further highlighting their potential as a sustainable energy solution.

Furthermore, biochar, another significant resource of forest-based biomass, offers a range of environmental benefits. When biochar is added to soil, it helps to improve soil structure, retain nutrients, and enhance water-retention capacity. This not only promotes healthier plant growth but also reduces the need for chemical fertilizers and irrigation, making it a sustainable and cost-effective solution for agriculture.

In addition to its agricultural applications, biochar has also gained attention as a potential tool for waste management and climate change mitigation. By utilizing locally available agricultural waste as a feedstock for biochar production, we can not only reduce waste but also minimize the environmental impacts associated with waste disposal. Moreover, studies have shown that biochar has the ability to remediate diverse pollutants of organic and inorganic origin, making it an effective tool for environmental remediation.

The objective of this chapter is to provide a comprehensive overview of the potential of forest resources for bioenergy production. It aims to explore the processes involved in the production of wood pellets, biochar, and heat generation and their respective applications in household and industrial sectors. This chapter explores the potential of forest-based biomass, specifically wood pellets and biochar, as renewable energy sources.

12.2 WOOD PELLETS

The term 'pellet' refers to a compact, spherical mass of a particular substance. A wood pellet is typically a compacted mass composed primarily of compressed wood material, taking on a spherical or cylindrical form. Typically, wood pellets are often used in plural form rather than singular, primarily when referring to them as a collective bulk material. Presently, a diverse range of feedstock materials has been transformed into pellets to facilitate their subsequent application. These include pellets created for poultry feed, biomass pellets designed for combustion purposes, hops pellets used in beer production, and catalytic pellets employed in catalytic chemical reactions. Wood pellets are commonly used as a form of biofuel for the purpose of power generation in numerous developed nations, including Japan, China, Germany, Austria, Holland, and the UK (Guo *et al.*, 2015). Wood pellets offer a potential pathway for transitioning from fossil fuels to renewable energy sources, thereby facilitating a reduction in carbon emissions. Wood pellets, derived from organic materials such as sawdust or tree waste, present a highly effective and environmentally sustainable energy option. Currently, nations and sectors are undergoing a shift from coal-based power generation facilities to bioenergy installations that use wood pellets as a fuel source. This transition is contributing to the mitigation of greenhouse gas emissions (Mola-Yudego *et al.*, 2014; Pradhan *et al.*, 2018b). Figure 12.1 shows the flow process of wood feedstock into wood pellets and its applications.

Using wood as an energy source is a logical choice due to the abundance of supporting evidence and the pressing need for alternative energy solutions. This holds particularly true in the current circumstances when financial resources are limited for a significant number of individuals, leading to the exploration of more effective strategies for future planning (Mathiesen *et al.*, 2011). The primary challenge in using these resources is the significant expense associated with their production, distribution, and standardization of usage, which may lead to complications in the different energy conversion technologies (Bishop *et al.*, 2012). Pelletizing is regarded as a crucial element in addressing this challenge, which is why the global wood pellet production sector has witnessed significant economic expansion. The production of pellets and the advancement of their technology are steadily expanding and transitioning toward complete automation on a daily basis.

12.2.1 Production of wood pellets

The production of wood pellets primarily relies on the kind of raw material used for the process of pelletization. The manufacturing of pellets involves several stages: preparing raw materials, the primary pelletization process, and post-pelletization activities.

12.2.1.1 Feedstock for wood pellets

Materials used for manufacturing wood pellets include sawdust, wood chips, logs, and wood dust. The wood pellets produced from these raw materials are very compact and have a low amount of moisture. This allows them to have a higher energy output compared to traditional biomass fuel (Xiao *et al.*, 2015). The choice of feedstock greatly impacts the production of wood pellets. Wood dust, being dry,

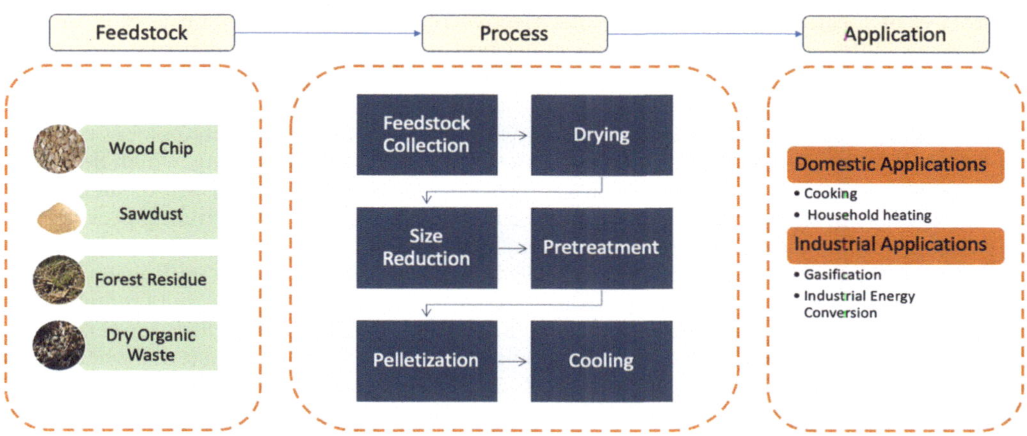

Figure 12.1 Wood pellets flow sheet.

Table 12.1. Feedstock required for production of 1 t wood pellets.

Feedstock Type	Quantity for 1 t Wood Pellets
Wood chips	5.3 lcm
Sawdust	7.5 lcm
Industrial wood chip	5.1 lcm
Log wood	2.2 scm

Source: Thek and Obernberger (2012).

does not require drying as a pretreatment process. However, sawdust, which has moisture content, needs to undergo a pretreatment process of drying before it can be pelletized. Table 12.1 presents the various types of raw materials and their respective quantities necessary for the manufacturing process of wood pellets.

Presently, there are various raw materials available for the manufacturing of wood pellets, including hardwoods such as beech, willow, and oak. Certain producers may have a preference for softwood over hardwood due to the fact that softwood has a lower density and is more easily processed into pellets compared to hardwood. Forest wastes, including roots and branches, can be used as a valuable resource for the production of wood pellets (Marinescu & Bush, 2009). Indeed, various kinds of wood (raw material) have the potential to undergo a conversion process into pellets. However, various factors such as usage standards, transportation and initial costs, and the accessibility of wood can impact the selection of feedstock.

12.2.1.2 Drying

Drying the wood feedstock involves removing excess moisture through a controlled drying process. This is typically carried out using industrial dryers or kilns. These dryers use heat and airflow to evaporate the moisture from the wood, reducing its moisture content to a desirable level. The purpose of drying the wood feedstock is to degrade the moisture content so that it can be removed through evaporation. Once the wood feedstock has been dried to the desired moisture content, it is ready for pelletization. During the drying process, the wood feedstock is typically exposed to heat and airflow in industrial dryers or kilns (Lam *et al.*, 2011).

12.2.1.3 Size reduction

The reduction in size of the raw material for wood pellets is a crucial stage in ensuring the effectiveness of the process of forming pellets. The main objective of this pretreatment process is to decrease the dimensions of the raw material to attain consistent particle size. The consistent size of the raw materials allows for the creation of wood pellets with a high density. As a result, these wood pellets exhibit efficient combustion. The initial stage of this preprocessing procedure involves the breaking down of the wooden raw material using chippers, which transform the elongated pieces of wood into smaller wood chips. Next, the wood chips are subjected to mechanical shredding, which helps to break down the material that may be more difficult to process compared to typical wood, such as bark or small branches. The primary purpose of the hammer mill is to facilitate the grinding process by using rapidly rotating hammers to break down the raw material into smaller particles. This reduction in size allows for the creation of finer particles from the feedstock. Additional types of milling machines, such as knife mills, linear knife grids, and disk mills, are also employed for the purpose of reducing the size of feedstock (Igathinathane *et al.*, 2009; Kirsten *et al.*, 2016). However, it was discovered that hammer mills demonstrated superior performance in reducing the size of wood raw material when compared to knife mills (Ghorbani *et al.*, 2010; Kirsten *et al.*, 2016; Miao *et al.*, 2011).

There are certain factors that influence the pretreatment of raw materials, such as the level of moisture present (Mani *et al.*, 2004). It was noted that wood with a high amount of moisture requires more mechanical energy compared to dry wood when undergoing size reduction pretreatment. The process of reducing the size of the feedstock also contributes to the supply of feedstock by facilitating its logistics (Miao *et al.*, 2011). The process of reducing the size is crucial to enhance the overall surface area of the feedstock, thereby improving its quality prior to the pelletization process (Mani *et al.*, 2004).

12.2.1.4 Pretreatment process

12.2.1.4.1 Steam explosion pretreatment

The process of steam explosion is widely recognized as a pretreatment method for preparing wood feedstock prior to pelletization. The steam explosion method involves exposing the feedstock to steam at a high temperature and then quickly reducing the pressure, causing the raw material to explode and break down. The wooden material is subjected to steam at a high pressure for a specific duration. This leads to the extraction of moisture from the wooden raw material. The material that goes through steam explosion pretreatment exhibits greater mechanical strength in comparison to other materials (Lam *et al.*, 2013; Tooyserkani *et al.*, 2012). The same pretreatment process can be used in other activities related to the production of biofuels, such as the production of bioethanol (Balat *et al.*, 2008). Research has indicated that feedstock subjected to steam explosion pretreatment has a greater density of pellets when compared to feedstock that has not undergone steam explosion (Shaw *et al.*, 2009). The application of a steam explosion pretreatment process has been noted to result in a reduction in the energy required for grinding. Nevertheless, it is crucial to take into account that the ultimate size of the feedstock, post-steam explosion, also plays a significant role in determining energy requirements (Adapa *et al.*, 2011).

12.2.1.4.2 Hydrothermal carbonization pretreatment

Hydrothermal carbonization pretreatment is a method used in the wood pellet pretreatment process to convert biomass into homogeneous chars at lower temperatures under self-generated pressures. It involves treating biomass with hot compressed water in the temperature range between 180 and 350°C (Lee *et al.*, 2018; Mumme *et al.*, 2011). During the hydrothermal carbonization (HTC) process, biomass is carbonized under mild temperature conditions using deionized water as a solvent. This technique helps to make the pelletized biomass more homogeneous and energy dense. The treatment temperature is carefully controlled to ensure that the water is in a supercritical condition. This hydrothermal carbonization pretreatment method enhances the grindability of the biomass and improves its composition in terms of the inorganic fraction (Jackowski *et al.*, 2019). Additionally, studies have shown that hydro chars produced through hydrothermal carbonization have relatively easy pelletizing properties. These properties make hydrothermal carbonization a promising process for valorizing low-quality solid biofuels, especially when dealing with wet biomass as a potential feedstock.

12.2.1.4.3 Biological pretreatment

Biological pretreatment is an essential step in the wood pellet production process. It involves the use of microorganisms or enzymes to break down the complex structure of lignocellulosic biomass and make it more accessible for further processing. This process is considered to be environmentally friendly and cost-effective compared to physical and chemical pretreatment methods.

The first step in biological pretreatment is the selection and cultivation of specific microorganisms or the production of enzymes that are capable of degrading the lignocellulosic material. These microorganisms, often cellulolytic, synthesize potent cellulolytic enzymes during hydrolysis, which help to break down the lignocellulosic structure. By using these enzymes, the biological pretreatment process disrupts the lignin fragments, making them more accessible for subsequent steps in the production of high-quality pellets (Canam *et al.*, 2011). One of the advantages of biological pretreatment is its low energy input and lack of chemical additions. Unlike conventional physicochemical methods, which require large amounts of energy and cause pollution, biological pretreatments offer a more sustainable and eco-friendly alternative (Sarkar *et al.*, 2012).

12.2.1.4.4 Torrefaction

The wood pellet production process hinges significantly on the crucial step of torrefaction. Torrefaction is a mild pyrolysis process wherein biomass undergoes thermal degradation, typically occurring at temperatures spanning from 200 to 300°C. This thermal treatment induces substantial alterations in the composition of the biomass (Chen *et al.*, 2015). In the course of torrefaction, there is a decrease in the hemicellulose content, along with reductions in the hydrogen to carbon and oxygen to carbon ratios. Conversely, the cellulose, lignin, and ash content of the biomass experience an increase. This transformation yields several enhancements, including improved grinding ability, heightened hydrophobicity, and an elevated calorific value in the biomass. These improved fuel properties render torrefied biomass a more suitable alternative for coal combustion. Following the torrefaction process, the torrefied biomass can be further processed into wood pellets (Safin *et al.*, 2017).

12.2.1.5 Pelletizing process

The next step after the drying and pretreatment processes is the pelletizing process. During the pelletizing process, pressure and heat are applied to the wood particles, causing lignin, a natural binder in wood, to soften and act as a glue, binding the particles together to form pellets. In the process of biomass pelletization, it is crucial to maintain the pressure within a specific range that yields high-quality pellets. Essentially, the goal is to optimize pressure so that the pellet production is energy-efficient, and the pellet mill operates at its maximum capacity, all while ensuring the quality of the pellets is preserved (Holm *et al.*, 2006). The primary equipment used for the process of pelletizing is a die and roller. Various roller–die combinations are being used for the manufacturing of wood pellets, including ring die pellet mills. These mills are comprised of a die ring that encircles stationary rollers. Flat die roller pellet mills are a type of machinery where the rollers are positioned above the horizontal die and revolve accordingly. The raw material is introduced into pellet mills and thereafter dispersed in a uniform manner (Tumuluru *et al.*, 2010). Table 12.2 shows the parameters and their effect associated with the manufacture of wood pellets within the pelletizing process.

Table 12.2. Parameters and their effect on pellet production.

Parameters	Effect	References
Moisture content	• Low moisture content during pelletization leads to the production of fragile pellets. • High moisture content requires a heating process to remove extra moisture. • Optimal moisture content is essential for pelletization as it works as a binding agent in pellet formation.	Carone *et al.* (2011); Li (2000); Tumuluru *et al.* (2011)
Biomass/ feedstock composition	• High lignin composition in feedstock increases durability of wood pellets. • Low extractive content of feedstock also enhances mechanical properties of wood pellets.	Castellano *et al.* (2015); Theerarattananoon *et al.* (2011)
Machine parameters	• Pellet milling machine specifications affect the shape of wood pellets. • Machine dies with thick characteristics increase the durability of the pellets produced.	Pradhan *et al.* (2018a); Theerarattananoon *et al.* (2011)
Particle size	• Fine particle size enhanced the physical properties of the pellets produced. • Feedstock with a mixture of small and large particles provides optimum quality during pelletization.	Kirsten *et al.* (2016); Stelte *et al.* (2011)

12.2.1.6 Cooling

One crucial step in the wood pellet production process is the cooling of the pellets after pelletization. This step is important because it helps to set and solidify the pellets, ensuring their durability and stability. After the pellets are formed through the pellet press, they come out at a high temperature due to the pressure applied during the process. The heat causes the lignin in the wood to plasticize slightly, acting as a natural binder that holds the pellets together. However, if the pellets are left to cool on their own without any intervention, they may become deformed or lose their shape. Therefore, the pellets are immediately transferred to a cooling system, typically a rotary cooler or a counterflow cooler. In this cooling system, the pellets are exposed to ambient air or cool water, which helps to rapidly lower their temperature. During the cooling phase, wood pellets undergo a crucial transformation. As they get cool, the pellets solidify, maintaining their structural integrity and becoming more resilient, thereby reducing the likelihood of breakage or crumbling. This cooling process plays a key role in stabilizing the pellets by firming up the lignin that melted on the pellet surfaces during earlier stages. This ensures that the pellets retain their original shape, contributing to their durability and overall quality (Börcsök & Pásztory, 2021).

12.2.2 Wood pellet applications

Wood pellets are multipurpose in nature and adaptable, finding employment in both commercial and industrial power plants as well as domestic settings. Cookstoves, boilers, and gasifiers are a few examples of biomass-based energy conversion devices that can readily employ wood pellets for power and heating applications.

12.2.2.1 Wood pellets for household use

Wood pellets have various applications in households. They can be used as a source of heat for stoves and fireplaces, providing efficient and clean burning that is both cost-effective and environmentally friendly. Additionally, wood pellets can also be used for cooking purposes in pellet grills or smokers, adding a unique smoky flavor to your food. These versatile fuel sources are easy to use and store, making them an ideal choice for both heating and cooking needs in residential settings. Recent studies have shown that a significant number of researchers have experimented with using wood pellets in a variety of household energy conversion applications (Cardozo *et al.*, 2014; Garcia-Maraver *et al.*, 2014; Wöhler *et al.*, 2017).

12.2.2.2 Wood pellets for industrial applications

Wood pellets have become increasingly popular in industrial applications due to their numerous benefits and versatility. These pellets, with a moisture content below 10%, meet the standard requirements for various industries. The technology used for their production is highly efficient and flexible, enabling the processing of different types of waste materials such as wood sawdust, straw, bark, and even coal dust and polyvinyl chloride (PVC) waste. In the industrial sector, wood pellets are primarily used for power generation and heating. The uniform shape and size of wood pellets, along with their low moisture content and high-energy density, make them an ideal choice for industrial-scale heat and power production plants.

One of the industries that can greatly benefit from wood pellet usage is the pharmaceutical industry. The consistently high-quality and heating properties of wood pellets make them ideal for use in pharmaceutical manufacturing processes, where precise and controlled heating is crucial. Another industry that can benefit from wood pellets is the chemical industry. The uniformity in water content and chemical composition of wood pellets, along with their high product density and heating value, make them a reliable and efficient source of solid biofuels for energy and heat production in industrial-sized plants within the chemical industry (Jarolin *et al.*, 2023).

12.3 BIOCHAR

Biochar is a carbon-rich substance generated through the biomass pyrolysis process (Chaves Fernandes *et al.*, 2020). It has gained significant popularity due to its various applications, such as waste management and climate change mitigation. One of the key advantages of biochar is its ability to be sourced from locally available agricultural waste, making it a cost-effective and environmentally friendly option. General biochar production and utilization flow pattern is shown in Figure 12.2. Many studies have shown that biochar is an effective tool for remediation of both organic and inorganic pollutants (Domene *et al.*, 2015; Liang *et al.*, 2021; Yazdani *et al.*, 2019). Its porous structure allows it to absorb and retain these pollutants, making it an ideal solution for soil and water treatment applications.

12.3.1 Biochar production technologies

Biochar is predominantly produced through thermochemical conversion methods, encompassing processes such as slow pyrolysis, fast pyrolysis, torrefaction, and gasification. Figure 12.3 illustrates the consideration of various process parameters in these methods. These thermochemical processes induce a permanent alteration in the physical properties and chemical structure of biomass, ultimately giving rise to the formation of biochar. This transformation occurs in an environment where oxygen is either absent or limited, and it is facilitated under specific temperature and pressure conditions (Leng *et al.*, 2019).

12.3.1.1 Slow pyrolysis

This method involves heating organic biomass, such as agricultural waste or forestry residues, in the absence or limited supply of oxygen. The forest waste is heated to temperatures of around 350–500°C, which causes it to break down and release volatile gases. During the slow pyrolysis process, the volatile gases are captured and can be used as a source of energy. This makes slow pyrolysis a sustainable

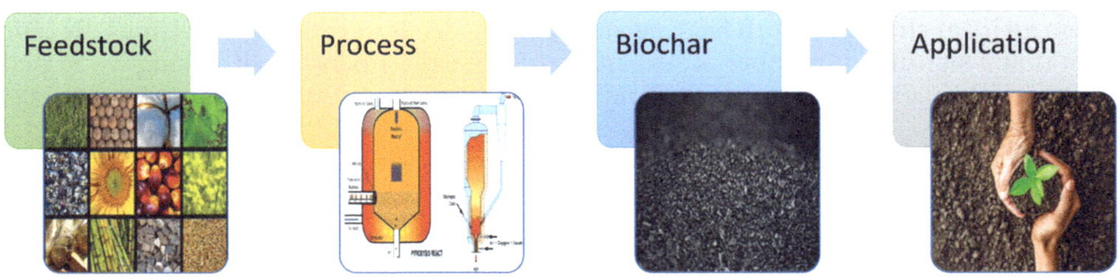

Figure 12.2 Biochar production pathway from feedstock to application.

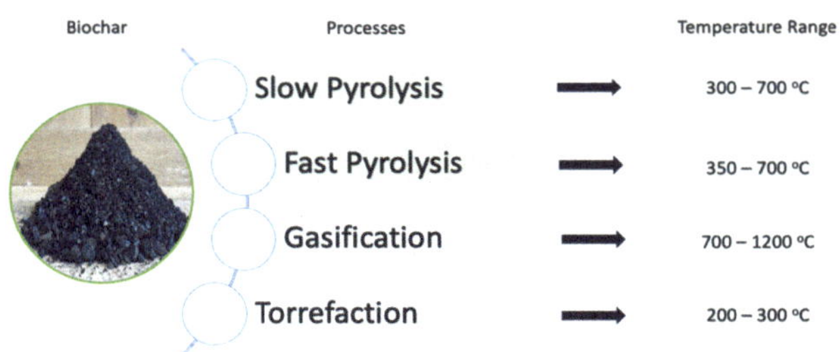

Figure 12.3 Biochar production technologies.

and efficient method for both biochar production and energy generation. Slow pyrolysis is a process that primarily results in the creation of char, with a smaller proportion of liquid and gas products also being produced (Demirbas & Arin, 2002). This process is characterized by a slower heating rate and a longer duration for the vapors to remain within the system. The specified conditions give rise to an environment favorable for the complete occurrence of secondary reactions. Moreover, the prolonged residence time of the vapors enables the effective removal of byproducts generated throughout these secondary reactions (Tripathi *et al.*, 2016).

12.3.1.2 Fast pyrolysis

Fast pyrolysis involves the rapid heating of biomass feedstock in the absence of oxygen at elevated temperatures. The temperature range for fast pyrolysis is between 900 and 1200°C, which is higher than the temperature range for slow pyrolysis at about 300°C (Demirbas & Arin, 2002). The higher temperature in fast pyrolysis facilitates a shorter residence time, leading to a more efficient process. With fast pyrolysis, the residence time is significantly reduced, allowing for a faster conversion of biomass into valuable products. The yield of biochar from fast pyrolysis is typically lower compared to slow pyrolysis. Fast pyrolysis produces around 15% biochar yield, whereas slow pyrolysis can yield up to 35% biochar (Adegoke *et al.*, 2021). Fast pyrolysis has some challenges for the production of biochar. Unlike slow pyrolysis, which has a long residence time and mainly produces biochar, fast pyrolysis produces biochar as a byproduct while focusing on obtaining bio-oil. The disadvantage of fast pyrolysis is that it yields less biochar as compared to slow pyrolysis.

12.3.1.3 Gasification

In the gasification process, biomass undergoes heating in an oxygen-limited environment, leading to the generation of a combustible gas referred to as syngas. This syngas is composed of carbon monoxide, hydrogen, and other gases, and it holds utility as a fuel for diverse applications (Han & Kim, 2008). The remaining solid residue after gasification is the biochar.

Gasification parameters significantly influence the quality of biochar produced. Among these parameters, the equivalence ratio holds particular importance. The equivalence ratio is defined as the ratio of the actual air–fuel ratio to the stoichiometric air–fuel ratio and is recognized as the most influential factor in the gasification process. The optimal energy ratio (ER) value for gasification is around 0.25–0.28, as it contributes to the production of biochar with desirable physical and chemical properties (Benedetti *et al.*, 2018). Other gasification parameters that affect the biochar quality include feedstock properties, gasifying agent, and pressure. Gasification of biomass for biochar production typically requires a high temperature ranging from 500 to 1000°C, depending on the desired properties of the biochar.

12.3.1.4 Torrefaction

To produce biochar through the torrefaction process, the forest waste undergoes a thermochemical transformation at temperatures between 200 and 300°C in the absence of oxygen (Wang *et al.*, 2017). This results in the formation of biochar with characteristics similar to bitumen and lignite coal (Barskov *et al.*, 2019). The benefits of torrefaction are twofold. First, it allows for the effective utilization and processing of waste materials, reducing the threat to the natural environment. Second, it produces a solid fuel that can be used for various purposes, such as fuel for vehicles or as a source of energy for combined heat and power systems.

The torrefaction process involves heating raw biomass in an inert nitrogen atmosphere for a period ranging from several minutes to several hours. During torrefaction, the biomass undergoes significant chemical and physical changes. The hemicellulose in the biomass degrades, resulting in a solid material with low moisture content and hydrophobic properties (Pilon & Lavoie, 2014). These methods for producing biochar work in different ways, and a comparison of some of the most important aspects of each method is provided in Table 12.3.

Table 12.3. Comparison of biochar production techniques.

	Temperature (°C)	Oxidizing Agent	Pretreatment	References
Slow pyrolysis	300–700	No	Required	Neves *et al.* (2011); Wang *et al.* (2020)
Fast pyrolysis	350–700	No	Required	Neves *et al.* (2011; Wang *et al.* (2020)
Gasification	700–1200	Controlled air	Required	Asadullah (2014)
Torrefaction	200–300	Excess air	Optional	Wang *et al.* (2020)

12.3.1.5 Challenges in biochar production technologies

Slow pyrolysis: The main challenge in slow pyrolysis is achieving high biochar yields while maintaining a slow heating rate. This interaction often requires a longer residence time and is energy-intensive in nature (Hu & Gholizadeh, 2019).

Fast pyrolysis: Understanding the rapid heating and cooling of the biomass feedstock is crucial in fast pyrolysis. This suggests a significant use of expensive and highly complex devices under strict regulations to prevent the accumulation of undesirable byproducts, such as tar (Rasaq *et al.*, 2021).

Gasification: The main obstacle in the gasification process involves managing the gasifying agent and maintaining control over the operating conditions of the system, while also achieving optimal biochar combustion without compromising the capture of syngas for energy utilization (Shahabuddin & Alam, 2022).

Torrefaction: Maintaining consistent heat distribution across biomass material and effectively managing the duration at high temperatures are the main challenges in producing biochar that is both thermally stable and energy dense (Ivanovski *et al.*, 2023).

12.3.2 Biochar applications

12.3.2.1 Biochar application for soil improvement

Biochar stands out as a versatile material, providing a multitude of benefits for soil enhancement in both agricultural and environmental contexts. Introducing biochar into the soil enables farmers to augment fertility, increase water-retention capacity, and enhance nutrient-retention capabilities. Moreover, biochar has been shown to boost soil microbial activity and regulate pH levels, establishing an advantageous environment conducive to optimal plant growth (Lehmann *et al.*, 2015). One of the key advantages of using biochar is its ability to reduce denitrification in the soil, leading to decreased emissions of N_2O, a potent greenhouse gas. This makes biochar application a valuable strategy for mitigating climate change. Moreover, biochar also plays a crucial role in carbon sequestration, helping to remove CO_2 from the atmosphere and store it in the soil for long term. The positive impacts of adding biochar to soil on its physical characteristics have been widely studied. For instance, incorporating biochar into the soil can enhance the overall surface area of the soil, promote better air circulation within the soil, and optimize the soil's bulk density, porosity, and compaction (Palansooriya *et al.*, 2019).

12.3.2.2 Biochar application for soil remediation and water treatment

The application of biochar has garnered substantial attention in recent years, attributed to its myriad benefits in soil remediation and water treatment. This charcoal-like material, derived through the pyrolysis process, has emerged as a valuable asset for enhancing agricultural practices and tackling environmental challenges effectively. One of the key applications of biochar is in agriculture, where it serves to enhance water retention and soil fertility. By adding biochar to the soil, farmers can reduce denitrification and subsequently decrease N_2O emissions. Furthermore, biochar has demonstrated its capacity to elevate crop yields by augmenting the soil's cation exchange capacity, water-retention ability, and nutrient-retention capability. Additionally, it contributes to enhanced soil microbial activity and regulation of pH levels (Elkhalifa *et al.*, 2022).

12.3.2.3 Biochar for carbon sequestration

The process of carbon sequestration involves capturing carbon and increasing the content of organic carbon in soil, which helps to enhance the soil's ability to store carbon and requires change in land management practices (Powlson *et al.*, 2011). Biochar has gained recognition as an effective material for increasing soil carbon content and enhancing the capacity of soil to store carbon is through the process of carbon sequestration.

12.3.2.4 Biochar for the mitigation of greenhouse gas emissions

The agricultural sector stands as a significant source of greenhouse gas emissions into the atmosphere (Lehmann *et al.*, 2015). The utilization of biochar in soil has demonstrated benefits, not only in terms of carbon sequestration but also in the reduction of gaseous emissions. Research indicates that specific biochar cycles, including those derived from crop residues, poultry litters, sewage sludge, cattle manure, and food waste, result in a net consumption of greenhouse gases rather than emission. This suggests that the application of biochar to soil can contribute to net negative greenhouse gas emissions, thereby exerting a positive impact on climate change (Elkhalifa *et al.*, 2022; Palansooriya *et al.*, 2019).

12.4 HEAT GENERATION

Wood pellets and biochar are both valuable sources of heat generation. Figure 12.4 illustrates the heat generation technologies. Wood pellets, produced from compressed sawdust and other biomass materials, are known for their high calorific value and consistent burn efficiency. They are widely used in residential and commercial heating systems, including pellet stoves and boilers. On the contrary, biochar, derived from the thermal decomposition of biomass in a torrefaction process, offers several benefits as a heat source. Biochar exhibits a higher calorific value and enhanced grindability in comparison to raw biomass. This characteristic renders biochar more efficient for advanced gasification and combustion technologies.

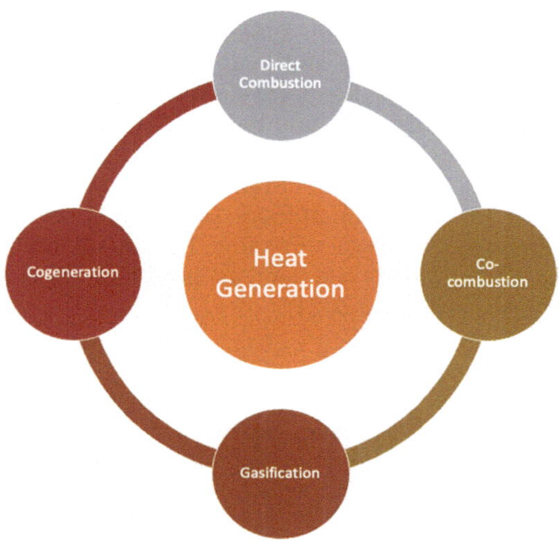

Figure 12.4 Heat generation technologies.

12.4.1 Direct combustion

Direct combustion is a commonly used method for heat generation using wood pellets and biochar (Mohammadi, 2021). By directly burning these biomass materials, heat is produced through the combustion process. This heat can then be used for various applications such as heating buildings, generating steam, or even powering engines. One of the advantages of direct combustion is its simplicity. It requires minimal equipment and can be easily implemented in various settings. Wood pellets and biochar, being renewable sources of fuel, also offer the advantage of being environmentally friendly. They release fewer greenhouse gases compared to fossil fuels, making them a more sustainable option for heat generation.

Direct combustion works by igniting the biomass material, whether it be wood pellets or biochar, and allowing it to burn in the presence of oxygen. As the biomass combusts, it releases heat energy in the form of flames and hot gases (Zakharevich *et al.*, 2015). This heat energy can then be harnessed and used for heating, cooking, or other industrial processes. The optimization of the combustion process involves controlling several parameters, including the quantity of supplied oxygen, temperature, and the residence time of the biomass material within the combustion chamber (Zakharevich *et al.*, 2015). These parameters play a pivotal role in influencing the efficiency of the combustion process and the composition of the resultant combustion products.

It can be challenging to maintain and minimize ash content and emissions while maximizing energy outputs when using direct combustion for heat generation. The complete combustion of biomass feedstock can be difficult as it requires control of oxygen or air supply and combustion temperature (Banerjee, 2023).

12.4.2 Co-combustion

Co-combustion, also recognized as biomass co-firing, involves the simultaneous burning of biomass alongside conventional fossil fuels such as coal in power plants. This process converts the stored chemical energy of biomass into heat energy, electricity, and mechanical power through combustion (Banerjee, 2023). An essential aspect to consider in co-combustion is the particle size of the pulverized mixture. According to available sources, particle size significantly influences combustion efficiency, unburned ash content, and overall combustion stability. Therefore, it is essential to ensure that the sizes of both the coal and biomass particles are suitable for their respective combustibility. To achieve efficient co-combustion, the biomass and coal need to be pulverized to sizes that allow for optimal combustion (Sakuragi & Otaka, 2020). Efficient co-combustion in coal-fired power plants involves the precise pulverization of biomass and coal to sizes that align with their combustibility. This pulverization enhances the calorific value per weight of the biomass, improves water repellency, and facilitates better grindability. It is crucial to emphasize that the effectiveness of the co-combustion process hinges on these factors. Additionally, the practice of co-combusting biomass and coal in the established power plants presents both environmental and economic advantages (Variny *et al.*, 2021).

Co-combustion is the simultaneous ignition of biomass with different fuels. This presents problems related to how well the different fuel sources work together and how to make the best combustion conditions to make energy production as efficient as possible while minimizing damage to the environment (Jiao *et al.*, 2023; Variny *et al.*, 2021).

12.4.3 Cogeneration

Cogeneration, or combined heat and power, is an exceptionally efficient approach to generating electricity and heat concurrently from a single fuel source (Song *et al.*, 2020). In the context of renewable energy, wood pellets and biochar are promising options for cogeneration. Wood pellets, which are small, cylindrical-shaped pieces of compressed wood, can be burned in a biomass-fired boiler to produce steam. This steam can then be used to generate electricity in a coal-fired power plant through a process called parallel co-firing. Moreover, the biochar production process can be seamlessly integrated into cogeneration systems. Biochar is typically produced through pyrolysis, a method that entails heating biomass in the absence of oxygen. This pyrolysis process not only yields biochar but also generates bio-oil and heat, establishing a practical and efficient approach to biochar production within cogeneration systems.

Combining biomass burning with power generation systems in cogeneration poses challenges in achieving high overall energy efficiency and cost-effectiveness. Cogeneration is a modern approach to power generation that involves harnessing energy from multiple sources, resulting in increased fuel utilization and associated costs (Ren *et al.*, 2024).

12.4.4 Gasification

Gasification is a highly efficient technology that has gained significant attention in recent years for the thermo-chemical conversion of biomass into useful energy. By heating the biomass to temperatures of over 700°C, a process known as gasification occurs, resulting in the production of a fuel gas called syngas (Seghman *et al.*, 2022). This syngas can be utilized in various downstream applications, making gasification an attractive option for heat generation.

Wood pellets and biochar are two commonly used biomass feedstocks in gasification processes. Wood pellets, which are obtained from compressed sawdust and other wood residues, provide a consistent and efficient fuel source. During the gasification process, the wood pellets are introduced into a gasifier along with a gasifying agent, such as air or steam. The high temperatures in the gasifier cause the biomass to undergo thermal decomposition, leading to the release of volatile gases and the conversion of carbon into syngas. This syngas can then be further processed and utilized for various energy applications, including heat production, electricity generation, or even as a feedstock for the production of biofuels and chemicals (Almpantis & Zabaniotou, 2021).

Understanding the properties of syngas, including its heating value and the presence of impurities such as tar, particulates, and heavy metals, poses a significant challenge. Construction and operation of gasification plants incur substantial initial costs. Gasification faces challenges, such as slower carbon conversion, which can affect the efficiency of the process and the quality of the syngas produced (Mohammadi & Anukam, 2023; Shahabuddin & Alam, 2022).

12.5 FUTURE PERSPECTIVES

The future of sustainable energy production and environmental conservation looks promising with the increased production of biochar. It offers crucial plant nutrients during crop cultivation while reducing nutrient loss. Exploring the use of biochar pellets as a potential strategy can help to lower handling costs and minimize the loss of biochar when it is incorporated into the soil. In the beginning, biochar pellets were utilized as a substitute for wood pellets. Nevertheless, they have demonstrated different benefits when compared to conventional household fuel usage and briquette technology. There are various advantages to consider, such as the ability to capture carbon, enhance soil quality, reduce pollution, and generate renewable energy.

12.6 CONCLUSION

In conclusion, the production of timber pellets and biochar, along with the utilization of heat generation technologies, presents a promising outlook for the advancement of environmentally friendly energy production and conservation. By utilizing biomass fuels, individuals can significantly reduce their reliance on fossil fuels while also contributing to carbon neutrality and waste management solutions. In addition, these techniques offer numerous advantages, such as reducing our reliance on non-renewable energy sources and minimizing CO_2 emissions. In addition, the production of wood pellets and biochar from various organic materials, such as crop residues, forest litter, sewage sludge, and animal manure, provides effective solutions for managing these types of waste. In addition, the introduction of biochar technology brings numerous benefits, including improved soil fertility, pollution control, and the generation of renewable energy.

REFERENCES

Adapa P., Tabil L. and Schoenau G. (2011). Grinding performance and physical properties of non-treated and steam exploded barley, canola, oat and wheat straw. *Biomass and Bioenergy*, **35**(1), 549–561, https://doi.org/10.1016/j.biombioe.2010.10.004

Adegoke I. A., Ogunsanwo O. Y. and Ige A. R. (2021). Bio-fuel properties and elemental analysis of bio-oil produced from pyrolysis of *Gmelina arborea*. *Acta Chemica Malaysia*, **5**(1), 38–41, https://doi.org/10.2478/acmy-2021-0006

Almpantis D. and Zabaniotou A. (2021). Technological solutions and tools for circular bioeconomy in low-carbon transition: simulation modeling of rice husks gasification for CHP by aspen PLUS V9 and feasibility study by aspen process economic analyzer. *Energies*, **14**(7), 2006, https://doi.org/10.3390/en14072006

Artemio C. P., Maginot N. H., Serafín C.-U., Rahim F. P., José Guadalupe R. Q. and Fermín C.-M. (2018). Physical, mechanical and energy characterization of wood pellets obtained from three common tropical species. *PeerJ*, **6**, e5504, https://doi.org/10.7717/peerj.5504

Asadullah M. (2014). Barriers of commercial power generation using biomass gasification gas: a review. *Renewable and Sustainable Energy Reviews*, **29**, 201–215, https://doi.org/10.1016/j.rser.2013.08.074

Balat M., Balat H. and Öz C. (2008). Progress in bioethanol processing. *Progress in Energy and Combustion Science*, **34**(5), 551–573, https://doi.org/10.1016/j.pecs.2007.11.001

Banerjee N. (2023). Biomass to energy – an analysis of current technologies, prospects, and challenges. *BioEnergy Research*, **16**(2), 683–716, https://doi.org/10.1007/s12155-022-10500-7

Barskov S., Zappi M., Buchireddy P., Dufreche S., Guillory J., Gang D., Hernandez R., Bajpai R., Baudier J., Cooper R. and Sharp R. (2019). Torrefaction of biomass: a review of production methods for biocoal from cultured and waste lignocellulosic feedstocks. *Renewable Energy*, **142**, 624–642, https://doi.org/10.1016/j.renene.2019.04.068

Benedetti V., Patuzzi F. and Baratieri M. (2018). Characterization of char from biomass gasification and its similarities with activated carbon in adsorption applications. *Applied Energy*, **227**, 92–99, https://doi.org/10.1016/j.apenergy.2017.08.076

Bishop J. D. K., Axon C. J., Tran M., Bonilla D., Banister D. and McCulloch M. D. (2012). Identifying the fuels and energy conversion technologies necessary to meet European passenger car emissions legislation to 2020. *Fuel*, **99**, 88–105, https://doi.org/10.1016/j.fuel.2012.04.045

Boquera L., Pons D., Fernández A. I. and Cabeza L. F. (2021). Characterization of supplementary cementitious materials and fibers to be implemented in high temperature concretes for thermal energy storage (TES) application. *Energies*, **14**(16), 5190, https://doi.org/10.3390/en14165190

Börcsök Z. and Pásztory Z. (2021). The role of lignin in wood working processes using elevated temperatures: an abbreviated literature survey. *European Journal of Wood and Wood Products*, **79**(3), 511–526, https://doi.org/10.1007/s00107-020-01637-3

Canam T., Town J. R., Tsang A., McAllister T. A. and Dumonceaux T. J. (2011). Biological pretreatment with a cellobiose dehydrogenase-deficient strain of *Trametes versicolor* enhances the biofuel potential of canola straw. *Bioresource Technology*, **102**(21), 10020–10027, https://doi.org/10.1016/j.biortech.2011.08.045

Cardozo E., Erlich C., Alejo L. and Fransson T. H. (2014). Combustion of agricultural residues: an experimental study for small-scale applications. *Fuel*, **115**, 778–787, https://doi.org/10.1016/j.fuel.2013.07.054

Carone M. T., Pantaleo A. and Pellerano A. (2011). Influence of process parameters and biomass characteristics on the durability of pellets from the pruning residues of *Olea europaea* L. *Biomass and Bioenergy*, **35**(1), 402–410, https://doi.org/10.1016/j.biombioe.2010.08.052

Castellano J. M., Gómez M., Fernández M., Esteban L. S. and Carrasco J. E. (2015). Study on the effects of raw materials composition and pelletization conditions on the quality and properties of pellets obtained from different woody and non woody biomasses. *Fuel*, **139**, 629–636, https://doi.org/10.1016/j.fuel.2014.09.033

Chaves Fernandes B. C., Mendes K. F., Júnior A. F. D., Caldeira V. P. da S., Teófilo T.M. da S., Silva T. S., Mendonça V., Souza M. de F. and Silva D. V. (2020). Impact of pyrolysis temperature on the properties of eucalyptus wood-derived biochar. *Materials*, **13**(24), 5841, https://doi.org/10.3390/ma13245841

Chen W.-H., Peng J. and Bi X. T. (2015). A state-of-the-art review of biomass torrefaction, densification and applications. *Renewable and Sustainable Energy Reviews*, **44**, 847–866, https://doi.org/10.1016/j.rser.2014.12.039

Criscuoli I., Ventura M., Sperotto A., Panzacchi P. and Tonon G. (2019). Effect of woodchips biochar on sensitivity to temperature of soil greenhouse gases emissions. *Forests*, **10**(7), 594, https://doi.org/10.3390/f10070594

Demirbas A. and Arin G. (2002). An overview of biomass pyrolysis. *Energy Sources*, **24**(5), 471–482, https://doi.org/10.1080/00908310252889979

Domene X., Enders A., Hanley K. and Lehmann J. (2015). Ecotoxicological characterization of biochars: role of feedstock and pyrolysis temperature. *Science of The Total Environment*, **512–513**, 552–561, https://doi.org/10.1016/j.scitotenv.2014.12.035

Elkhalifa S., Parthasarathy P., Mackey H. R., Al-Ansari T., Elhassan O., Mansour S. and McKay G. (2022). Biochar development from thermal TGA studies of individual food waste vegetables and their blended systems. *Biomass Conversion and Biorefinery*, **12**, 118, https://doi.org/10.1007/s13399-022-02441-0

Garcia-Maraver A., Zamorano M., Fernandes U., Rabaçal M. and Costa M. (2014). Relationship between fuel quality and gaseous and particulate matter emissions in a domestic pellet-fired boiler. *Fuel*, **119**, 141–152, https://doi.org/10.1016/j.fuel.2013.11.037

Ghorbani Z., Masoumi A. A. and Hemmat A. (2010). Specific energy consumption for reducing the size of alfalfa chops using a hammer mill. *Biosystems Engineering*, **105**(1), 34–40, https://doi.org/10.1016/j.biosystemseng.2009.09.006

Grodsky S. M., Moorman C. E., Fritts S. R., Castleberry S. B. and Wigley T. B. (2016). Breeding, early-successional bird response to forest harvests for bioenergy. *PLoS ONE*, **11**(10), e0165070, https://doi.org/10.1371/journal.pone.0165070

Guo M., Song W. and Buhain J. (2015). Bioenergy and biofuels: history, status, and perspective. *Renewable and Sustainable Energy Reviews*, **42**, 712–725, https://doi.org/10.1016/j.rser.2014.10.013

Han J. and Kim H. (2008). The reduction and control technology of tar during biomass gasification/pyrolysis: an overview. *Renewable and Sustainable Energy Reviews*, **12**(2), 397–416, https://doi.org/10.1016/j.rser.2006.07.015

Holm J. K., Henriksen U. B., Hustad J. E. and Sørensen L. H. (2006). Toward an understanding of controlling parameters in softwood and hardwood pellets production. *Energy & Fuels*, **20**(6), 2686–2694, https://doi.org/10.1021/ef0503360

Hu X. and Gholizadeh M. (2019). Biomass pyrolysis: a review of the process development and challenges from initial researches up to the commercialisation stage. *Journal of Energy Chemistry*, **39**, 109–143, https://doi.org/10.1016/j.jechem.2019.01.024

Igathinathane C., Womac A. R., Sokhansanj S. and Narayan S. (2009). Size reduction of high- and low-moisture corn stalks by linear knife grid system. *Biomass and Bioenergy*, **33**(4), 547–557, https://doi.org/10.1016/j.biombioe.2008.09.004

Ivanovski M., Petrovič A., Goričanec D., Urbancl D. and Simonič M. (2023). Exploring the properties of the torrefaction process and its prospective in treating lignocellulosic material. *Energies*, **16**(18), 6521, https://doi.org/10.3390/en16186521

Jackowski M., Semba D., Trusek A., Wnukowski M., Niedzwiecki L., Baranowski M., Krochmalny K. and Pawlak-Kruczek H. (2019). Hydrothermal carbonization of brewery's spent grains for the production of solid biofuels. *Beverages*, **5**(1), 12, https://doi.org/10.3390/beverages5010012

Jarolin K., Jarolin K., Wang S., Dymala T., Song T., Heinrich S., Shen L. and Dosta M. (2023). Characterizing devolatilized wood pellets for fluidized bed applications. *Biomass Conversion and Biorefinery*, **13**(6), 5133–5150, https://doi.org/10.1007/s13399-021-01486-x

Jiao Y., Tian L., Yu S., Song X., Wu Z., Wei J. and Xu J. (2023). AAEM species migration/transformation during co-combustion of carbonaceous feedstocks and synergy behavior on co-combustion reactivity: a critical review. *Energies*, **16**(22), 7473, https://doi.org/10.3390/en16227473

Kirsten C., Lenz V., Schröder H.-W. and Repke J.-U. (2016). Hay pellets – the influence of particle size reduction on their physical–mechanical quality and energy demand during production. *Fuel Processing Technology*, **148**, 163–174, https://doi.org/10.1016/j.fuproc.2016.02.013

Lam P. S., Sokhansanj S., Bi X., Lim C. J. and Melin S. (2011). Energy input and quality of pellets made from steam-exploded Douglas fir (*Pseudotsuga menziesii*). *Energy & Fuels*, **25**(4), 1521–1528, https://doi.org/10.1021/ef101683s

Lam P. S., Lam P. Y., Sokhansanj S., Bi X. T. and Lim C. J. (2013). Mechanical and compositional characteristics of steam-treated Douglas fir (*Pseudotsuga menziesii* L.) during pelletization. *Biomass and Bioenergy*, **56**, 116–126, https://doi.org/10.1016/j.biombioe.2013.05.001

Lee J., Lee K., Sohn D., Kim Y. M. and Park K. Y. (2018). Hydrothermal carbonization of lipid extracted algae for hydrochar production and feasibility of using hydrochar as a solid fuel. *Energy*, **153**, 913–920, https://doi.org/10.1016/j.energy.2018.04.112

Lehmann J., Kuzyakov Y., Pan G. and Ok Y. S. (2015). Biochars and the plant–soil interface. *Plant and Soil*, **395**(1–2), 1–5, https://doi.org/10.1007/s11104-015-2658-3

Leng L., Huang H., Li H., Li J. and Zhou W. (2019). Biochar stability assessment methods: a review. *Science of The Total Environment*, **647**, 210–222, https://doi.org/10.1016/j.scitotenv.2018.07.402

Li Y. (2000). High-pressure densification of wood residues to form an upgraded fuel. *Biomass and Bioenergy*, **19**(3), 177–186, https://doi.org/10.1016/S0961-9534(00)00026-X

Liang M., Lu L., He H., Li J., Zhu Z. and Zhu Y. (2021). Applications of biochar and modified biochar in heavy metal contaminated soil: a descriptive review. *Sustainability*, **13**(24), 14041, https://doi.org/10.3390/su132414041

Mani S., Tabil L. G. and Sokhansanj S. (2004). Grinding performance and physical properties of wheat and barley straws, corn stover and switchgrass. *Biomass and Bioenergy*, **27**(4), 339–352, https://doi.org/10.1016/j.biombioe.2004.03.007

Marinescu M. and Bush T. (2009). Wood to Energy: Use of the Forest Biomass for Wood Pellets. Florida Cooperative Extension Service, Gainesville, University of Florida, United States.

Mathiesen B. V., Lund H. and Karlsson K. (2011). 100% Renewable energy systems, climate mitigation and economic growth. *Applied Energy*, **88**(2), 488–501, https://doi.org/10.1016/j.apenergy.2010.03.001

Miao Z., Grift T. E., Hansen A. C. and Ting K. C. (2011). Energy requirement for comminution of biomass in relation to particle physical properties. *Industrial Crops and Products*, **33**(2), 504–513, https://doi.org/10.1016/j.indcrop.2010.12.016

Mohammadi A. (2021). Overview of the benefits and challenges associated with pelletizing biochar. *Processes*, **9**(9), 1591, https://doi.org/10.3390/pr9091591

Mohammadi A. and Anukam A. (2023). The technical challenges of the gasification technologies currently in use and ways of optimizing them: a review. In: Latest Research on Energy Recovery (chapter 5), P. Vizureanu (ed.), IntechOpen, London, UK. https://doi.org/10.5772/intechopen.102593

Mola-Yudego B., Selkimäki M. and González-Olabarria J. R. (2014). Spatial analysis of the wood pellet production for energy in Europe. *Renewable Energy*, **63**, 76–83, https://doi.org/10.1016/j.renene.2013.08.034

Mumme J., Eckervogt L., Pielert J., Diakité M., Rupp F. and Kern J. (2011). Hydrothermal carbonization of anaerobically digested maize silage. *Bioresource Technology*, **102**(19), 9255–9260, https://doi.org/10.1016/j.biortech.2011.06.099

Neves D., Thunman H., Matos A., Tarelho L. and Gómez-Barea A. (2011). Characterization and prediction of biomass pyrolysis products. *Progress in Energy and Combustion Science*, **37**(5), 611–630, https://doi.org/10.1016/j.pecs.2011.01.001

Palansooriya K. N., Ok Y. S., Awad Y. M., Lee S. S., Sung J. K., Koutsospyros A. and Moon D. H. (2019). Impacts of biochar application on upland agriculture: a review. *Journal of Environmental Management*, **234**, 52–64, https://doi.org/10.1016/j.jenvman.2018.12.085

Pilon G. and Lavoie J.-M. (2014). Biomass Char Production at Low Severity Conditions under CO_2 and N_2 Environments, **1**, pp. 67–78, WIT Press, Southampton, UK. https://doi.org/10.2495/978-1-78466-034-5/007

Powlson D. S., Whitmore A. P. and Goulding K. W. T. (2011). Soil carbon sequestration to mitigate climate change: a critical re-examination to identify the true and the false. *European Journal of Soil Science*, **62**(1), 42–55, https://doi.org/10.1111/j.1365-2389.2010.01342.x

Pradhan P., Arora A. and Mahajani S. M. (2018a). Pilot scale evaluation of fuel pellets production from garden waste biomass. *Energy for Sustainable Development*, **43**, 1–14, https://doi.org/10.1016/j.esd.2017.11.005

Pradhan P., Mahajani S. M. and Arora A. (2018b). Production and utilization of fuel pellets from biomass: a review. *Fuel Processing Technology*, **181**, 215–232, https://doi.org/10.1016/j.fuproc.2018.09.021

Rasaq W. A., Golonka M., Scholz M. and Białowiec A. (2021). Opportunities and challenges of high-pressure fast pyrolysis of biomass: a review. *Energies*, **14**(17), 5426, https://doi.org/10.3390/en14175426

Ren J., Qian Z., Wang X., Huang W. and Wang B. (2024). Investigation of a biomass-driven cogeneration system integrated with an externally fired gas turbine, organic Rankine cycle, and absorption refrigeration cycle: thermodynamic and exergoeconomic analyses and optimization. *Sustainability*, **16**(11), 4495, https://doi.org/10.3390/su16114495

Safin R. R., Safiullina A. K. and Nazipova F. V. (2017). Increase in energy efficiency of use of vegetable waste. *IOP Conference Series: Earth and Environmental Science*, **87**, 032025, https://doi.org/10.1088/1755-1315/87/3/032025

Sakuragi K. and Otaka M. (2020). Effect of biomass carbonization on the grinding of coal/biomass mixtures. *ACS Omega*, **5**(34), 21722–21727, https://doi.org/10.1021/acsomega.0c02629

Sarkar N., Ghosh S. K., Bannerjee S. and Aikat K. (2012). Bioethanol production from agricultural wastes: an overview. *Renewable Energy*, **37**(1), 19–27, https://doi.org/10.1016/j.renene.2011.06.045

Seghman P., Krátký L. and Jirout T. (2022). Effect of membrane separation process conditions on the recovery of syngas components. *Acta Polytechnica*, **62**(3), 394–399, https://doi.org/10.14311/AP.2022.62.0394

Shahabuddin M. and Alam T. (2022). Gasification of solid fuels (coal, biomass and MSW): overview, challenges and mitigation strategies. *Energies*, **15**(12), 4444, https://doi.org/10.3390/en15124444

Shaw M. D., Karunakaran C. and Tabil L. G. (2009). Physicochemical characteristics of densified untreated and steam exploded poplar wood and wheat straw grinds. *Biosystems Engineering*, **103**(2), 198–207, https://doi.org/10.1016/j.biosystemseng.2009.02.012

Song G., Zhao L., Zhao H., Xiao J., Wang H. and Guo S. (2020). Design and assessment of a novel cogeneration process of synthetic natural gas and char via biomass pyrolysis-coupled hydrothermal gasification. *Industrial & Engineering Chemistry Research*, **59**(51), 22205–22214, https://doi.org/10.1021/acs.iecr.0c04504

Stelte W., Holm J. K., Sanadi A. R., Barsberg S., Ahrenfeldt J. and Henriksen U. B. (2011). Fuel pellets from biomass: the importance of the pelletizing pressure and its dependency on the processing conditions. *Fuel*, **90**(11), 3285–3290, https://doi.org/10.1016/j.fuel.2011.05.011

Theerarattananoon K., Xu F., Wilson J., Ballard R., Mckinney L., Staggenborg S., Vadlani P., Pei Z. J. and Wang D. (2011). Physical properties of pellets made from sorghum stalk, corn stover, wheat straw, and big bluestem. *Industrial Crops and Products*, **33**(2), 325–332, https://doi.org/10.1016/j.indcrop.2010.11.014

Thek G. and Obernberger I. (2012). The Pellet Handbook – The Production and Thermal Utilization of Biomass Pellets. Routledge, New York, USA.

Tooyserkani Z., Sokhansanj S., Bi X., Lim C. J., Saddler J., Lau A., Melin S., Lam P. S. and Kumar L. (2012). Effect of steam treatment on pellet strength and the energy input in pelleting of softwood particles. *Transactions of the ASABE*, **55**(6), 2265–2272, https://doi.org/10.13031/2013.42484

Tripathi M., Sahu J. N. and Ganesan P. (2016). Effect of process parameters on production of biochar from biomass waste through pyrolysis: a review. *Renewable and Sustainable Energy Reviews*, **55**, 467–481, https://doi.org/10.1016/j.rser.2015.10.122

Tumuluru J. S., Wright C. T., Kenney K. L. and Hess R. J. (2010). A Technical Review on Biomass Processing: Densification, Preprocessing, Modeling and Optimization. 2010 Pittsburgh, Pennsylvania, June 20–June 23, 2010, St. Joseph, MI, American Society of Agricultural and Biological Engineers, https://doi.org/10.13031/2013.29874

Tumuluru J. S., Wright C. T., Hess J. R. and Kenney K. L. (2011). A review of biomass densification systems to develop uniform feedstock commodities for bioenergy application. *Biofuels, Bioproducts and Biorefining*, **5**(6), 683–707, https://doi.org/10.1002/bbb.324

Variny M., Varga A., Rimár M., Janošovský J., Kizek J., Lukáč L., Jablonský G. and Mierka O. (2021). Advances in biomass co-combustion with fossil fuels in the European context: a review. *Processes*, **9**(1), 100, https://doi.org/10.3390/pr9010100

Wang L., Barta-Rajnai E., Skreiberg Ø., Khalil R., Czégény Z., Jakab E., Barta Z. and Grønli M. (2017). Impact of torrefaction on woody biomass properties. *Energy Procedia*, **105**, 1149–1154, https://doi.org/10.1016/j.egypro.2017.03.486

Wang D., Jiang P., Zhang H. and Yuan W. (2020). Biochar production and applications in agro and forestry systems: a review. *Science of The Total Environment*, **723**, 137775, https://doi.org/10.1016/j.scitotenv.2020.137775

Wöhler M., Jaeger D., Reichert G., Schmidl C. and Pelz S. K. (2017). Influence of pellet length on performance of pellet room heaters under real life operation conditions. *Renewable Energy*, **105**, 66–75, https://doi.org/10.1016/j.renene.2016.12.047

Xiao Z., Yuan X., Jiang L., Chen X., Li H., Zeng G., Leng L., Wang H. and Huang H. (2015). Energy recovery and secondary pollutant emission from the combustion of co-pelletized fuel from municipal sewage sludge and wood sawdust. *Energy*, **91**, 441–450, https://doi.org/10.1016/j.energy.2015.08.077

Yazdani M. R., Duimovich N., Tiraferri A., Laurell P., Borghei M., Zimmerman J. B. and Vahala R. (2019). Tailored mesoporous biochar sorbents from pinecone biomass for the adsorption of natural organic matter from lake water. *Journal of Molecular Liquids*, **291**, 111248, https://doi.org/10.1016/j.molliq.2019.111248

Zakharevich A. V., Velikanov V. D. and Vympin E. A. (2015). The ignition fragment wood heated to a high temperature of the plate. *MATEC Web of Conferences*, **23**, 01018, https://doi.org/10.1051/matecconf/20152301018

doi: 10.2166/9781789064100_0233

Chapter 13

A review of extraction of biodiesel from non-edible oils and microalgae using nanocatalysts

Sindhu Subramanian[1], V. M. Vijay Kirushnan[2], A. Amritha[3], V. S. Sanjay Kumaar[4], N. V. Shobika[5], P. Dhamothar[6] and Swetha S.[7]*

[1,2,3,4,5,6,7]*Department of Chemical Engineering and Materials Science, Amrita School of Engineering, Amrita Vishwa Vidyapeetham, Coimbatore, India.*

**Corresponding author. E-mail: s_sindhu@cb.amrita.edu*

ABSTRACT

With rapidly on-going energy demand and fast depletion of fossil fuels, the option of choosing biodiesel holds the upper-hand. Biodiesel production from edible oils was observed to be not feasible by industry due to the usage of feedstocks as food resources. Recently research using heterogeneous catalysts for biodiesel production from the non-edible oils and algal oil is considered way superior to homogeneous catalysts due to the lower selectivity, activity, recovery, and reusability of catalysts. This chapter aims to review critically a portion of the usage of non-edible oils and algal oils for biodiesel production via heterogeneous catalysts and nanoheterogeneous catalysts. It includes the preparation and characterization of catalysts, optimized conditions for alkyl ester production, and characterization of biodiesel. It also includes the selection of suitable catalysts for industrial application.

Keywords: Algal oil, biodiesel, characterization, heterogeneous catalyst, nanoheterogeneous catalyst, non-edible oil, properties, transesterification.

13.1 INTRODUCTION

In today's world, the major sources of energy are petrol, natural gas, and coal. As there is limited availability of non-renewable energy, it takes thousands of years to restore once it is exhausted. Researchers are searching for the best renewable energy alternatives to meet this demand because they preserve resources, do not harm the environment, and lessen reliance on imported fuels. Biodiesel has been found to be one of the best alternatives because it is environmentally friendly and easily affordable. Currently, there are four generations of biofuel manufacturing. Food, edible oils such as soybean, sunflower, oil palm, and so on are the main feedstocks in the first generation for producing different biofuels such as ethanol, biodiesel, and vegetable oil. It was challenging to deal with rising food prices and increased competition for arable land in the first generation due to population growth and rising consumption levels. As a result, the cost of food on the market rose. For this reason, second-generation biofuels are produced from non-edible plant materials such as wood, husk, jatropha, and so on. The third-generation biofuels from algae and microbes have the potential to yield 15–300 times more than the yield from the first and second generations. Algae also exhibit higher growth rate tendencies and an accumulation of high cell lipids. Therefore, third-generation fuels are beneficial for the production and use of biodiesel, biomethane, and biohydrogen (Aniruddha *et al.*, 2022; Mofijur *et al.*, 2021; Sree *et al.*, 2021).

The following procedures are used to produce biodiesel from non-edible oil and microalgae: the oil extracted from non-edible oils or algae will undergo a transesterification process with the help of catalysts. Here, esters are created as the biofuel is the product of a straightforward displacement reaction, with the byproduct being glycerol. In this chapter, we elaborate about the procedures for creating different types of heterogeneous catalysts, their characteristics, and how they affect the production of biodiesel.

13.2 PREPARATION AND CHARACTERIZATION OF CATALYSTS

Microalgae and non-edible oil are used to produce biodiesel, but the process is extremely slow. To speed up the reaction, a catalyst is used. Catalysts are of two types, namely homogeneous and heterogeneous. Homogenous catalysts, which provide a high yield and are also commercially feasible, are those that are in the same phase as a reactant. With respect to selectivity and reactivity, a homogenous catalyst is highly active. Examples of homogenous catalysts are NaOH, KOH, and so on. The primary drawback of homogenous catalysts is their extreme difficulty in separating from the mixture of reactants. Catalysts that exhibit distinct phases based on reactants are known as heterogeneous catalysts. Heterogeneous catalysts, such as CaO, MgO, and so on, readily recover from the reaction mixture and yield a high yield (Shobana *et al.*, 2020). Heterogeneous catalysts are classified as either base or acid. Base heterogenous catalysts can be either carbonates or transitional, alkali, mixed metal oxides. Anionic resins, zeolitic materials, and sulfated zirconia are used as acid heterogeneous catalysts. A heterogeneous catalyst that was created artificially has a high yield and is frequently used in the manufacturing of biodiesel (Akubude *et al.*, 2019).

13.2.1 Preparation of heterogeneous catalysts for second-generation feedstocks

The heterogeneous CaO–Bi_2O_3 catalyst, which is used to produce biodiesel from *Monotheca buxifolia*, was obtained by impregnation using leftover egg shells and bismuth nitrate pentahydrate ($Bi(NO_3)_3{\cdot}5H_2O$) (Farooq *et al.*, 2022). Biodiesel is extracted from royna oil using KOH that is created using a traditional method (Rahman *et al.*, 2022). Biodiesel is produced from silk-cotton seeds and CaO, which is obtained from $CaCO_3$ (Rahul Soosai *et al.*, 2022). A wet impregnation method was used for preparing $Na/SiO_2/TiO_2$ that serves as a catalyst to produce biodiesel from non-edible wild olive oil (Khan *et al.*, 2022).

13.2.2 Preparation of heterogeneous catalysts for third-generation feedstocks

Sweet potato leaves can be utilized as a heterogeneous catalyst to extract biodiesel from *Scenedesmus obliquus*. The unexploited sweet potato leaves were pretreated and calcinated to obtain the desired nature of catalysts (Eldiehy *et al.*, 2022). In a two-step process, the de-oiled microalgal biomass is carbonized, and then sulfuric acid is added to the carbonized biochar to yield biodiesel from *Tetradesmus obliquus* KMC24. This is accomplished by means of a heterogeneous catalyst produced from de-oiled microalgal biomass (Roy & Mohanty, 2021).

13.2.3 Preparation of heterogeneous nanocatalysts for second-generation feedstocks

Biodiesel is produced from neem oil using Fe, Ni, and Cu-doped ZnO prepared through the co-precipitation method (Noreen *et al.*, 2021). The CaO nanocatalyst, derived from calcium nitrate tetrahydrate and ethylene glycol solution, is employed in the production of biodiesel from custard apple seeds (Sivanesh *et al.*, 2022). *Prunus bokhariensis* can be utilized for biodiesel production with the assistance of Ag_2O nanocatalyst, which is created using a green method involving the combination of *M. buxifolia* leaves extract and $AgNO_3$ (Dawood *et al.*, 2022). The synthesis involves hydrothermal preparation of calcium chloride with $FeCl_3 \cdot 6H_2O$, $CoSO_4$, and water, followed by calcination to produce a bi-functional Co-doped Fe_2O_3–CaO nanocatalyst used for converting soybean oil into methyl ester (Xia *et al.*, 2022). Caper seeds can be utilized in biodiesel production with a copper nickel-doped zirconium dioxide catalyst. This catalyst is created using the in-situ wet impregnation method, where zirconia is prepared and doping salts are added (Munir *et al.*, 2021b). The zinc-doped calcium oxide and eggshell nanocatalyst can serve as a catalyst for the production of biodiesel from castor oil (Dalvand & Mahdavian, 2022; Naveenkumar & Baskar, 2021). *M. buxifolia* oil can serve as a raw material for producing green nanoparticles of CaO derived from leaves of *Boerhavia procumbens* and $CaSO_4$ (Rozina *et al.*, 2022). *Celastrus paniculatus* wild seed oil is utilized for biodiesel production with the assistance of a trimetallic (Ce, Cu, La) loaded montmorillonite clay catalyst that is created through an in-situ wet impregnation method (Munir *et al.*, 2021a). *Brachychiton populneus* can be utilized for biodiesel production using an NiO nanocatalyst derived from nickel sulfate and latex extract (Dawood *et al.*, 2021).

13.2.4 Preparation of heterogeneous nanocatalysts for third-generation feedstocks

The nano-CaO produced using the sol–gel technique is suitable for converting *Euglena sanguinea* into biodiesel through transesterification (Sivanesh *et al.*, 2022). *Chlorella vulgaris* was used to produce fatty acid methyl ester (FAME) through transesterification with catalysts TiO_2 and $SrTiO_3$. The strontium titanate is synthesized by a photochemical method. The highest yield of FAME was obtained from $SrTiO_3$ (Aghilinategh *et al.*, 2020). *C. vulgaris* can undergo transesterification using CaO derived from eggshell calcination to produce a high yield of FAME. The CaO catalyst can be produced using various concentrations of $Ca(NO_3)_2$ (0.1, 0.5, 1.0, and 1.5 M), with the optimal catalyst being derived from a concentration of 0.5 M of $Ca(NO_3)_2$ (Davoodbasha *et al.*, 2021). Transesterification of *Coelastrella* M-60 was conducted using three catalysts: conventional acid catalyst, TiO_2, $CaCO_3$, and CaO individually. The $CaCO_3$ catalyst was synthesized through a hydrothermal process. CaO produced the highest FAME yield, approximately 95% (Karpagam *et al.*, 2020). The bio-iron nanocatalyst which was extracted from the seaweed was used for the transesterification of *Coelastrella* M-60 and *Dictyococcus* sp. VSKA18. In total, 90% of catalyst was recovered in the *Dictyococcus* sp. VSKA18 (Sakthi Vignesh *et al.*, 2020). *Nannochloropsis oculata* undergoes transesterification in the presence of NiO prepared by the sol–gel method, NaOH, and HCl separately. Among the catalysts used for transesterification of *N. oculata*, the NiO catalyst gave high yield (Purkan *et al.*, 2021). The study utilized BG-11 for transesterification with a SO_4^{2-}/Fe_3O_4–Al_2O_3 catalyst prepared through the wet-impregnation method. Three different magnetite contents were tested: 0.15, 0.25, and 0.35. The optimal catalyst content was determined to be 0.25 (Safakish *et al.*, 2020).

13.2.5 Characterization of heterogeneous catalysts for second-generation feedstocks

Scanning electron microscopy (SEM) analysis was conducted on CaO–Bi_2O_3, revealing irregularly structured particles with a heterogeneous distribution (Farooq *et al.*, 2022). The SEM image of CaO shows the presence of porous surface (Rahul Soosai *et al.*, 2022). When SEM was conducted on Na/SiO_2/TiO_2, the particles exhibited a smooth, spherical configuration and were agglomerated with irregular shapes (Khan *et al.*, 2022).

In X-ray diffraction (XRD) analysis, the 2θ peaks for CaO–Bi_2O_3 were observed between 20° and 46° (Farooq *et al.*, 2022). The 2θ values of CaO were present between 17° and 55° (Rahul Soosai *et al.*, 2022). The 2θ peaks appeared at 23.6° and 48° for Na/SiO_2/TiO_2 in the XRD analysis (Khan *et al.*, 2022).

The energy dispersive X-ray (EDX) analysis results of CaO–Bi_2O_3 indicated that the Bi_2O_3 was well distributed in CaO (Farooq *et al.*, 2022). The EDX analysis of CaO shows the sample is mostly composed of Ca and O compounds (Rahul Soosai *et al.*, 2022). The EDX confirms the successful fabrication of oxygen, silicon, and titanium on the Na/SiO_2/TiO_2 catalyst (Khan *et al.*, 2022). The Fourier-transform infrared (FT-IR) spectra of CaO were found at 3641.40 cm^{-1} (Rahul Soosai

Table 13.1 Properties of catalyst.

Catalyst	Surface Area (m^2/g)	References
K/Al_2O_3	20.46	Kazemifard *et al.* (2019)
K/Fe_2O_3–Al_2O_3	18.43	Zürich (2016)
$Ca(OCH_3)_2$	30	Teo *et al.* (2016)
CaO	19.810	Sivanesh *et al.* (2022)
$CaCO_3$	73.519	Karpagam *et al.* (2020)
Iron nanoparticle	15.542	Sakthi Vignesh *et al.* (2020)
2S Fe–Al (0.25)	11.79 ± 0.07	Safakish *et al.* (2020)
CaO	16.4	Pandit and Fulekar (2017)
KF/KOH–Fe_3O_4	315.613	Farrokheh *et al.* (2022)
CaO–Bi_2O_3	16.7726	Farooq *et al.* (2022)

et al., 2022). A rare catalyst combination of both calcium and bismuth oxides gave a surface area of 16.772 m^2/g (Farooq *et al.*, 2022). Table 13.1 illustrates the surface area of various nanocatalysts used in the production of biodiesel from non-edible oils.

13.2.6 Characterization of heterogeneous catalysts for third-generation feedstocks

The SEM image of calcinated sweet potato leaves displayed a crystalline structure with indeterminate shapes and sizes (Eldiehy *et al.*, 2022). The SEM analysis revealed a non-porous structure in the de-oiled biomass (Roy & Mohanty, 2021). In XRD, the 2θ peaks of the calcinated sweet potato leave green catalyst were observed between 25° and 60° which indicated the respective crystals (Eldiehy *et al.*, 2022). The 2θ peaks of the catalyst peaks were in the range of 35–50° for the de-oiled biomass catalyst (Roy & Mohanty, 2021). The EDX analysis showed an abundance of alkaline and alkali earth metal viz. potassium, calcium, magnesium, silica, chlorine, and sodium (Eldiehy *et al.*, 2022). The carbon, oxygen, nitrogen, and sulfur were observed in the de-oiled biomass catalyst (Roy & Mohanty, 2021). The FT-IR peaks at 1600 and 1376 cm^{-1} indicated the presence of polyaromatic compounds, peaks at 1027 cm and 1200 cm^{-1} confirmed the presence of $-SO_3H$ group, and peaks at 1722 cm^{-1} indicated the presence of the C=O group in the de-oiled biomass catalyst (Roy & Mohanty, 2021).

13.2.7 Characterization of heterogeneous nanocatalysts for second-generation feedstocks

An SEM analysis was conducted on the Ag_2O nanocatalyst, revealing a hexagonal structure with minor clustering. The hexagonal structure indicates the presence of polymeric and magnetic interfaces on the catalytic surface (Dawood *et al.*, 2022). The SEM revealed a particle size between 40 and 80 nm for Co/Fe_2O_3–CaO. The Fe_2O_3–CaO particles were irregular in shape and had pores within the grains (Xia *et al.*, 2022). The SEM image shows green CaO nanoparticles with a square-shaped structure, having an average particle size ranging from 22 to 26 nm in diameter (Rozina *et al.*, 2022). The SEM image of the unmodified clay displayed a fluffy and densely packed layered structure. Surface morphology results revealed a spherical and cubic structure with a particle size of 32 nm for green nickel oxide (Dawood *et al.*, 2021).

An XRD analysis confirmed the crystalline structure of zirconium dioxide by detecting diffraction peaks at 0.13°, 35.33°, 49.89°, and 50.87°. The copper–nickel oxide spectra displayed peaks at 2θ values of 37.18°, 43.19°, and 62.86°. The Debye–Scherrer equation was used to determine the diameter to be 39.28 nm (Munir *et al.*, 2021b). The crystal structure of nano-CaO was verified using the XRD analysis. Sharp peaks were detected at 31°, 32.34°, 34°, 37°, 54°, and 63.47° within the scanning range of 10–90° (Rozina *et al.*, 2022; Sivanesh *et al.*, 2022). The nickel oxide nanocatalyst displayed three distinct peaks at 37.42°, 43.4°, and 62.89° in the XRD analysis, indicating its crystalline structure

(Dawood *et al.*, 2021). Distinct diffraction peaks were observed for the montmorillonite clay at 5.7°, 19.91°, 29.21°, and 34.46°. Peaks at 25.3°, 29.2°, and 44.6° are associated with La_2O_3, whereas peaks at 33.6°, 47.5°, and 56.4° are linked to CeO_2. The total intensity of the sharp peaks in these metal oxides is significantly lower than what was expected based on the pattern. The metals lanthanum, cerium, and copper oxide were found to have settled in the spaces (Rozina *et al.*, 2022).

An EDX analysis reveals the catalyst's atomic composition, which consists of oxygen, iron, carbon, molybdenum, sodium, cobalt, and sulfur (Munir *et al.*, 2021b). Only the atomic percentages of oxygen and calcium were detected in the CaO nanocatalyst (Rozina *et al.*, 2022). Cerium had a loading of approximately 7.47% and lanthanum had a loading of around 7.98% in the catalyst, whereas copper had a lower loading of 2.62% in montmorillonite clay. Elements such as sodium, potassium, calcium, silicon, and oxygen are detected in clay following modification through EDX analysis. The detection of zirconium and oxygen peaks confirms the existence of these elements in the catalyst (Munir *et al.*, 2021a). The EDX analysis of the nickel oxide nanocatalyst reveals only the atomic percentages of oxygen and nickel (Dawood *et al.*, 2021).

The FT-IR peaks at 111 and 200 cm^{-1} indicate the existence of the CaO lattice in the Co/Fe_2O_3–CaO catalyst. The 3400–3300 cm^{-1} peak in Cu/Ni-doped ZrO indicates the –OH stretching, whereas the 500–400 cm^{-1} peak is attributed to metal oxides (Munir *et al.*, 2021b). Various peaks were detected in the FT-IR analysis of the CaO nanocatalyst, indicating the Ca–O stretch, hydroxyl group, CO_3, and the C=C stretch (Rozina *et al.*, 2022). The peaks at 1020 and 796 cm^{-1} in both the unmodified and modified clay catalysts indicate the presence of silicon–oxygen–silicon and silicon–oxygen–aluminum bonds interacting. At 694 cm^{-1}, there was also a small peak of aluminum–magnesium $(OH)_2$ (Munir *et al.*, 2021a). In the nickel oxide nanocatalyst, metal oxide stretching was shown in FT-IR (Dawood *et al.*, 2021).

13.2.8 Characterization of heterogeneous nanocatalysts for third-generation feedstocks

Nanoparticles in the form of spheres, rods, and sticks were observed in the SEM image of calcium oxide derived from eggshell, seashell, and calcium nitrate. The calcium oxide catalyst derived from calcium nitrate exhibited a particle size ranging from 200 to 400 nm. The nanoparticle size was confirmed based on the surface morphology of KF/KOH, strontium titanate, $Ca(OCH_3)_2$, lanthanum oxide, Fe_3O_4, and CaO (Davoodbasha *et al.*, 2021; Farrokheh *et al.*, 2022; Karpagam *et al.*, 2020; Pandit & Fulekar, 2017; Sivanesh *et al.*, 2022; Teo *et al.*, 2016; Vinoth Arul Raj *et al.*, 2021). The XRD analysis confirmed the crystal structure of CaO, KF/KOH, and Fe_3O_4 (Farrokheh *et al.*, 2022; Karpagam *et al.*, 2020; Sivanesh *et al.*, 2022; Vinoth Arul Raj *et al.*, 2021). The 2θ values ranging from 31° to 55°, 32.3° to 72.5°, 15.2° to 34.4°, 30.2° to 62.9°, 18.16° to 37.34°, and 37.22° to 94.92° were identified for CaO, strontium titanate, SO_4^{2-}/Fe_3O_4–Al_2O_3, Fe_3O_4, *Spirulina*, and *N. Oculata*, respectively (Aghilinategh *et al.*, 2020; Farrokheh *et al.*, 2022; Joseph *et al.*, 2017; Safakish *et al.*, 2020; Sivanesh *et al.*, 2022; Vinoth Arul Raj *et al.*, 2021; Wu & Miao, 2014). Strontium titanate shows a pattern between 32.397° and 72.567°, CaO between 10° and 90°, and KOH and KF_3 crystals between 15.2° and 34.4° (Aghilinategh *et al.*, 2020; Farrokheh *et al.*, 2022; Sivanesh *et al.*, 2022). The composition of $Ca_xMg_yCO_3$ was verified through XRD analysis. The Brunauer–Emmett–Teller (BET) analysis revealed the specific surface area of all catalysts. The average pore sizes of the materials mentioned were as follows: KF/KOH–Fe_3O_4 (2.5 nm), CaO (3.05 nm), iron nanoparticles (2.3 nm), $Ca(OCH_3)_2$ (3.347 nm), and $Ca(OCH_3)_2$ (1.18 nm) (Farrokheh *et al.*, 2022; Sakthi Vignesh *et al.*, 2020; Teo *et al.*, 2016; Vinoth Arul Raj *et al.*, 2021). The average pore volume of CaO, iron nanoparticles, and $Ca(OCH_3)_2$ were determined to be 0.09, 0.03, and 0.02 cm^3/g, respectively. EDX analysis detected the elements strontium, titanium, and oxygen in the strontium titanate nanocatalyst; calcium and oxygen in the CaO nanocatalyst; iron, oxygen, and carbon in the iron nanoparticles; and lanthanum and oxygen in the lanthanum oxide nanocatalyst (Aghilinategh *et al.*, 2020; Farrokheh *et al.*, 2022; Sivanesh *et al.*, 2022; Vinoth Arul Raj *et al.*, 2021). Properties of catalysts such as surface area and pore size are crucial in influencing the reaction rate.

The surface area of the Cao nanocatalyst was determined to be 19.810 m^2/g using BET spectrum analysis (Sivanesh *et al.*, 2022). The BET analysis indicated that the bio-nano-$CaCO_3$ had a specific surface area of 73.519 m^2/g (Karpagam *et al.*, 2020). The surface area of heterogeneous iron nanoparticles synthesized with *Sargassum polycystum* was determined to be 15.542 m^2/g (Sakthi Vignesh *et al.*, 2020). A mixture of microalgae containing various species is treated with an $SO_4^{2-}/Fe_3O_4–Al_2O_3$ magnetic nanocatalyst with a surface area of 11.79 ± 0.07 m^2/g (with a margin of error of 0.07) (Safakish *et al.*, 2020). The synthesized nanocatalyst CaO used for *Acutodesmus obliquus* resulted in a surface area of 16.4 m^2/g (Pandit & Fulekar, 2017). The surface area of the calcium methoxide catalyst was determined to be 30 m^2/g (Teo *et al.*, 2016). The BET analysis revealed that the KF/KOH–Fe_3O_4 catalyst has a surface area of 315.613 m^2/g, indicating that a large surface area enhances the transesterification process and boosts the catalyst's activity (Farrokheh *et al.*, 2022). The surface areas of K/Al_2O_3 and K/Fe_2O_3–Al_2O_3 were found to be 20.46 and 18.43 m^2/g (Vinoth Arul Raj *et al.*, 2021). Table 13.1 shows the surface area of different nanocatalysts utilized in the manufacture of biodiesel from algae.

13.3 FEEDSTOCK AVAILABILITY

Non-edible seeds were collected and lipids were extracted using a Soxhlet extractor. The lipid is extracted, pretreated, and then subjected to transesterification to produce biodiesel. Figure 13.1 illustrates the process of producing biodiesel from non-edible seeds.

Algae were gathered from various sources. The algae are cultivated in a photo-bioreactor with appropriate incubation time and intermittent flashing light. The light effect is created by strategically positioning a series of LEDs to evenly distribute light. The distributed light aids in the generation of a substantial amount of algae with a high cell density. The algae were centrifuged and the algal biomass was collected. The collected algal biomass was washed and dried at a suitable temperature. The algal oil was extracted from biomass with a suitable solvent. This algal oil is used for transesterification to produce biodiesel. The algal oil can also be extracted from a Folch extractor and Bligh and Dyer method. A schematic representation of the production of biodiesel from algal biomass is shown in Figure 13.2.

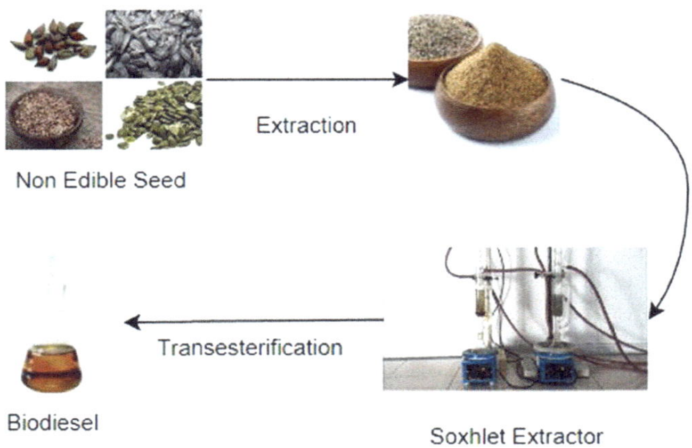

Figure 13.1 Schematic representation of production of biodiesel from non-edible seeds.

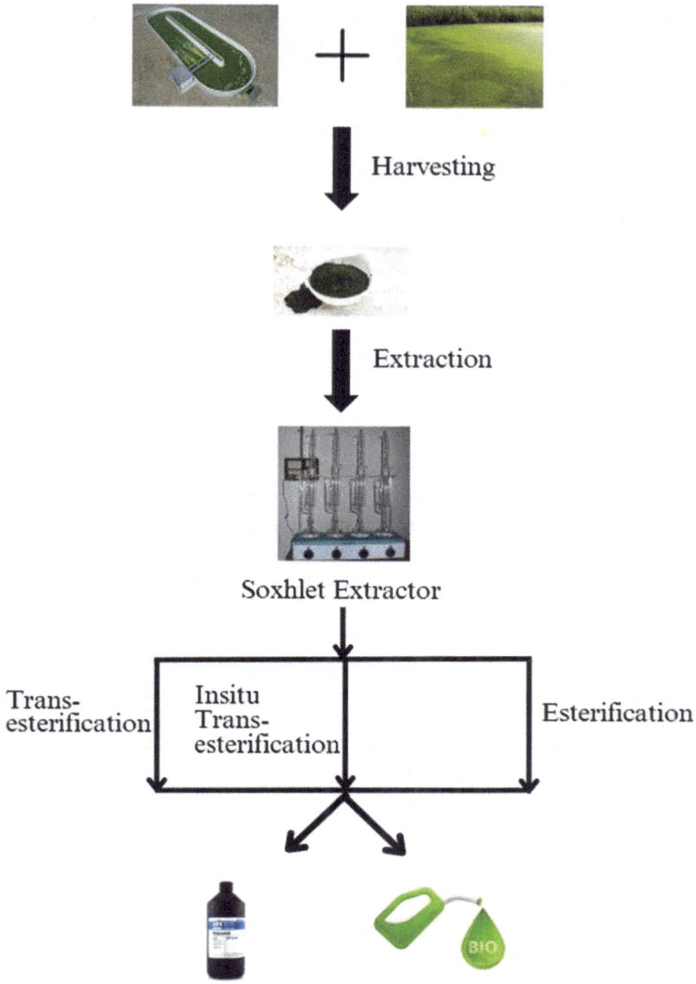

Figure 13.2 Schematic representation of production of biodiesel from microalgae.

13.3.1 Non-edible seeds

M. buxifolia is primarily found in Asia, Africa, and certain regions of Europe. *M. buxifolia* is a medicinal shrub used for treating gastro-urinary disorders (Farooq *et al.*, 2022; Rozina *et al.*, 2022). Royna seeds are cultivated in Asia. The royna plants thrive exclusively in fertile soil and are utilized as a food preservative (Rahman *et al.*, 2022). Neem oil and custard apple are predominantly found in Asia, Australia, and America. Animals do not consume custard apple seeds, and neem oil has numerous medicinal uses (Noreen *et al.*, 2021; Sivanesh *et al.*, 2022). *P. bokhariensis* is indigenous to Australia and certain regions of Asia. *P. bokhariensis* is a deciduous tree with extremely hard, inedible seeds. These are not consumed by animals (Dawood *et al.*, 2022). *Capparis spinosa* L. is a shrub that typically grows in rocky areas, old walls, and hilly slopes. These are used medicinally for gout and hemorrhoids (Munir *et al.*, 2021b). Southeast Asia, Australia, and numerous Pacific islands are the main habitats of *C. paniculatus*. The seed oil is used as a lamp fuel, in soap making, and as a lubricant.

It has numerous medical uses, including as an antibacterial, analgesic, antioxidant, and antimalarial agent (Munir *et al.*, 2021a).

13.3.2 Microalgae

C. vulgaris is a commonly used algae for the production of biodiesel and is easily cultivable in both fresh water and saltwater. Its lipid content is estimated to be between 28 and 32% (Aghilinategh *et al.*, 2020). *Coelastrella* sp. M-60 is a type of green, unicellular microalgae that grows in fresh water and is thermotolerant (Karpagam *et al.*, 2020). Freshwater algae *Coelastrella* sp. M-60 and *Dictyococcus* sp. VSKA18 were both utilized because they were readily available and economically viable for biodiesel (Sakthi Vignesh *et al.*, 2020). Another source for biodiesel production is the green freshwater algae *A. obliquus* (Pandit & Fulekar, 2017). A marine algae called *N. oculata*, which is mostly found in seas and has a lipid content of between 31 and 68%, is another source (Teo *et al.*, 2016). It is also possible to produce biodiesel from the two most common marine algae, *C. vulgaris* and *Spirulina platensis* (Farrokheh *et al.*, 2022).

13.4 BIODIESEL PRODUCTION

13.4.1 Biodiesel production from second-generation feedstocks using heterogeneous catalysts

With the help of CaO–Bi_2O_3, *M. buxifolia* is transesterified. For the transesterification of *M. buxifolia*, the optimal conditions are 65°C, 6% ZnO catalyst concentration, and a methanol-to-oil ratio of roughly 8:1 (Farooq *et al.*, 2022). Through a transesterification reaction, high yield of biodiesel is extracted from royna seed oils at a temperature of 70°C for 1 h, using a 1:7 alcohol-to-oil ratio and 1% catalyst (Rahman *et al.*, 2022).

13.4.2 Biodiesel production from third-generation feedstocks using heterogeneous catalysts

S. obliquus is transesterified for approximately 90 min at 60°C using a methanol-to-oil ratio of 15:1 and a 7 wt% catalyst. A yield of roughly 99.50% was obtained from the *S. obliquus* oil (Eldiehy *et al.*, 2022). *T. obliquus* is transesterified for approximately 6 h at 80°C using a 5 wt% catalyst and an alcohol-to-oil ratio of roughly 10:1. A yield of approximately 96.25% was obtained from the *T. obliquus* oil (Roy & Mohanty, 2021).

13.4.3 Biodiesel production from second-generation feedstocks using nanocatalysts

The non-edible oil is extracted from its respective sources and then it is subjected to transesterification. The optimal conditions for transesterification of neem oil are 65°C with 2% of ZnO catalyst concentration and a methanol-to-oil ratio of about 5:1 (Noreen *et al.*, 2021). Using the nano-CaO, the oil extracted from the custard apple seeds is transesterified to methyl ester (Sivanesh *et al.*, 2022). An alcohol-to-oil ratio of 12:1 and 3.5 wt% Ag_2O nanocatalyst continuously stirred for 2.5 h at 130°C reaction temperature are the suitable conditions for transesterification of 50 ml of *P. bokhariensis* seed oil, resulting in an 89% yield (Dawood *et al.*, 2022). An oil yield of 90.2% was obtained by using *C. spinosa* L. seed oil with an oil-to-methanol ratio of 1:6 at 70°C for 1.5 h, with copper–nickel-doped zirconium as the catalyst (Munir *et al.*, 2021b). The *M. buxifolia* oil undergoes transesterification with green CaO nanoparticles under a 9:1 methanol-to-oil molar ratio, 0.83 wt% catalyst, a 3-h reaction time, and a temperature of 85°C, resulting in a 95% yield. Transesterification of soybean oil with Co-doped Fe_2O_3–CaO resulted in a 98.2% yield. This was achieved using a methanol-to-oil molar ratio of 16:1, a catalyst percentage of 3 wt%, and a reaction time of 150 min at 70°C (Xia *et al.*, 2022). The *C. paniculatus* seed oil containing 3.5 wt% of montmorillonite clay catalyst is reacted at 120°C under optimal conditions with an oil-to-methanol ratio of 12:1 for 120 min to obtain a yield of 89.42% (Munir *et al.*, 2021a). The *B. populneus* undergoes transesterification with a green nickel oxide nanocatalyst under specific conditions: a methanol-to-oil ratio of 9:1, catalyst of 2.5 wt%, and a temperature of 85°C, resulting in a maximum yield of 97.5% (Dawood *et al.*, 2021).

13.4.4 Biodiesel production from third-generation feedstocks using nanocatalysts

Biodiesel can be produced by transesterification of lipids from algae with the use of a nanocatalyst. *C. vulgaris* undergoes transesterification with KF/KOH–Fe_3O_4 under various conditions. Optimal yield was achieved by adjusting the methanol-to-oil ratio, reaction time, and catalyst weight. The optimum conditions for the reaction were a 2-h duration with 1.5 wt% catalyst and a 6:1 methanol-to-oil ratio (Farrokheh *et al.*, 2022; Vinoth Arul Raj *et al.*, 2021). *Nannochloropsis salina* undergoes transesterification with nano-CaO. The maximum yield was obtained by varying the catalyst weight, methanol-to-oil ratio, temperature, and reaction time. The optimum conditions for the reaction were 3% of catalyst, methanol-to-oil ratio of 6:1, and temperature of 60°C for 55 min (Purkan *et al.*, 2021). *S. platensis* undergoes transesterification with $Ca(OCH_3)_2$ by changing the catalyst weight, methanol-to-oil ratio, temperature, and reaction time. The optimum conditions were 3% catalyst weight, 30:1 methanol-to-oil ratio, temperature of 80°C, and 3 h (Farrokheh *et al.*, 2022). The BG-11 medium (*C. vulgaris*, *Chlamydomonas debaryana*, *Euglena* sp., *Oscillatoria quadripunctata*, *Microcystis aeruginosa*, and *Scenedesmus bijuga*) undergoes transesterification with SO_4^{2-}/Fe_3O_4–Al_2O_3. The maximum yield was obtained by varying the temperature, catalyst weight, methanol-to-oil ratio, and reaction time. The optimum conditions for transesterification were 8 wt% of catalyst, 6 h of reaction time, methanol-to-algae of 9 ml/g, and temperature of 120°C (Safakish *et al.*, 2020). *Naviculla* sp. undergoes transesterification in the presence of lanthanum oxide to produce methyl esters. The yield of biodiesel is also affected by the nature of the catalyst. The nanocatalyst prepared from the algal mass is more effective than the waste cooking oil catalyst (Joseph *et al.*, 2017). The *Chlorella pyrenoidosa* undergoes transesterification in the presence of the Fe_3O_4 nanocatalyst to yield biodiesel at 45°C, 400 rpm, and methanol-to-oil ratio of about 4:1 (Wu & Miao, 2014). The production of biodiesel from *E. sanguinea* using the nano-calcium oxide catalyst takes place by a transesterification reaction for about 2 h and 10 min with methanol-to-oil ratio of 6:1 and 1 wt% catalyst (Sivanesh *et al.*, 2022).

In-situ transesterification is a process that involves combining the extraction of lipids from microalgae with the transesterification reaction, and performing them simultaneously. It decreases the duration required for lipid extraction and transesterification reaction (Safakish *et al.*, 2020). In this process, methanol plays an important dual role, as a solvent for lipid extraction from microalgae and as a reactant in transesterification reactions. Electrolysis is a process by which electric current is passed between electrodes to break down ionic substances resulting in chemical change. The FAME was produced through the transesterification of triglycerides using electrolysis. This method is cost-effective and has many advantages over reflux methods (Farrokheh *et al.*, 2022).

13.5 CHARACTERIZATION OF BIODIESEL

13.5.1 Characterization of biodiesel from second-generation feedstocks using heterogeneous catalysts

The FT-IR analysis of FAMEs obtained from the non-edible silk cotton seed oil confirmed the C=O group, C–O stretching, and C=C bending at 1737.86, 1172.72 and 721.38 cm^{-1} (Rahul Soosai *et al.*, 2022).

Gas chromatography coupled with mass spectrometry (GC-MS) was used to analyze the biodiesel extracted from *M. buxifolia*, revealing the presence of 19.50% palmitic acid, 17.28% tridecylic acid, 14.76% lauric acid, and 11.83% capric acid (Farooq *et al.*, 2022). GC-MS analysis of the biodiesel obtained from the non-edible silk cotton seed oils confirmed the presence of 24.77% of hexadecenoic acid, 25.25% of octadecenoic acid methyl ester and 34.53% of 9,12-octadecenoic acid (*Z,Z*) methyl ester (Rahul Soosai *et al.*, 2022). Non-edible wild olive oil FAME contains 61.40% 8-octadecenoic acid methyl ester, 22.43% hexadecanoic acid methyl ester, 13.627% oxiraneoctanoic acid methyl ester, and 1.115% docosanoic acid methyl ester (Khan *et al.*, 2022).

Proton nuclear magnetic resonance (^{1}H-NMR) analysis of the biodiesel obtained from the royna seed oils showed peaks at 3.7 and 2.3 ppm (Rahman *et al.*, 2022). ^{1}H-NMR and carbon-13 nuclear magnetic resonance confirmed the presence of methyl esters at the respective peaks in biodiesel of non-edible silk cotton seed oils (Rahul Soosai *et al.*, 2022).

13.5.2 Characterization of biodiesel from third-generation feedstocks using heterogeneous catalysts

GC-MS analysis confirmed the presence of 9-octadecenoic acid methyl ester, hexadecanoic acid methyl ester, and 9-octadecenoic methyl ester in FAME obtained from *S. obliquus* (Eldiehy *et al.*, 2022). The FAME of *T. obliquus* composed of palmitic acid (40.96%), arachidic acid (22.39%), tetracosanoic acid (12.09%), palmitoleic acid (8.17%), oleic acid (7.28%), erucic acid (2.12%), and linolenic acid (1.51%) (Roy & Mohanty, 2021).

^{1}H-NMR analysis of the biodiesel obtained from the *S. obliquus* revealed that the FAME formation was confirmed at 3.6 ppm (Eldiehy *et al.*, 2022).

13.5.3 Characterization of biodiesel from second-generation feedstocks using nanocatalysts

FT-IR study of biodiesel obtained from neem oil had peaks at 1159 cm^{-1} for stretching of C–O absorption as well as the presence of esters, 1459 cm^{-1} for methylene bending vibrations, 1377 cm^{-1} for the O–CH_2 bond, 1742 cm^{-1} for C=O methyl esters. The ester group peak was indicated at 1745.64 cm^{-1}. The C–H stretching for sp^2 and sp^3 occurs at 2924.18 and 2852.81 cm^{-1}, respectively. Bending vibrations of –CH_3, –CH_2, and O–H appear at 1464.02, 1377.22, and 3471.98 cm^{-1}, respectively (Noreen *et al.*, 2021). The methyl esters from *C. spinosa* L. seed oil show stretchings at 2922, 2853, and 1743 cm^{-1} for sp^2 –C–H, sp^3 –C–H, and C–O bonds in FT-IR. The peak at 1462 cm^{-1} relates to the –CH_2 vibration. The C–O stretching was observed at two places which occur due to the methoxycarbonyl group (Munir *et al.*, 2021b). The peaks for FAME from non-edible *C. paniculatus* wild seed oil are found at 1740, 2853, 2922, 1458, 1158, 1025, and 721.20 cm^{-1} that indicate the triglycerides in the oil. The CH_2 and CH_3 bonds were shown at 1362 and 1434 cm^{-1} peaks. The most prominent peak for FAME appears at 1435 cm^{-1} (Munir *et al.*, 2021a). The formed FAME from soya bean oil showed peaks at 2919.57 and 2850.58 cm^{-1} indicating unsaturated hydrocarbons and saturated methane group. The vibrations at 1460.52, 722.16, 1740.58, and 1162.79 cm^{-1} indicate the existence of in-plane with external bending vibration, C=O triglyceride esters, and a C–O bond (Dawood *et al.*, 2021).

GC-MS was performed on the biodiesel obtained from the custard apple seeds where the presence of 9-octadecenoic acid and tetradecyl ester of 282 and 286 g/mol molecular weights were confirmed (Sivanesh *et al.*, 2022). *M. buxifolia* products contain hexadecenoic acid methyl ester and octadecanoic acid methyl ester, with a high concentration of 9-octadecenoic acid methyl ester (Dawood *et al.*, 2022). The analyzed GC-MS results of caper seed oil biodiesel revealed the presence of saturated FAME fragments such as methyl tetra decanoate acid, hexadecenoic acid methyl ester, 7-hexadecenoic acid methyl ester, and 9-octadecanoic acid methyl ester (Munir *et al.*, 2021b). Biodiesel produced from *C. paniculatus* wild seed oil is found to have methyl esters of 10-octadecenoic acid, as well as notable fragmentation ions (Munir *et al.*, 2021a). The gas chromatogram of *B. populneus* non-edible oil showed the occurrence of hexadecanoic acid methyl ester, pentadecanoic acid, heptadecanoic ethyl ester, 9,12-octadecanoic acid, 9-octadecanoic acid, methyl stearate, oleic acid, linoleic acid, ethyl oleate, and *cis*-10-nonadecanoic acid (Dawood *et al.*, 2021). GC-MS of biodiesel obtained from mahua oil reveals the presence of palmitic acid, myristic acid, stearic acid, oleic acid, linoleic acid, linolenic acid, and arachidic acid fatty acids (Tukaram Bai *et al.*, 2021).

The chemical shifts in the H-NMR spectra of CPBD (4,4'-(9H-carbazole-2,7-diyl)bis(1-butylpyridin-1-ium) dibromide) reveal the long signals of biodiesel derived from *C. paniculatus* wild seed oil. The chemical shifts and signals in the carbon nuclear magnetic resonance (C-NMR) spectra of CPBD were identified (Munir *et al.*, 2021a). *C. spinosa* L. seed oil found the peaks corresponding to (–COO–), (–CH=CH–), (–C–O), and (–CH_2–)$_n$, respectively (Munir *et al.*, 2021b). H-NMR of biodiesel from *M.*

buxifolia indicates the signal peak of olefinic hydrogen (–CH=CH–) positioned at 5.277–5.394 ppm, whereas the peak of allylic hydrogen ($-CH_2-$) appeared at 2.799 ppm. In C-NMR analysis of the carbonyl (C–O) group in methyl ester was detected at 77.48 ppm, whereas the signal for long chain of ethylene carbon $(-CH_2-)_n$ appeared at 29.08–29.75 ppm. The signal for the carbonyl group (–COOH) was spotted at 174.21 ppm. The signals for unsaturation (–CH=CH–) in the biodiesel spectrum was positioned at 127.70 ppm corresponding to inner non-conjugated carbon (Rozina *et al.*, 2022).

The properties of the biodiesel obtained from the second-generation feedstocks are shown in Table 13.2.

Table 13.2 Properties of biodiesel from the second-generation feedstocks.

Source for Biodiesel	Properties of Biodiesel	Standards (ASTM and British)	References
Neem oil	Density: 0.892 g/cm³ Viscosity: 5.55 mm²/s Calorific value: 40.1 kJ/g Acid value: 0.11 mg KOH/g Cetane number: 50.63	ASTM D-6751	Noreen *et al.* (2021)
P. bokhariensis	Flash point: >93°C Density at 15°C: <120 kg/l Kinematic viscosity at 40°C: 1.9–6.0 Sulfur: <0.5 wt%	ASTM	Dawood *et al.* (2022)
Mahua oil	Density: 896 Kinematic viscosity at 40°C: 1.9–6 Flash point: min 130 Cetane number: min 47	ASTM	Tukaram Bai *et al.* (2021)
C. paniculatus wild seed oil	Density at 15°C: 0.86 Flash point: 7°C Pour point: −10°C Cloud point: −8°C	ASTMD-951, ASTM D-6751, GB-T, and EN-14214	Munir *et al.* (2021a)
Custard apple seeds	Density: 0.875 g/cm³ Viscosity: 5 mm²/s Acid value: 0.34 KOH/g Flash point: 158°C Pour point–: −15 to 10°C	ASTM D-6751	Sivanesh *et al.* (2022)
M. buxifolia seed oil	Cloud point: −8°C Pour point: −9°C Flash point: 95°C Density: 0.821 g/cm³		Dawood *et al.* (2022)
B. populneus oil	Density at 15°C: 0.876 kg/l Flash point: 100°C Pour point: −15°C Cloud point: −14°C	ASTM D-1298 ASTM D-93 ASTM D-97 ASTM D-2500	Dawood *et al.* (2021)

13.5.4 Characterization of biodiesel from third-generation feedstocks using nanocatalysts

From the FT-IR results of nano-CaO, the wavelength of 300–1000, 1000–1340, 1600–1900, and 3000–3700 cm^{-1} which indicates the presence of Ca–O, C–O, C=O, and O–H functional groups and C–H, Ca–O lattice, and C=O vibrations are observed at 632.65, 447, and 875.68 cm^{-1} (Sivanesh *et al.*, 2022). For nano-iron particles, sulfate, and CO_2 groups peaks are observed at 1085.9 and 1118.71 cm^{-1} and O–H, C–H, C=C, C–I stretching were also found (Sakthi Vignesh *et al.*, 2020). For $SO_4^{2-}/Fe_3O_4–Al_2O_4$, Al–O–H, AlO_6, and AlO_4 vibrations were observed at 3560, 400–650, and 650–850 cm^{-1} (Safakish *et al.*, 2020). For KF/KOH–Fe_2O_3, FT-IR shows peaks at 1164, 1171, and 1197 cm^{-1} that were observed due to the presence of triglyceride spectrum, methyl ester spectrum, and O–CH_3 vibration (Farrokheh *et al.*, 2022; Vinoth Arul Raj *et al.*, 2021). For $Ca(OCH_3)_2$

Table 13.3 Properties of biodiesel from the third-generation feedstocks.

Source for Biodiesel	Properties of Biodiesel	Standards (ASTM and British)	References
C. vulgaris	Viscosity at −40°C: 4.0 mm^2/s Density: 0.86 g/cm^3 Pour point: 13°C Cloud point: 2°C Flash point: 143°C	ASTM D-975	Farrokheh *et al.* (2022)
Coelastrella M-60	Iodine value: 42.295 Cetane number: 66.094 Degree of unsaturation: 38.124 Saponification value: 186.214 Cold filter plugging point: 13.948 Long chain saturation factor: 9.684	ASTM and EN14214	Sakthi Vignesh *et al.* (2020)
Dictyococcus sp. VSKA18	Iodine value: 27.97 Cetane number: 126.14 Degree of unsaturation: 20.44 Saponification value: 63.37 Cold filter plugging point: 7.06 Long chain saturation factor: 7.49	ASTM and EN14214	Sakthi Vignesh *et al.* (2020)
N. salina	Cetane number: 86.9 Boiling point: 360°C Kinematic viscosity: 5.85 mm^2/s Heat of combustion: 2696.12 kcal/mol	ASTM	Teo *et al.* (2016)
Naviculla sp.	Cetane number: 50.20 Heat of combustion: 44.34 kcal/mol	EN14214	Joseph *et al.* (2017)
C. pyrenoidosa	Cetane number: 58.42 Iodine value: 69.33 Degree of unsaturation: 79.85 Saponification value: 196.9 Long chain saturated factor: 5.73 Cold filter plugging point: 1.52	ASTM D-6751	Wu and Miao (2014)
S. platensis	Density: 860 kg/m^3 Viscosity: 5.66 mm^2/s Pour point: −18°C Flash point: 130°C	ASTM	Farrokheh *et al.* (2022)

nanocatalyst, CH_3 nanocatalyst is observed at 3650 cm^{-1} and C–O stretching at 1470 cm^{-1} (Teo *et al.*, 2016).

To confirm the FAME, GC-MS was used. The lipid extracted from *C. vulgaris* undergoes transesterification in the presence of catalysts such as strontium titanate, KF/KOH–Fe_3O_4, or SO_4^{2-}/Fe_3O_4–Al_2O_3 to produce FAMEs such as myristoleic acid, *n*-hexadecanoic acid methyl ester, and dodecanoic acid. These methyl esters have molecular weight ranging from 170 to 400 g/mol (Aghilinategh *et al.*, 2020; Farrokheh *et al.*, 2022; Safakish *et al.*, 2020; Vinoth Arul Raj *et al.*, 2021). The *Coelastrella* M-60 undergoes transesterification in the presence of bio-nano-CaO, bio-nano-$CaCO_3$, or bio-iron nanoparticles to yield lauric acid, myristric acid, palmitic acid, and so on. The GC-MS found the molecular weight of these methyl esters ranging from 200 to 300 g/mol (Karpagam *et al.*, 2020; Sakthi Vignesh *et al.*, 2020). *Dictyococcus* sp. VSKA18 undergoes transesterification in the presence of bio-iron nanoparticles to yield linolelaidic acid, linolenic acid, and so on, which have a molecular weight ranging from 200 to 300 g/mol (Sakthi Vignesh *et al.*, 2020). *N. oculata* undergoes transesterification with a yield of 89.72% in the presence of NiO (Purkan *et al.*, 2021). When the *E. sanguinea* undergoes transesterification in the presence of nano-CaO, it yields FAME which contains dodecanoic acid with a molecular weight of 186 g/mol and this was confirmed by GC-MS (Sivanesh *et al.*, 2022). GC-MS verified the production of FAME during the transesterification of *S. platensis* in the presence of $Ca(OCH_3)_2$ (Nautiyal *et al.*, 2014). *N. salina* undergoes transesterification in the presence of CaO to yield stigmast-4-*en*-3-one and 2(3*H*)-benzothiazolethione. The molecular weights of these were found to be 412.7 and 190.2 g/mol, respectively (Vinoth Arul Raj *et al.*, 2021). The properties of the biodiesel obtained from the third-generation feedstocks are shown in Table 13.3.

13.6 REUSABILITY OF CATALYSTS

Iron, nickel, copper-doped zinc oxide nanocatalysts are reused for six times where the yield gets reduced to 73% (Farooq *et al.*, 2022). Calcium oxide catalyst was used for about eight cycles in which a drastic reduction of 30% occurred in the last cycle (Rahman *et al.*, 2022). The copper-doped Fe_2O_3–CaO catalyst showed better yield percentage and re-usage than the normal Fe_2O_3–CaO catalyst which was around 78.8% in the fifth cycle (Khan *et al.*, 2022). Similarly, the Cu–Ni-doped zirconium oxide catalyst used for the biodiesel production from caper seed oils gave around 77.4% even in the sixth cycle, which is found to be the most effective in terms of the highest yield obtained even after six cycles of catalyst being reused (Eldiehy *et al.*, 2022). Another doped nanocatalyst is the zinc-doped calcium oxide, where it is reused at most for four times and the yield being 75.03% in fourth cycle itself, which is not preferable much (Roy & Mohanty, 2021). The green nanocatalyst of Cao being reused for seven times, where the lowest yield of 65% was obtained (Noreen *et al.*, 2021). In the case of the trimetallic-based montmorillonite nanocatalyst, the yield started to decrease to 75% after reusing it for fifth and sixth times (Dawood *et al.*, 2022). The SO_4^{2-}/Fe_3O_4–Al_2O_3 nanocatalyst is reused five times. When the catalyst is reused for five times, the yield obtained was about 87.6%. From the sixth time yield decreased to 75.2% (Safakish *et al.*, 2020). The KF/KOH–Fe_3O_4 can be reused for 10 times and the yield obtained was about 94% (Farrokheh *et al.*, 2022).

Table 13.4 demonstrates the catalyst's reusability and the resulting yield.

13.7 ECONOMIC ANALYSIS OF CATALYSTS

Table 13.5 shows the latest price of the catalysts being used widely (Zürich, 2016). Analysis shows that CaO, Fe_3O_4–CaO, KOH, and KOH/Al_2O_3 are found to be economical with the least price.

Table 13.4 Reusability of catalyst.

Sl. No.	Name of the Catalyst	Reuse Time	Yield (%)	References
1	Fe, Ni, and Cu-doped ZnO nanocatalyst	6	3rd cycle: 91.5 4th cycle: 85 5th cycle: 79 6th cycle: 73	Farooq *et al.* (2022)
2	Cao nanocatalyst	8	30% reduction in yield was observed from the 1st cycle to last cycle	Rahman *et al.* (2022)
3	Co-doped Fe_2O_3–Cao nanocatalyst		4th cycle: 87 5th cycle: 78.8	Khan *et al.* (2022)
4	Cu–Ni-doped ZrO_2 catalyst		1st three cycles: around 90% 4th cycle: 88.4 5th cycle: 82.1 6th cycle: 77.4	Eldiehy *et al.* (2022)
5	Zn-doped Cao nanocatalyst	1st to 4th cycle	Ranges from 83.6 to 75.03	Roy and Mohanty (2021)
6	Green nanocatalyst of Cao	7	5th cycle: maximum of 88% maintained (optimum conditions) 6th cycle: 75 7th cycle: 65	Noreen *et al.* (2021)
7	Trimetallic-based montmorillonite nanocatalyst	6	1–4 cycles: almost the same yield of 86%; not much decline in yield was observed 5th and 6th cycles: started to decrease to 75%	Dawood *et al.* (2022)
8	SO_4^{2-}/Fe_3O_4–Al_2O_3	5	1st to 5th cycle: 87.6 6th cycle: 75.2	Safakish *et al.* (2020)
9	KF/KOH–Fe_3O_4	10	94	Farrokheh *et al.* (2022)

Table 13.5 Economic analysis of catalysts.

Catalysts	Weight (g)	Price (Rs.)	Environmental Regulations	Price for 1 kg Catalyst
CaO (90%)	500	200	High basicity, high catalytic activity, eco-friendly	400
ZnO (99%)	500	860	Aquatic toxicity, irritant, highly flammable	1720
$(Bi(NO_3)_3{\cdot}5H_2O$	500	3850	–	7700
Ag_2O (97%)	25	8100	Strong oxidizer, irritant, acute and chronic aquatic toxicity, light sensitive, dangerous to environment	324,000
NiO	250	2850	Water insoluble, carcinogen, might be an irritant	11,400
CaO–Bi_2O_3	600	1400	–	2333
Fe_2O_3–CaO	1000	600	–	600
MgO heavy (98%)	500	1500	Air and moisture sensitive	3000
Lanthanum oxide (99.9%)	100	1300	Hygroscopic, air sensitive, water insoluble, irritant(eyes)	13,000
KOH	500	400	Air and moisture sensitive, irritant	800
KOH–Al_2O_3	1000	950	–	950

13.8 CONCLUSION AND FUTURE WORK

An alternative energy source is required due to the depletion of fossil fuels. Biodiesel is proven to be a viable substitute for fossil fuels. It can be produced by extracting oil from non-edible seeds and microalgae. This chapter explored the production of biodiesel from different non-edible seeds and microalgae using various catalysts. The inedible seeds are more abundant than the first-generation seeds and also provide a high yield. Microalgae are advantageous for biodiesel production due to its ease of cultivation and high lipid content. We propose that for the second generation, *B. populneus* seed oil and for the third generation, *C. vulgaris* would be the appropriate sources due to the availability of feedstocks. Green nickel oxide catalyst and KF/KOH–Fe_3O_4 catalyst were identified as suitable for biodiesel production based on their recyclability, surface area, and yield. Researchers are attempting to demonstrate that genetically modified microalgae can also generate a high yield. This will serve as the feedstock for future biodiesel production.

ACKNOWLEDGMENTS

The authors acknowledge the Department of Chemical Engineering and Material Science, Amrita School of Engineering, Amrita Vishwa Vidyapeetham, Coimbatore, India for the support. We thank all the researchers who conducted experiments for biodiesel production using nanocatalysts that provide the base for this chapter.

REFERENCES

Aghilinategh M., Barati M. and Hamadanian M. (2020). The modified supercritical media for one-pot biodiesel production from *Chlorella vulgaris* using photochemically-synthetized $SrTiO_3$ nanocatalyst. *Renewable Energy*, **160**, 176–184, https://doi.org/10.1016/j.renene.2020.06.081

Akubude V. C., Nwaigwe K. N. and Dintwa E. (2019). Production of biodiesel from microalgae via nanocatalyzed transesterification process: a review. *Materials Science for Energy Technologies*, **2**, 216–225, https://doi.org/10.1016/j.mset.2018.12.006

Aniruddha R., Rajendran A. and Sindhu S. (2022). A study on biofuel generation from microalgae species. *Materials Today: Proceedings*, **57**, 1660–1665, https://doi.org/10.1016/j.matpr.2021.12.269

Dalvand P. and Mahdavian L. (2022). Biodiesel production in the presence of eggshell nano-catalyst. *Chemistry and Technology of Fuels and Oils*, **58**, 55–62, https://doi.org/10.1007/s10553-022-01351-1

Davoodbasha M. A., Pugazhendhi A., Kim J. W., Lee S. Y. and Nooruddin T. (2021). Biodiesel production through transesterification of *Chlorella vulgaris*: synthesis and characterization of CaO nanocatalyst. *Fuel*, **300**, 121018, https://doi.org/10.1016/j.fuel.2021.121018

Dawood S., Koyande A. K., Ahmad M., Mubashir M., Asif S., Klemeš J. J., Bokhari A., Saqib S., Lee M., Qyyum M. A. and Show P. L. (2021). Synthesis of biodiesel from non-edible (*Brachychiton populneus*) oil in the presence of nickel oxide nanocatalyst: parametric and optimisation studies. *Chemosphere*, **278**, 130469, https://doi.org/10.1016/j.chemosphere.2021.130469

Dawood S., Ahmad M., Zafar M., Asif S., Klemeš J. J., Bokhari A., Mubashir M., Han N., Ibrahim M. M., El-Bahy Z. M. and Khoo K. S. (2022). Biodiesel synthesis from *Prunus bokhariensis* non-edible seed oil by using green silver oxide nanocatalyst. *Chemosphere*, **291**, 132780, https://doi.org/10.1016/j.chemosphere.2021.132780

Eldiehy K. S. H., Daimary N., Borah D., Sarmah D., Bora U., Mandal M. and Deka D. (2022). Towards biodiesel sustainability: waste sweet potato leaves as a green heterogeneous catalyst for biodiesel production using microalgal oil and waste cooking oil. *Industrial Crops and Products*, **187**, 115467, https://doi.org/10.1016/j.indcrop.2022.115467

Farooq M., Ramli A., Gul M., Naeem A., Perveen F., Khan I. W., Saeed S., Sahar J. and Abid G. (2022). Utilization of indigenous gurgure (*Monotheca buxifolia*) waste seeds as a potential feedstock for biodiesel production using environmentally benign bismuth modified CaO catalyst. *Chemical Engineering Research and Design*, **183**, 67–76, https://doi.org/10.1016/j.cherd.2022.04.022

Farrokheh A., Tahvildari K. and Nozari M. (2022). Biodiesel production from the *Chlorella vulgaris* and *Spirulina platensis* microalgae by electrolysis using CaO/KOH–Fe_3O_4 and KF/KOH–Fe_3O_4 as magnetic nanocatalysts. *Biomass Conversion and Biorefinery*, **12**, 403–417, https://doi.org/10.1007/s13399-020-00688-z

Joseph M. M., Renjith K. R., John G., Nair S. M. and Chandramohanakumar N. (2017). Biodiesel prospective of five diatom strains using growth parameters and fatty acid profiles. *Biofuels*, **8**, 81–89, https://doi.org/10.1080/17597269.2016.1204585

Karpagam R., Rani K., Ashokkumar B., Ganesh Moorthy I., Dhakshinamoorthy A. and Varalakshmi P. (2020). Green energy from *Coelastrella* sp. M-60: bio-nanoparticles mediated whole biomass transesterification for biodiesel production. *Fuel*, **279**, 118490, https://doi.org/10.1016/j.fuel.2020.118490

Kazemifard S., Nayebzadeh H., Saghatoleslami N. and Safakish E. (2019). Application of magnetic alumina-ferric oxide nanocatalyst supported by KOH for in-situ transesterification of microalgae cultivated in wastewater medium. *Biomass and Bioenergy*, **129**, 105338, https://doi.org/10.1016/j.biombioe.2019.105338

Khan I. W., Naeem A., Farooq M., Ghazi Z. A., Saeed T., Perveen F. and Malik T. (2022). Biodiesel production by valorizing waste non-edible wild olive oil using heterogeneous base catalyst: process optimization and cost estimation. *Fuel*, **320**, 123828, https://doi.org/10.1016/j.fuel.2022.123828

Mofijur M., Siddiki S. Y. A., Shuvho M. B. A., Djavanroodi F., Fattah I. M. R., Ong H. C., Chowdhury M. A. and Mahlia T. M. I. (2021). Effect of nanocatalysts on the transesterification reaction of first, second and third generation biodiesel sources – a mini-review. *Chemosphere*, **270**, 128642, https://doi.org/10.1016/j.chemosphere.2020.128642

Munir M., Ahmad M., Mubashir M., Asif S., Waseem A., Mukhtar A., Saqib S., Munawaroh H. S. H., Lam M. K., Khoo K. S., Bokhari A. and Show P.L. (2021a). A practical approach for synthesis of biodiesel via non-edible seeds oils using trimetallic based montmorillonite nano-catalyst. *Bioresource Technology*, **328**, 124859, https://doi.org/10.1016/j.biortech.2021.124859

Munir M., Ahmad M., Saeed M., Waseem A., Nizami A. S., Sultana S., Zafar M., Rehan M., Srinivasan G. R., Ali A. M. and Ali M. I. (2021b). Biodiesel production from novel non-edible caper (*Capparis spinosa* L.) seeds oil employing Cu–Ni doped ZrO_2 catalyst. *Renewable and Sustainable Energy Reviews*, **138**, 110558, https://doi.org/10.1016/j.rser.2020.110558

Nautiyal P., Subramanian K. A. and Dastidar M. G. (2014). Kinetic and thermodynamic studies on biodiesel production from *Spirulina platensis* algae biomass using single stage extraction–transesterification process. *Fuel*, **135**, 228–234, https://doi.org/10.1016/j.fuel.2014.06.063

Naveenkumar R. and Baskar G. (2021). Process optimization, green chemistry balance and technoeconomic analysis of biodiesel production from castor oil using heterogeneous nanocatalyst. *Bioresource Technology*, **320**, 124347, https://doi.org/10.1016/j.biortech.2020.124347

Noreen S., Khalid K., Iqbal M., Baghdadi H. B., Nisar N., Siddiqua U. H., Nisar J., Slimani Y., Khan M. I. and Nazir A. (2021). Eco-benign approach to produce biodiesel from neem oil using heterogeneous nano-catalysts and process optimization. *Environmental Technology & Innovation*, **22**, 101430, https://doi.org/10.1016/j.eti.2021.101430

Pandit P. R. and Fulekar M. H. (2017). Egg shell waste as heterogeneous nanocatalyst for biodiesel production: optimized by response surface methodology. *Journal of Environmental Management*, **198**, 319–329, https://doi.org/10.1016/j.jenvman.2017.04.100

Purkan P., Safa A., Abdulloh A., Baktir A., Arissirajudin R., Hermansyah H. and Kim S. W. (2021). In-situ biodiesel synthesis from microalgae *Nannochloropsis oculata* using nio nanocatalyst. *Rasayan Journal of Chemistry*, **14**, 2097–2103, https://doi.org/10.31788/RJC.2021.1436317

Rahman M. W., Mondal A. K., Hasan M. S. and Sultana M. (2022). Biodiesel production from a non-edible source of royna (*Aphanamixis polystachya*) oil. *Energy, Sustainability and Society*, **12**, 1–11, https://doi.org/10.1186/s13705-022-00360-6

Rahul Soosai M., Moorthy I. M. G., Varalakshmi P. and Yonas C. J. (2022). Integrated global optimization and process modelling for biodiesel production from non-edible silk-cotton seed oil by microwave-assisted transesterification with heterogeneous calcium oxide catalyst. *Journal of Cleaner Production*, **367**, 132946, https://doi.org/10.1016/j.jclepro.2022.132946

Roy M. and Mohanty K. (2021). Valorization of de-oiled microalgal biomass as a carbon-based heterogeneous catalyst for a sustainable biodiesel production. *Bioresource Technology*, **337**, 125424, https://doi.org/10.1016/j.biortech.2021.125424

Rozina A. M., Elnaggar A. Y., Keat Teong L., Sultana S., Zafar M., Ahmad M., Munir M., Hussein E. E. and Ul Abidin S. Z. (2022). Sustainable and eco-friendly synthesis of biodiesel from novel and non-edible seed oil of *Monotheca buxifolia* using green nano-catalyst of calcium oxide. *Energy Convers Manag X*, **13**, 100142, https://doi.org/10.1016/j.ecmx.2021.100142

Safakish E., Nayebzadeh H., Saghatoleslami N. and Kazemifard S. (2020). Comprehensive assessment of the preparation conditions of a separable magnetic nanocatalyst for biodiesel production from algae. *Algal Research*, **49**, 101949, https://doi.org/10.1016/j.algal.2020.101949

Sakthi Vignesh N., Vimali E., Sangeetha R., Arumugam M., Ashokkumar B., Ganeshmoorthy I. and Varalakshmi P. (2020). Sustainable biofuel from microalgae: application of lignocellulosic wastes and bio-iron nanoparticle for biodiesel production. *Fuel*, **278**, 118326, https://doi.org/10.1016/j.fuel.2020.118326

Shobana M., Prasad R. K., Reddy Ragula U. B. and Kumaresan D. (2020). Kinetics and characterization of transesterification of cottonseed oil to biodiesel using calcined clam shells as catalyst. *Biofuels*, **11**, 597–605, https://doi.org/10.1080/17597269.2017.1378994

Sivanesh S., Aswin K. N., Antony A., Surya Varma M., Lekshmi A., Kamalesh K., Naageshwaran M., Soundarya S. and Subramanian S. (2022). Biodiesel production from custard apple seeds and *Euglena sanguinea* using CaO nano-catalyst. *Bioresource Technology*, **344**, 126418, https://doi.org/10.1016/j.biortech.2021.126418

Sree J. V., Chowdary B. A., Kumar K. S., Anbazhagan M. P. and Subramanian S. (2021). Optimization of the biodiesel production from waste cooking oil using homogeneous catalyst and heterogeneous catalysts. *Materials Today: Proceedings*, **46**, 4900–4908, https://doi.org/10.1016/j.matpr.2020.10.332

Teo S. H., Islam A. and Taufiq-Yap Y. H. (2016). Algae derived biodiesel using nanocatalytic transesterification process. *Chemical Engineering Research and Design*, **111**, 362–370, https://doi.org/10.1016/j.cherd.2016.04.012

Tukaram Bai M., Swarna U., Mahalakshmi R. C. and Sridevi V. (2021). Production of methyl ester from mahua oil: characterization and optimization by using RSM. *Materials Today: Proceedings*, **44**, 1609–1616, https://doi.org/10.1016/j.matpr.2020.11.815

Vinoth Arul Raj J., Praveen Kumar R., Vijayakumar B., Gnansounou E. and Bharathiraja B. (2021). Modelling and process optimization for biodiesel production from *Nannochloropsis salina* using artificial neural network. *Bioresource Technology*, **329**, 124872, https://doi.org/10.1016/j.biortech.2021.124872

Wu H. and Miao X. (2014). Biodiesel quality and biochemical changes of microalgae *Chlorella pyrenoidosa* and *Scenedesmus obliquus* in response to nitrate levels. *Bioresource Technology*, **170**, 421–427, https://doi.org/10.1016/j.biortech.2014.08.017

Xia S., Li J., Chen G., Tao J., Li W. and Zhu G. (2022). Magnetic reusable acid-base bifunctional Co doped Fe_2O_3–CaO nanocatalysts for biodiesel production from soybean oil and waste frying oil. *Renewable Energy*, **189**, 421–434, https://doi.org/10.1016/j.renene.2022.02.122

Zürich E. T. H. (2016). Magic of chemistry. *Loba Chemie – Laboratory Reagents and Fine Chemicals*, 1–84.

IWA Publishing's authorised EU representative for General Product Safety Regulations is Diane D'Arras, 15 rue Duret, 75116 Paris, France, e-mail: safety@iwap.co.uk.

Printed and bound by CPI Group (UK) Ltd, Croydon, CR0 4YY
12/05/2026
02108869-0011